6th International Conference on Advanced Engineering and Technology (ICAET 2019)

AA007015

IOP Conference Series: Materials Science and Engineering
Volume 811

Incheon, South Korea
14 - 16 December 2019

ISBN: 978-1-7138-3605-6
ISSN: 1757-8981

Printed from e-media with permission by:

Curran Associates, Inc.
57 Morehouse Lane
Red Hook, NY 12571

Some format issues inherent in the e-media version may also appear in this print version.

Additional copies of this publication are available from:

Curran Associates, Inc.
57 Morehouse Lane
Red Hook, NY 12571 USA
Phone: 845-758-0400
Fax: 845-758-2633
Email: curran@proceedings.com
Web: www.proceedings.com

TABLE OF CONTENTS

IOP Conf. Series: Materials Science and Engineering **811** (2020) 011001 doi:10.1088/1757-899X/811/1/011001

Preface

The 6th ICAET 2019 is the 6th International Conference on Advanced Engineering and Technology (6th ICAET 2019) took place in Incheon, Korea during December 13-15, 2019.

The conference program covered invited, oral, and poster presentations from scientists working in similar areas to establish platforms for collaborative research projects in this field. This conference will bring together leaders from industry and academia to exchange and share their experiences, present research results, explore collaborations and to spark new ideas, with the aim of developing new projects and exploiting new technology in this field.

The committee of ICAET expresses their sincere thanks to all authors for their high-quality research papers and careful presentations. All reviewers are also thanked for their careful comments and advices. Thanks are finally given to IOP Publication as well for producing this volume.

The Organizing Committee of ICAET 2019

Committee Chair

Prof. Dongyan Shi

Harbin Engineering University

China

IOP Conf. Series: Materials Science and Engineering **811** (2020) 011001 doi:10.1088/1757-899X/811/1/011001

Conference Photograph

IOP Conf. Series: Materials Science and Engineering **811** (2020) 011001 doi:10.1088/1757-899X/811/1/011001

Sponsors

Incheon Disaster Prevention Research Center (IDPRC)	IDPRC 인천대학교 방재연구센터 Incheon Disaster Prevention Center

Recently, various efforts to prevent and prepare are vitally needed for prevention of Disasters and calamities. So we understand the necessity for technology of disaster and we built up the Incheon Disaster Prevention Research Center (IDPRC) in Incheon National University (INU) in 1997.

Accordingly, Incheon Disaster Prevention Research Center (IDPRC) in Incheon National University has progressed of research on the prevention of disasters and calamities through the various seminars, conference and lectures. These research could be conducted cooperation with Incheon National University (INU) in various fields' structure, soil, hydraulic and environment.

Incheon Disaster Prevention Research Center (IDPRC) will try to be a leader in the disaster of industry through the various research activities and global conference.

Incheon National University	INU 인천대학교 INCHEON NATIONAL UNIVERSITY

Incheon National University established a global campus by integrating two city council funded colleges into one in March 2010 and became a national university operated by the legal entity. With the start of attracting the branch of Lawrence Berkeley Research Center that is a world famous US national policy research center for education and research, Incheon National University will attract St. Petersburg University in Russia, Polymers University in Britain, and Kent University in Belgium and other foreign universities and research centers. These efforts will make INU spring to a world-wide competitive university.

As a local development leading university, Incheon National University established INU VISON 2020 and improved its competition actively through investment in selection and concentration. In compliance with the geographic feature and areal specialty, it will concentrate on the international trade, goods distribution, applied technology convergence, BNT-based life science, urban science and the sinology- based local humanity fields.

IOP Conf. Series: Materials Science and Engineering **811** (2020) 011001 doi:10.1088/1757-899X/811/1/011001

Lists of Committees

Committee name	Organizer name	Organization
Honorary Chair	Dongyan Shi	Harbin Engineering University
Conference General Chairs	Dongkeon kim	Dong-A University
Honorary Chair	Zhixun Yang	Harbin Engineering University
Technical Committees	J.W. Jung	Harbin Engineering University
	J.W. Hu	Incheon National University

First name	Surename	Title	E-mail address
A K M	Sipi	Extraction of methylene blue from aqueous solutions by waste cooking oil and its back-extraction	shchang@uitm.edu.my
Roland T.	Loto	Effect of repetitive heat treatment process on the localized corrosion resistance of 409 ferritic, 430 ferritic, 3163 austenitic and 444 ferritic stainless steels for high temperature variation	tolu.loto@gmail.com
Yue-Shi	Lee	The Maintenance of Customer Purchasing Behaviours	sjyen@mail.mcu.edu.tw
Sulaiha	Ali	Thermogravimetric analysis and pyrolysis kinetics of Malaysian wood species	sulaihaali2013@gmail.com
V V	Romanovskiy	Analysis of circuit design of a rowing electric installation with a distributed DC bus of a port icebreaker	butsanetsaa@gumrf.ru
V	Karetnikov	Experience of use high-precise position systems for inland waters	butsanetsaa@gumrf.ru
Nur Khairunnisa	Abd Halim	Eulerian-Eulerian Approach to CFD Simulation of Two-Phase Bubble Column	nisaahalim38@gmail.com
David	Park	Unidirectional heat transfer into a rotating cylinder containing R-134a refrigerant	swkim@alaska.edu
S C	Zhang	A method of selection the textile materials for virtual reconstruction	569642835@qq.com
Noppakun	Sanpo	Silica Aerogel Thermal Insulation Coating as Commodity Usage	darapond.tri@mahidol.edu
Sireethorn	Termkleebbuppa	Copper-Zinc Oxide Synergistic Approach as Low-Emissivity Material for Energy-Saving Windows	darapond.tri@mahidol.edu
N.	Yuvaraj	EDM Process by using Copper Electrode with INCONEL 625 Material	prakash.palanivel@skillteq.com
Xiaoyan	Teng	Damage Mode of Long Tube Structure at Different Water Depths under Strong Impact	sunyuxinsunny@163.com
Xuanxin	Tian	Numerical Simulation of Explosive Welding of Marine Aluminum Steel Transition Joint Based on AUTODYN	1055067077@qq.com
Subpa-asa	Prang	The Study on Effect of Lightweight Concrete Block by Water Hyacinth Adding	prangsub@gmail.com
H Y	Setyawan	Progress on pine derivative products as fuel source in Indonesia	hendrix@ub.ac.id
Yu	Shao	Research on Modular Coordination of Indemnificatory Finished Product Housing in Severe Cold Region of China	2486058209@qq.com
P.	Yu	Global existence of the 3-D Heat-conductive incompressible viscous fluid	yp9106@gmail.com
A.	Rattanapan	The Potential use of Waste Cement as a Filler in Polymer Composites	apaipan.r@cit.kmutnb.ac.th
P.	Sapsrithong	Morphology and Mechanical Properties of Polyamide 6 and Polybutylene Terephthalate Blends Compatibilized with Epoxidized Natural Rubber	pornsri.s@cit.kmutnb.ac.th
I N	Budiarsa	Identification Plastic Properties of Spot Welded Joints Using the Instrumented Indentation Technique	nyoman.budiarsa@unud.ac.id
J.	Kumbunleu	The Study on Chitosan Coating on Poly(Lactic Acid) Film Packaging to Extend Vegetable and Fruit Life	thritima.s@cit.kmutnb.ac.th
S.	Tuampoemsab	Unvulcanized Rubber Tape Made From Natural Rubber and Synthetic Polyisoprene Modified with Rosin as a Tackifier	surakit.t@cit.kmutnb.ac.th
Muamer	Abuzwidah	Traffic Safety Benefits of Using Highway Lighting Technology	mabuzwidah@sharjah.ac.ae
A	Fadlilah	Chemical Quality of Fresh New Zealand White Rabbit Meat in Batu, Indonesia	anik.fadlilah04@gmail.com
M	Muchlas	Evaluation of Chemical Composition and Lipid Components in Maize Straw based complete feed diets Supplemented by Conden Tannin and Myristic acid	muchamad.muchlas@gmail.com
Adil	Fahada	A Semi-Stack Approach for Accurate Network Traffic Classification Using Multi-View Stacking	afalharthi@bu.edu.sa
A. A.	Jaafar	Improvement of heat sink performance by using Graphene nanosheets coated by chemical spray method	salam.waheed@uokufa.edu.iq
Yungchiang	Lin	Evaluation of Repair Quality for Concrete Structure by Non-destructive Technique	yungchaing.lin@gmail.com
Meryem EL	Badaoui	Modeling of a parabolic trough using two heat transfer fluids and an economic estimation in the Moroccan dairy industry	meryem.elbadaoui12@gmail.com
M A	Fajobi	Steel Corrosion behaviour in acidic solution for application in petrochemical distillation systems	fajobimuyiwa@gmail.com
O A	Odunlami	Corrosion Inhibition of Rosemary oil on carbon steel in dilute Sulphuric acid	oboloms@gmail.com
S	Rusdi	Amine-based Carbon Dioxide Absorption: Evaluation of Species Activity Coefficient for MEA-H2O System from 303 to 333K	sholeh.mamun@uii.ac.id
S	Ma'mun	Amino-Acid-Salt-based Carbon Dioxide Capture: Precipitation Behavior of Potassium Sarcosine Solution	sholeh.mamun@uii.ac.id
I	Odun-Ayo	A Systematic Mapping Study of Innovative Cloud Applications	isaac.odun-ayo@covenantuniversity.edu.ng
M A	Fajobi	Development of a Human powered pedal washing Machine	fajobi.muyiwa@covenantuniversity.edu.ng
S L	Annisa	Contribution of Nano Crumb Rubber on Recycled Asphalt Concrete As Pedestrian Paving Block	sigit@eng.ui.ac.id
S P	Hadiwardoyo	Temperature effect on the deformation of the recycled hot-mix asphalt concrete with Nano Crumb Rubber as an added material with Wheel Tracking Machine	sigit@eng.ui.ac.id
J	Sarasook	A Using of Solar Roof Top for Electrical Energy Conservation in Educational Organization	jassadat77@gmail.com
S	Charoenchit	A Study of Inspection for Electrical Safety Systems Case Study Four Locations in Thailand	jassadat77@gmail.com
Xuyang	Liu	Analysis of Characteristics and Research on Processing Method of the High Viscosity DW-1 Colloid	1493658835@qq.com
Haiyang	Xue	Analysis and Research on the Working Parameters of Colloidal Spraying for the Thermal Insulation Layer of Rocket Tank	1079032509@qq.com
M	Afifuddin	Experimental Research on Ferro Foam Concrete I beams with different number of wire mesh layer	m.afifuddin@unsyiah.ac.id
Yuanlong	Chen	Design and Analysis of Flow Field in Electrochemical Machining of Micro Grooves on Gear Tooth Surface	fs1831766842@gmail.com
Ruth María	Grajeda Rosado	Thermal analysis of the urban canyon based on the variables: orientation, wind and vehicular anthropogenic heat	hchavezenator@gmail.com
Ruth María	Grajeda Rosado	Mapping the LST (land surface temperature) with satellite information and software ArcGis	hchavezenator@gmail.com
	HUSAINI	Prediction of the Fatigue Life of the SAE 5160 Carbon Steel Lower Arm Based on Strain-Life Approach	edi@unsyiah.ac.id
Norfarah Nadia	Ismail	Higher Modes and Superposed Mode Behavior for Flexible Pavement Layer System	norfarahnadia@iium.edu.my
Dwinita	Apritasari	Effect of Hinge Properties on Nonlinear Analysis of Eccentrically Steel Braced Frames	mulia@eng.ui.ac.id
Kazuma	Tanaka	Dynamic behavior analysis of feed drive system using linear guideways with different ball diameters	yamazaki@mail.dendai.ac.jp
Akinori	Tsuzura	Vibration compensation of checkweigher using Acceleration sensor	yamazaki@mail.dendai.ac.jp
Ahmad Zaki	Risadi	Damage Assessment Using Natural Frequency on Concentric Braced Frame Steel Structures	jrastandi@gmail.com
Daisheng	Zhang	Simulation Analysis and Optimization of a Passenger Car Ride Comfort	785453623@qq.com

IOP Conf. Series: Materials Science and Engineering **811** (2020) 011003 doi:10.1088/1757-899X/811/1/011003

Peer review statement

All papers published in this volume of *IOP Conference Series: Materials Science and Engineering* have been peer reviewed through processes administered by the proceedings Editors. Reviews were conducted by expert referees to the professional and scientific standards expected of a proceedings journal published by IOP Publishing.

IOP Conf. Series: Materials Science and Engineering **811** (2020) 012001 doi:10.1088/1757-899X/811/1/012001

Extraction of Methylene Blue from Aqueous Solutions by Waste Cooking Oil and its Back-Extraction

A K M Sipi[1], S H Chang[1*]

[1]Faculty of Chemical Engineering, Universiti Teknologi MARA, 13500 Permatang Pauh, Penang, Malaysia

shchang@uitm.edu.my

Abstract. This work aimed to use waste cooking oil as a greener replacement for the petroleum-based solvents in removing methylene blue (MB) from aqueous solutions by solvent extraction. The effects of different extraction parameters such as equilibrium pH, aqueous to organic phase ratio, and initial MB concentration on the extraction of MB from aqueous solutions by waste cooking oil were investigated. The back-extraction of MB from MB-loaded waste cooking oil was also explored. The maximum extraction (~90%) and back-extraction (>90%) of MB were attained at equilibrium pH of 8, aqueous to organic phase ratio of 1:1, initial MB concentration of 20 mg/L, and with 1 M of sodium bicarbonate as the back-extraction agent.

1. Introduction

Dyestuffs are organic or inorganic materials that could absorb and reflect light to show colors. They are categorized into acid, basic, reactive, direct, vat, mordant, disperse, metal complex, and sulfur dyestuffs and are widely used as colorants in textile, leather, paper printing, food, cosmetics, and pharmaceutical industries. However, the release of a large number of dyestuffs into the environment during dyeing processes from numerous industries (textiles, dyeing, paper and pulp, tannery and paint, and dye manufacturing) is of global concern due to their toxicity, carcinogenicity, mutagenicity, and non-biodegradability [1].

With the rise in demand for dyestuffs following the rapid population growth, industrialization, and urbanization, an assortment of dyestuffs has been synthesized. Methylene Blue (MB) is one of the dyestuffs that has been synthesized and used extensively in various industries. Despite the fact that MB shows superior stability than its natural counterparts, its complex aromatic molecular structure has resulted in its recalcitrant nature which poses severe environmental, esthetical and medical issues when it is released into the environment [2]. Hence, the removal of MB from wastewater prior to discharge is indispensable.

Today, a number of physicochemical and biological technologies have been used to remove dyestuffs from wastewater which encompass adsorption, coagulation-flocculation, activated sludge, membrane bioreactor, advanced oxidation, and electrochemical processes [1]. Among the various treatment technologies, solvent extraction appears as one of the promising technologies for treating wastewater containing dyestuffs owing to its ability to remove and recover dyestuffs from wastewater through the extraction and back-extraction processes, respectively [3]. However, the conventional solvents used in solvent extraction are mostly petroleum-based which are harmful, non-biodegradable, and non-renewable [4]. Hence, this work attempted to use eco-friendly and renewable biomass, namely, waste cooking oil, as a greener replacement for the petroleum-based solvents in removing MB from aqueous solutions by solvent extraction. The effects of different extraction parameters such as equilibrium pH (pHeq), solvent type, aqueous to organic (A:O) phase ratio, and initial MB concentration on the extraction of MB from aqueous solutions by waste cooking oil were investigated. The back-extraction of MB from MB-loaded waste cooking oil was also explored.

2. Materials and Method

2.1. Materials

The waste cooking oil was collected from a local restaurant which was filtered with a cheesecloth before use, while the fresh cooking oil was bought from a local supermarket which was used as received. MB (Merck, > 82% purity), sodium hydroxide (NaOH) (Merck, > 97% purity), hydrochloric acid (HCl) (Qrec, 37% purity), tributylphosphate (TBP) (Sigma-Aldrich, > 99% purity) and sodium bicarbonate ($NaHCO_3$) (Sigma-Aldrich, > 99% purity) were used as received.

2.2 Method

The extraction of MB from aqueous solutions was carried out based on the procedure reported in our previous work [4]. Firstly, an organic solvent made up of waste or fresh cooking oil and 80 mM of TBP (phase modifier) was added to an aqueous solution containing a specific amount of MB (20, 40 or 60 mg/L) in a conical flask at an A:O phase ratio (1:1, 2:1 or 1:2). An orbital shaker (Lab Companion, SK-300) was then used to agitate the mixture at 150 rpm for 7 minutes after which it was allowed to stand to separate for about 5 min. Next, some aqueous sample was taken from the flask with a syringe and its pH_{eq} was measured with a pH meter (Eutech, pH 700). If the pH_{eq} measured varied from the desired pH_{eq}, the sample was returned to the flask and its pH_{eq} was adjusted with 1 M NaOH or 1 M HCl. The mixture was allowed to separate and the pH_{eq} was measured and adjusted again. This continued until the desired pH_{eq} was obtained. Next, the mixture was moved to a separating funnel for phase separation and some aqueous sample was collected for MB concentration analysis by using a UV-visible spectrophotometer (PerkinElmer, LAMBDA 365) at a wavelength of 662 nm [5]. The percentage of extraction (%E) of MB is calculated by [5]:

$$\%E = \frac{[MB]_{o,aq} - [MB]_{f,aq}}{[MB]_{o,aq}} \text{ x } 100\% \quad (1)$$

where $[MB]_{o,aq}$ and $[MB]_{f,aq}$ are the initial and final concentrations of MB in the aqueous feed phase. Lastly, MB was back-extracted from the MB-loaded waste cooking oil by using a $NaHCO_3$ solution (0.2-1.0 M) following the same procedure as that of MB extraction but without the pH_{eq} adjustment step. The % of back-extraction (BE) of MB is given by [5]:

$$\%BE = \frac{[MB]_f}{[MB]_{o,aq} - [MB]_{f,aq}} \text{ x } 100\% \quad (2)$$

where $[MB]_f$ is the final dye concentration in the $NaHCO_3$ solution.

3. Result and Discussion

3.1. Effect of pH_{eq} on MB extraction

Figure 1 shows the effect of pH_{eq} on the %E of MB by using waste and fresh cooking oils. Other parameters like A:O phase ratio and initial MB concentration were fixed at 1:1 and 20 mg/L, respectively. As can be seen in figure 1, the highest %E of MB by both waste (93.4%) and fresh (73%)

cooking oil were achieved at pH_{eq} of about 8 but decreased with pH_{eq} at pH_{eq} below and above 8. The latter may be attributed to the increasing competition between H^+ ions and MB cations to bind with the active components in the cooking oil as pH_{eq} decreased below 8 [6] and the increasing negative charge surrounding MB cations which repelled them from the active components in the cooking oil as pH_{eq} increased above 8 [7]. Meanwhile, the greater performance of waste cooking oil in MB extraction than its fresh counterpart could be explained from its larger amount of acidic proton-bearing active components such as free fatty acids that could react with MB cations through a cation exchange process [3, 5]. Since pH_{eq} of 8 achieved the highest %E of MB, it was selected for further studies.

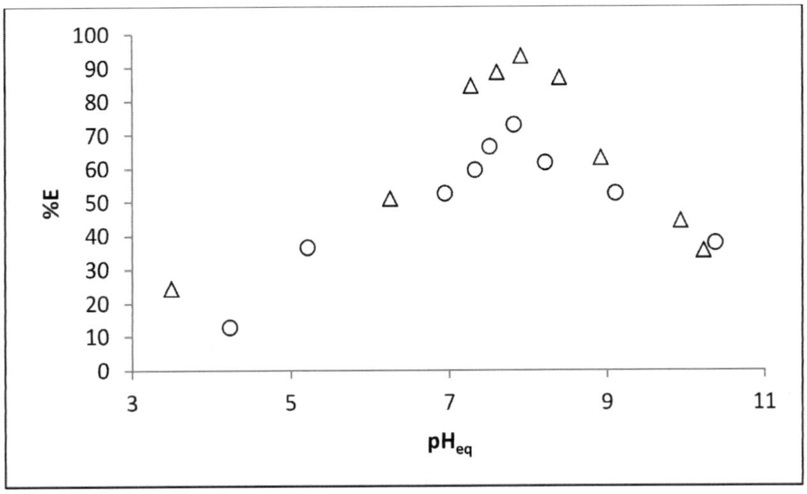

Figure 1. Effect of pH_{eq} on %E of MB by using waste (Δ) and fresh (○) cooking oils.

3.2. Effect of A:O phase ratio on MB extraction

Figure 2 shows the effect of A:O phase ratio on %E of MB by using waste cooking oil. Other parameters like pH_{eq} and initial MB concentration were fixed at 8 and 20 mg/L, respectively. It was found that A:O of 1:2 attained the highest %E (91%), followed closely by A:O of 1:1 (89%), and the lowest %E (66%) was achieved by A:O of 2:1. This was because a higher amount of organic phase than that of aqueous phase implied a greater amount of active components available for MB extraction, and thus favored the MB extraction, while the opposite disfavored it due to the lack of active components for MB extraction [5]. Since there was no significant difference in the %E of MB attained between A:O of 1:1 and 1:2, the A:O of 1:1 was selected for further studies.

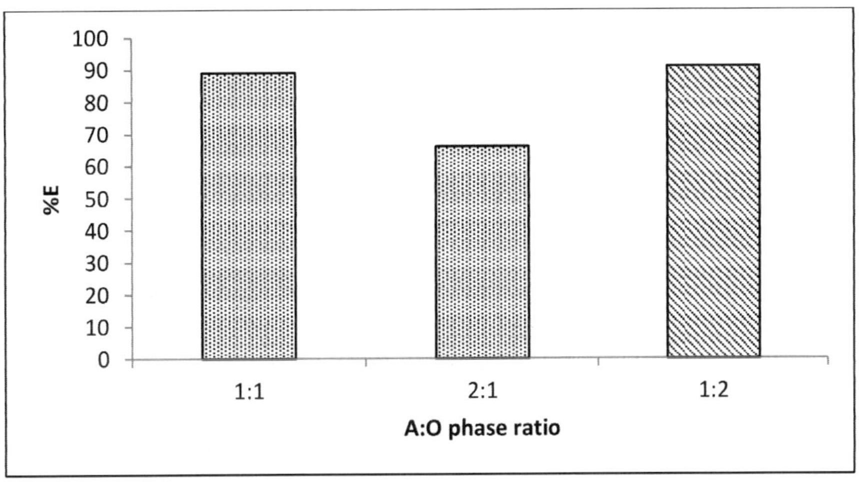

Figure 2. Effect of A:O phase ratio on %E of MB.

3.3. Effects of initial MB concentration on MB extraction

Figure 3 shows the effect of initial MB concentration on %E of MB by using waste cooking oil. Other parameters like pH_{eq} and A:O phase ratio were fixed at 8 and 1:1, respectively. As shown in figure 3, the %E of MB decreased gradually from 87 to 80% with the initial MB concentration. This could be ascribed to the insufficient active components for MB extraction at high initial MB concentrations and this finding is consistent with other similar previous works [3, 5]. Since the maximum %E of MB was achieved at an initial MB concentration of 20 mg/L, it was selected for further studies.

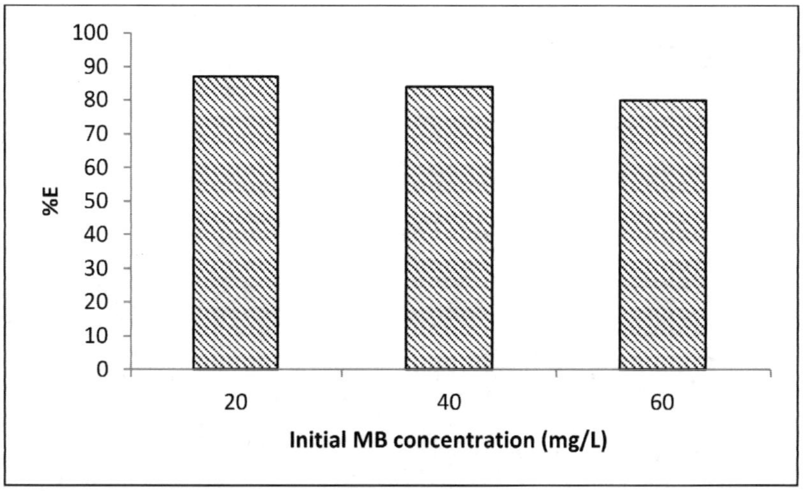

Figure 3. Effect of initial MB concentration on %E of MB.

3.4. Effect of $NaHCO_3$ concentration on MB back-extraction

Figure 4 shows the effect of $NaHCO_3$ concentration on %BE of MB from MB-loaded waste cooking oil which was obtained from the extraction experiments conducted under the optimum conditions (pH_{eq} of 8 (figure 1), A:O of 1:1 (figure 2), and initial MB concentration of 20 mg/L (figure 3)). As can be seen in figure 4, %BE of MB increased with $NaHCO_3$ concentration and the maximum %BE was accomplished with 1 M $NaHCO_3$. The was due to the increasing amount of $NaHCO_3$ molecules available for reaction with MB as the concentration of $NaHCO_3$ increased. This is in good agreement with similar previous works [6].

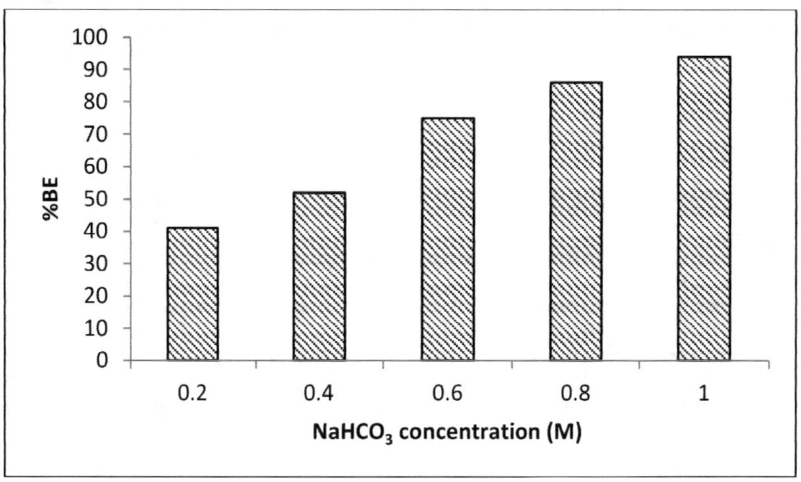

Figure 4. Effect of $NaHCO_3$ concentration on %BE of MB.

4. Conclusions

All extraction parameters studied were found to affect MB extraction significantly. The maximum extraction (~90%) and back-extraction (>90%) of MB were attained at equilibrium pH of 8, aqueous to organic phase ratio of 1:1, initial MB concentration of 20 mg/L, and with 1 M of sodium bicarbonate as the back-extraction agent. Hence, waste cooking oil is a potential greener replacement for petroleum-based solvents in MB removal by solvent extraction.

Acknowledgments

The authors thanked UiTM for funding this work.

References

[1] Katheresan V, Kansedo J, Lau S Y 2018 Efficiency of various recent wastewater dye removal methods: A review, *J. Environ. Chem. Eng.* **6** 4676-97.

[2] Muthuraman G, Teng T T, Leh C P, Norli I 2009 Extraction and recovery of methylene blue from industrial wastewater using benzoic acid as an extractant, *J. Hazard. Mater.* **163** 363-9.

[3] Soniya M, Muthuraman G 2015 Recovery of methylene blue from aqueous solution by liquid-liquid extraction, *Desalin. Water Treat.* **53** 2501-9.

[4] Chang S H, Teng T T, Ismail N 2010 Extraction of Cu(II) from aqueous solutions by vegetable oil-based organic solvents, *J. Hazard. Mater.* **181** 868-72.

[5] Soniya M, Muthuraman G 2015 Recovery of methylene blue from aqueous solution by liquid-liquid extraction, *Desalin. Water Treat.* **53** 2501-9.

[6] Othman N, Fatiha N, Noah M, Zing O, Bukhari M, Rahman H A, Rashid R 2017 Kinetic extraction of basic dye using vegetable oil as a solvent, *Mal. J. Fund. Appl. Sci.* **13** 685-9.

[7] Wahab A A A, Chang S H, Som A M 2016 Stoichiometry of Cu (II) ion extraction with di-2-ethylhexylphosphoric acid dissolved in waste palm cooking oil, *Int. J. Technol.* **7(5)** 778-5.

IOP Conf. Series: Materials Science and Engineering **811** (2020) 012002 doi:10.1088/1757-899X/811/1/012002

Effect of Cyclic Heat Treatment Process on the Localized Corrosion Resistance of 409 Ferritic, 430 Ferritic, 3163 Austenitic and 444 Ferritic Stainless Steels for High Temperature Applications

Roland T. Loto[1]

[1]Department of Mechanical Engineering, Covenant University, Ogun state, Nigeria

tolu.loto@gmail.com

Abstract. The corrosion resistance of as-received 409 ferritic, 430 ferritic, 444 ferritic and 316 austenitic stainless steels was studied in 0.05M H_2SO_4/3.5% NaCl solution by potentiodynamic polarization and potentiostatic measurement and compared to their counterparts subjected to repetitive temperature variation at 1000°C. As-received 409 ferritic steel exhibited the highest corrosion rate at 8.406 mm/y compared to as-received 316 austenitic steel with the lowest corrosion rate at 3.130 mm/y. Corrosion rate of heat treated 409, 430 and 444 ferritic stainless steels significantly increased to 10.766 mm/y, 5.694 mm/y and 6.096mm/y while the corrosion rate of as-received 316 austenitic steel decreased to decreased to 0.409 mm/y. As-received 444 ferritic steel exhibited the weakest resistance to pitting corrosion while as-received 409 ferritic steel exhibited the most significant metastable pitting activity. Pitting corrosion susceptibility of heat treated 409 ferritic, 316 austenitic and 444 ferritic steel decreased significantly to 0.137 V, 0.384 V and 0.096 V while heat treated 430 ferritic steel exhibited significant improvement in pitting corrosion resistance.

1. Introduction

Vehicular exhaust system is made up of piping linkages responsible for the transport of hot reaction fluids from the combustion section of the engine to the exhaust manifold and through the exhaust muffler to the tail end of the exhaust pipe [1]. Corrosion of the exhaust system results in breakdown of the metallic parts causing leakage of internal parts, components, vibration, air pollution and noise issues. The operating conditions of exhaust systems in addition to liquefied corrosive ions in the exhaust fluid requires strong precautions in material selection for the manufacture of exhaust system parts and components to ensure optimum performance and long operating lifespan. Stainless steels are used in the manufacture of exhaust mufflers due to their proven excellent corrosion resistance, heat resistance and aesthetic properties. However, corrosion constitutes the major problem for the accelerated degradation of exhaust mufflers and other components of the exhaust system leading to loss of operating performance [2]. Carbon steels are mostly used in the manufacture exhaust systems. However, their weak corrosion resistance is responsible for their short operating lifespan. Aluminised carbon steel is more applicable than carbon steel with strong pitting resistance, but is weak under high temperature operating conditions. Specialty stainless steels offer higher performance but are generally costly [3]. 409 ferritic stainless steels has become the conventional replacement for carbon steel grades. However, they have low coefficient of thermal expansion, coarsening at welding temperatures and are susceptible to localized corrosion deteriorations [4]. Some grades of stainless steels are capable of

IOP Conf. Series: Materials Science and Engineering **811** (2020) 012002 doi:10.1088/1757-899X/811/1/012002

withstanding the deteriorating actions of corrosive anions under high temperature. However, there are cases where corrosion resistance of the steel is enhanced, decreases or remain unchanged [5-7]. Previous research on the passive film stability and pitting corrosion of S40900 stainless steel in 0.1 M NaCl and chloride/sulphate solutions show the passive film presents n-type semi conductive behavior with grain boundaries playing an important role as pitting initiation sites. Results also showed S40900 corrosion rate is proportional to the concentration of SO_4^{2-} and Cl^- anions in the electrolyte [8, 9]. 304L steel showed no passivation and resistance to pitting after 0% NaCl concentration coupled with increase in corrosion rate [10]. This research focuses on the effect of repetitive high temperature conditions on the corrosion resistance of 430 ferritic, 444 ferritic and 316 austenitic stainless steel in sulphate-chloride solution in comparison to the corrosion resistance of the conventional 409 ferritic stainless steels.

2. Material and Methods

As-received 409 ferritic stainless steel (409AS) was sourced from a vehicle exhaust muffler. 430 ferritic stainless steel (430AS), 316 austenitic stainless steel (316AS) and 444 ferritic stainless steel (444AS) were sourced from Vienna, Austria. The elemental compositions of the steels were confirmed with PhenomWorld scanning electron microscope (Model No. MVE0224651193). One test for each stainless steel specimen were cut and mounted in resin mounts with visible surface areas of 1 cm^2 after which the exposed surfaces of the steels were metallographically prepared. 800 mL of 0.05 M H_2SO_4 solution at 3.5% NaCl concentration was formulated from standard grade concentration of the acid (98% purity) and NaCl (recrystallized). A second set of prepared stainless steel specimens (409HT, 430HT, 316HT and 444HT) were subjected to heat treatment within a muffle furnace at 1000°C for 30 minutes and subsequently cooled to room temperature. Electrochemical test of 409AT, 430AT, 316AT and 444AT stainless steels was done with Digi-Ivy potentiostat at 30 °C ambient temperatures interfaced with a computer. Polarization curves were plotted at scan rate of 0.0015 V/s from -1.5 V to +1.5 V.

3. Results and discussion

3.1 Potentiodynamic polarization studies

Potentiodynamic polarization data for as-received and heat treated stainless steel corrosion in 0.05M H_2SO_4/3.5% NaCl concentration are shown in table 1. 409AS exhibited the highest corrosion rate value of 8.406 mm/y among the as-received steels while 316AS had the lowest corrosion rate at 1.581 mm/y. 444AS and 430AS showed less susceptibility to H_2 embrittlement from observation of its cathodic Tafel slope. The relatively high anodic Tafel slope of 430ST confirms it is susceptibility to anodic degradation. Data for cathodic Tafel slopes of the as-received steel shows the reduction mechanisms is under activation control. However, the anodic reaction mechanisms control the redox reaction mechanism based on the anodic Tafel slope values. Heat treatment significantly increased the corrosion resistance of the alloy specimens. 409HT exhibited a corrosion rate value of 10.766 mm/y. Corrosion rate of 430HT and 444HT increased to 5.694 mm/y and 6.096 mm/y. Difference in corrosion potential and corrosion rate between the as-received and heat treated 430 ferritic steel confirms the deteriorating effect of heat treatment. The corrosion potential difference between as-received and heat treated 409 and 444 ferritic steels from -0.499 V and -0.454 V to -0.502 V coupled in addition to increased corrosion rate also shows heat treatment enhances the initiation and growth of localized corrosion reactions. Heat treatment increased the corrosion resistance of 316 austenitic steel from 1.581 mm/y to 0.490 mm/y while its corrosion potential shifted from -0.411 V to -0.323 V.

IOP Conf. Series: Materials Science and Engineering **811** (2020) 012002 doi:10.1088/1757-899X/811/1/012002

Table 1. Polarization data for as-received and heat-treated 409 ferritic, 430 ferritic, 316 austenitic and 444 ferritic stainless steel corrosion in 0.05 M H_2SO_4/3.5% NaCl solution.

A-received Steels							
Steel	**Corrosion Rate (mm/y)**	**Corrosion Current (A)**	**Corrosion Current Density (A/cm^2)**	**Corrosion Potential (V)**	**Polarization Resistance, R_p (Ω)**	**Cathodic Potential, B_c (V/dec)**	**Anodic Potential, B_a (V/dec)**
409ST	8.406	7.82E-04	7.82E-04	-0.499	32.8	-6.109	-1.211
430ST	3.504	3.26E-04	3.26E-04	-0.502	78.8	-6.445	4.976
316ST	1.581	1.51E-04	1.51E-04	-0.411	170.1	-6.060	7.317
444ST	3.130	2.95E-04	2.95E-04	-0.454	87.1	-6.250	4.550
Heat Treated Steels							
Steel	**Corrosion Rate (mm/y)**	**Corrosion Current (A)**	**Corrosion Current Density (A/cm^2)**	**Corrosion Potential (V)**	**Polarization Resistance, R_p (Ω)**	**Cathodic Potential, B_c (V/dec)**	**Anodic Potential, B_a (V/dec)**
409ST	10.766	1.00E-03	1.00E-03	-0.502	25.6	-6.798	-1.193
430ST	5.694	5.30E-04	5.30E-04	-0.463	48.5	-6.537	1.050
316ST	0.490	4.68E-05	4.68E-05	-0.323	1169.0	-7.535	11.070
444ST	6.096	5.75E-04	5.75E-04	-0.502	44.7	-6.008	0.252

3.2 Potentiostatic studies

The metastable pitting and stable pitting graphical portions for as-received and heat treated 409 ferritic, 430 ferritic, 316 austenitic and 444 ferritic steel are shown from figure 1(a) to figure 2(b). Table 2 shows the potentiostatic data of the steels obtained. 409AS displayed the widest metastable pitting portion for the as-received steels [11]. The metastable pitting current of 409AS at onset is 2.58 x 10^{-3} A which is greater than the values obtained for 430AS, 316AS and 444AS. 430AS metastable pitting current initiated at 2.14 x 10^{-3} A and passivated at 5.91 x 10^{-5} A compared to 409AS which passivated at 2.90 x 10^{-5} A. 316AS and 444AS underwent early metastable pitting at 4.68 x 10^{-4} A and 3.60 x 10^{-4} A, and passivated at 3.97 x 10^{-4} A and 6.10 x 10^{-5} A. The passivation range and pitting potential of 316ST are 0.145 V and -0.031 V due to weak resistance to localized corrosion deterioration. 444AS and 409AS had passivation range values of 0.592 V and 0.296 V, and pitting corrosion potential of 0.087 V and 0.296 V due to strong resistance to localized corrosion compared to 430ST and 409ST. The passivation range of 409HT, 316HT and 444HT (heat treated steels) [table 2] shifted from 0.296 V, 0.145 V and 0.592 V to 0.137 V, 0.384 V and 0.096 V due to decreased resistance to localized corrosion [figure 1(b)]. 444HT is relatively more susceptible to high temperature variation with due to differences in its corrosion rate and visible decrease in passivation plot. Absence of metastable pitting activity on the plot of 316HT in addition to significantly lower corrosion rate shows it is applicable for high temperature application. The steel passivated at -0.210 V (1.68 x 10^{-4} A). 430HT exhibited passivation range of 0.384 V making it applicable vehicular mufflers. The short metastable pitting plot of 430HT initiated at 7.88 x10^{-4} A and passivated at 1.15 x 10^{-4} A is due to resistance to oxidation. The pitting potential values of 409HT, 316HT and 444HT shows decrease to localized corrosion.

Table 2. Potentiostatic data for untreated and heat treated 409ST, 430ST, 316ST and 444ST corrosion in 0.05 M H_2SO_4/3.5% NaCl solution.

As-received Steel					
Steel	Metastable Pitting Current (A)	Passivation Current (A)	Passivation Potential (V)	Pitting Potential (V)	Passivation Range (V)
409	2.58E-03	2.90E-05	-0.209	0.087	0.296
430	2.14E-03	5.91E-05	-0.214	-0.038	0.176
316	4.68E-04	3.97E-04	-0.176	-0.031	0.145
444	3.60E-04	6.10E-05	-0.332	0.260	0.592
Heat Treated Steel					
Steel	Metastable Pitting Current (A)	Passivation Current (A)	Passivation Potential (V)	Pitting Potential (V)	Passivation Range (V)
409	3.24E-03	1.07E-04	-0.297	-0.160	0.137
430	7.88E-04	1.15E-04	-0.326	0.058	0.384
316	-	1.68E-04	-0.210	-0.174	0.036
444	2.66E-03	1.66E-03	-0.294	-0.198	0.096

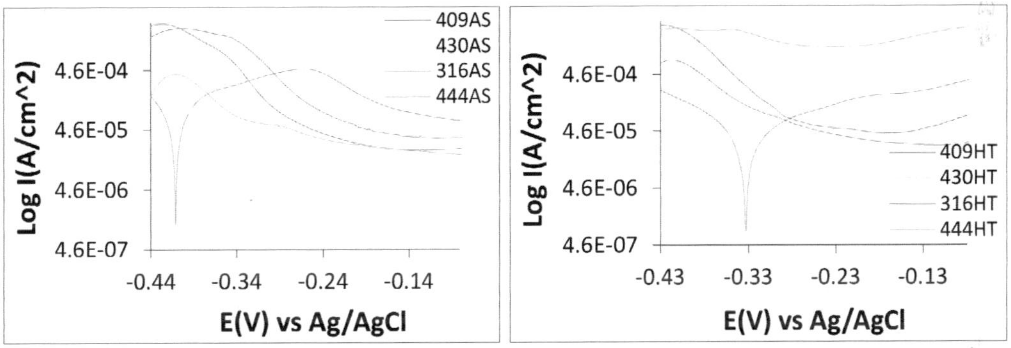

(a) (b)

Figure 1. Metastable pitting plot of (a) as-received (b) heat treated 409 ferritic, 430 ferritic, 316 austenitic and 444 ferritic stainless steel corrosion.

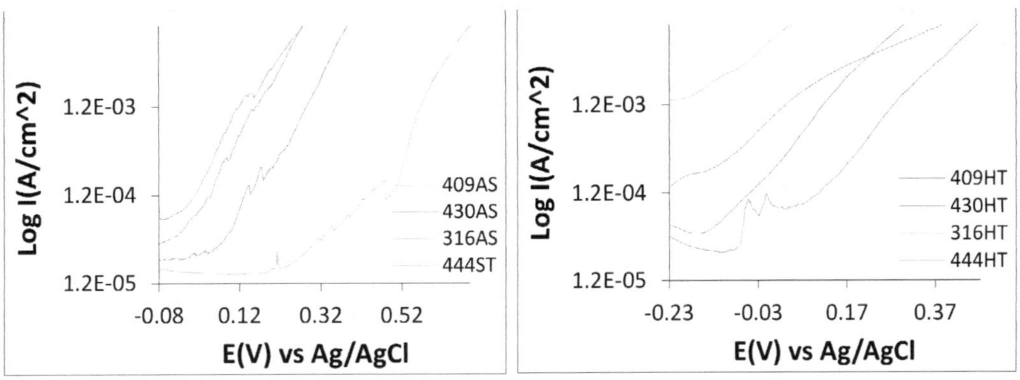

(a) (b)

Figure 2. Stable pitting plot of (a) as-received (b) heat treated 409 ferritic, 430 ferritic, 316 austenitic and 444 ferritic stainless steel corrosion.

4. Conclusion

Repetitive temperature variation significantly influenced the corrosion resistance of as received and heat-treated 430 ferritic, 316 austenitic and 444 ferritic stainless steel when compared to the corrosion resistance of 409 ferritic steel in chloride-sulphate solution. Corrosion rate of as-received 430 ferritic, 316 austenitic and 444 ferritic steel were significantly lower than the value for as-received 409 ferritic steel. Application of heat treatment processes increased the corrosion rate of the heat-treated stainless steel counterparts with heat-treated 409 ferritic steel having the highest value. As-received 444 ferritic steel exhibited the lowest vulnerability to pitting corrosion compared to as-received 409 steel. The pitting resistance of heat-treated 409 ferritic, 316 austenitic and 444 ferritic were significantly lower than their as-received counterparts while 430 ferritic steel showed significant improvement in pitting corrosion resistance.

Acknowledgement
The author is grateful to Covenant University Ota, Ogun State, Nigeria for the sponsorship and provision of research facilities for this project.

References

[1] Ikpe A E, Abdulsamad G 2017 Engineering material selection for automotive exhaust systems using CES software, *Int. J. Eng. Sci. Technol.* **3(2)** 50.

[2] Hoffmann C, Gümpel P 2009 Pitting corrosion in the wet section of the automotive exhaust systems, *J. Achiev. Mater. Manuf. Eng.* **34** 115.

[3] Heck R M, Farrauto R J 2001 Automobile exhaust catalysts, *Appl. Catal. A: Gen.* **221** 443.

[4] Han P H, Xu Z H, Wang C P, Li M C, Bi H Y 2014 Condensate corrosion behavior of Type 409 stainless steel in simulated automotive muffler environments, *Int. J. Electrochem. Sci.* **9** 3784.

[5] Loto R T, Aiguwurhuo O, Evana U 2016 Corrosion resistance study of heat-treated 420 martensitic stainless steel and 316 austenitic stainless steel in dilute acid concentrations, *Rev. Téc. Ing. Univ. Zulia* **39(7)** 35.

[6] Loto R T, Loto C A, Ohijeagbon I 2018 Effect of heat treatment processes on the localized corrosion resistance of austenitic stainless steel type 301 in chloride/sulphate solution, *Results Phys.* **11** 570.

[7] Loto R T 2019 Electrochemical data on the effect of extreme temperatures on 2101 duplex steel corrosion performance in chloride-sulphate environment, *Orient. J. Chem.* **35(2)** 531.

[8] De Souza J S, De Oliveira L A, Sayeg I J, Antunes R A 2017 Electrochemical study of the AISI 409 ferritic stainless steel: Passive film stability and pitting nucleation and growth, *Mat. Res.* **20(6)**.

[9] Loto R T 2019 Effect of SO_4^{2-} and Cl^- anionic attack on the localized corrosion resistance and morphology of 409 ferritic stainless steel, *Results Phys.* **12** 738.

[10] Loto R T 2017 Study of the corrosion resistance of type 304L and 316 austenitic stainless steels in acid chloride solution, *Orient. J. Chem.* **33(3)** 1090.

[11] AlHazaa A N, El-Sayed M S, Hany S A 2015 Galvanic corrosion in 3.5 wt% NaCl solutions of magnesium alloy AZ31 coupled with Ni after different bonding periods of time, *Int. J. Electrochem. Sci.* **10** 5420.

IOP Conf. Series: Materials Science and Engineering **811** (2020) 012003 doi:10.1088/1757-899X/811/1/012003

Thermogravimetric Analysis and Pyrolysis Kinetics of Malaysian Wood Species

Sulaiha Ali[1, a], Siti Aslina Hussain[1, 2, *], Mohd. Zahirasri Mohd Tohir[1, 2, c], Ahmad Ainuddin Nuruddin[3, d]

[1]Department of Chemical and Environmental Engineering, Faculty of Engineering, Universiti Putra Malaysia, 43400 UPM Serdang, Selangor, Malaysia
[2]Safety Engineering Interest Group (SEIG), Department of Chemical and Environmental Engineering, Faculty of Engineering, Universiti Putra Malaysia, 43400 UPM Serdang, Selangor, Malaysia
[3]Institute of Tropical Forestry and Forest Products (INTROP), Universiti Putra Malaysia, 43400 UPM Serdang, Selangor, Malaysia

[a]sulaihaali2013@gmail.com, [b]aslina@upm.edu.my, [c]zahirasri@upm.edu.my, [d]ainuddin@upm.edu.my

Abstract: The mass loss of three wood species in Malaysia, Shorea laevis, Koompassia malaccensis, and Shorea parvifolia were measured against time and temperature at three different heating rates (5, 10 and 20 °C/min) in nitrogen environment, through thermogravimetry analysis (TGA) technique. This measurement revealed that the decomposition of these wood generally occurred in three stages, the dehydration stage, the active stage, and; the passive stage. The maximum decomposition peak in derivative thermogravimetric (DTG) analysis curve shifted towards higher temperature range with the expedition in heating rate. The kinetic decomposition parameters of the timbers were determined by adopting thermal kinetic method of Coats and Redfern. Principally, the density and moisture content of timber does not influence the kinetic parameters studied in this work. Overall, the thermal decomposition in the second-stage (active stage) was dominated by the hemicellulose and cellulose while the decomposition of lignin was apparent in the third-stage (passive stage) of pyrolysis. The higher activation energy in the second-stage indicates that more cellulose is contained in the sample. The activation energy of lignin is observed to be much lesser compared to the activation energy of the hemicellulose and cellulose for two species tested except for the Shorea parvifolia.

1. Introduction

The wood industry is one of the significant contributors to Malaysia's economy, wherein 2018, the timber industry recorded RM22.3 billion in exports, contributing 1.6% to the country's gross domestic product and 2.2% to total exports [1]. Due to their qualities, Malaysian wood is highly favored to be used as materials for furniture such as sofa, dining table, and cabinet. However, timber is considered as combustible material, which these surfaces may notably add to the fire loading of a compartment, promote the spread of flame, and speed up the outbreak of flashover. The thermogravimetric analysis (TGA) is the simplest way of investigating the thermal property of degradation for wood as it provides quantitative information on the decomposition of polymeric substance as it only requires a small amount of sample, which is less in cost. Hostikka et al. [2] evaluated the performance of three reaction schemes of wood pyrolysis in reproducing thermogravimetric experiments of birch wood and concluded that the first-order single-step reaction scheme provides good prediction for the heat release rate. Sharma et al. [3] studied the thermal decomposition and the reaction mechanism of the eucalyptus wood and revealed that the reaction mechanism of wood pyrolysis overlaps the diffusion mechanism at lower conversion value (α = 0.05- 0.4). Lu et al. [4] examined the pyrolysis characteristics of raw/torrefied C. japonica and anthracite coal and found that the pyrolysis behavior of fuel blends can be determined in terms of the weight percentages of biomass and coal. The results obtained from TGA conversion curves are essential to develop kinetics models that are further used for completing the fire prediction models of the materials. Although many works have been done on fire properties of wood, the archival material related to the wood in Malaysia was found to be very limited in scope and incomplete [5]. Therefore, there is a reasonable need for the study of thermal degradations and pyrolysis kinetics of the local wood species in Malaysia. Thus, this work investigates the thermal degradations of three selected Malaysian solid wood material based on their application indoor and outdoor. Furthermore, the decomposition kinetic parameters are also determined by adopting thermal kinetics method, the Coats and Redfern (CR)[6].

2. Materials and methods

2.1. Samples preparation

Three Malaysian timber species were selected namely the Shorea laevis(Balau), the Koompassia malaccensis (Kempas) and the Shorea parvifolia(Light red meranti) as these species were listed by the Malaysian Timber Council (MTC) as 20 Malaysian commercial timbers based on the application, demand, and availability [7]. The specimens were prepared to be in fine fragment form in order to conduct the analysis.

2.2. Thermogravimetric analysis (TGA) technique

The thermogravimetric analysis of the three Malaysian timber species was performed using the Mettler Toledo equipment in Nitrogen atmosphere with the flow rate of 50 ml/min from 30 to 900°C. The samples were put on the platinum pan, then placed on a ceramic crucible and heated under three different heating rates of 5, 10 and 20°C/min. The mass loss of wood samples was recorded as a function of temperature and analyzed with STARe SW 12.10 thermal analysis software.

2.3 Kinetics theory

The reasons for the kinetics analysis of wood pyrolysis were eventually to evaluate the activation energy, the pre-exponential factor, and the reaction mechanism. The kinetics parameters in this study were determined by embracing the CR method, assuming n = 1 and the value of 2RT/E is minimal which can be expressed as:

$$ln\left[\frac{-ln(1-X)}{T^2}\right] = ln\frac{AR}{\beta E_a} - \frac{E_a}{RT} \quad (1)$$

Where X is the fuel conversions, T is the absolute temperature in K, A is the pre-exponential factor, E_a is the activation energy, and R is the gas constant (8.314 J/K mol). The plot of $\ln[-\ln(1-X)/T^2]$ versus 1/T becomes a linear line where the E_a and the A are to be determined from the slope and the intercept of the regression line, respectively.

3. Results and discussions

The thermal degradations and the pyrolysis kinetics characteristic for three species of Malaysian wood were evaluated in this work. The characteristic reaction temperatures, devolatilization rates, and mass fractions were used to describe the TGA and DTG analysis curves at various heating rates. For the determination of kinetics parameters, the first-order reactions of the CR method were calculated, and the discussion of each result is provided in the following sections.

3.1 Thermal decompositions of Malaysian wood species

3.1.1 Thermal decomposition of heavy hardwood: Shorea laevis. The mass fraction and the time derivative of the mass fraction showed as TGA and DTG curves for Shorea laevis were as displayed in Figure 1 (a), (b), and (c). In concurrence with previous findings [8], generally, the DTG curves show two primary regions. The determination of wood pyrolysis stage is based on the mechanism of superposition of the individual wood structure components; starting with the decomposition of hemicellulose at 180-350°C (approximate range for first-stage of wood pyrolysis), followed by the decomposition of cellulose at 275-350°C (approximate range of second-stage of wood pyrolysis), and lastly the decomposition of lignin at 250-500°C (approximate range of third-stage of wood pyrolysis) [3, 4]. In this work, the Shorea laevis encountered the loss of moisture content at ~100 - 240°C. The main decompositions mostly take place at the second-stage of ~ 200°C-440°C due to the depletion of the cellulose [9]. The peak temperature at which the decomposition intensity is the highest or where the primary degradation occurred were at 358 (intensity rate = 0.9435 wt%/°C), 367 (intensity rate = 0.17 35 wt%/°C) and 378 °C (intensity rate = 0.35 wt%/°C) due to heating rates of 5, 10 and 20°C/min respectively. It is also observed that the higher the heating rate, the higher the decomposition intensity rate and the peak shifted to higher temperature. The decomposition of lignin occurred at the third stage of pyrolysis, at ~390 to 600 °C without showing any characteristic peak.

3.1.2 Thermal decomposition of medium hardwood: Koompassia malaccensis. The TGA and DTG curves of the Koompassia malaccensis were as disclosed in Figure 1 (d), (e) and (f). The decompositions of the Koompassia malaccensis was paramount at the second-stage (~200°C-480°C) due to the depletion of cellulose. The peak temperatures for Kempas were 341 (intensity rate:0.0896 wt%/°C), 363 (intensity rate:0.1755 wt%/°C) and 362 °C (intensity rate:0.3378 wt%/°C) at 5, 10 and 20 °C/min heating rates respectively. The decomposition of lignin samples demonstrated a volatile evolution due to decomposition of lignin at the temperature of 573 °C/min at the heating rate of 5°C/min.

3.1.3 Thermal decomposition of light hardwood: Shorea parvifolia. The TGA and DTG curves of the Shorea parvifolia were as exhibited in Figure 1 (g) (h) and (i). At the second-stage (~220 – 500 °C), this species demonstrated its peak temperature of cellulose decomposition particularly at 354 (intensity rate: 0.11wt%/°C), 374 (intensity rate: 0.20wt%/°C) and 385 °C (intensity rate: 0.38wt%/°C) under the heating rates of 5,10 and 20°C/min. It is noted that the expedition of heating rates accompanied the acceleration of peak temperatures. The third-stage of Shorea parvifolia pyrolysis does not indicate characteristic peak at all three different heating rates tested.

3.2 Pyrolysis Kinetics

In this work, the E_a and the A of all species were determined by adopting the iso-conversional kinetics method of Coats and Redfern (CR). The results were as tabulated in Table 1. The values of R^2 were found to be between 0.81 to 0.99, which reflects that the pyrolysis processes are well correlated whether in the second or third-stage.

During the second-stage of pyrolysis, only the E_a of the Koompassia malaccensis increased from 30 to 49.05 kJmol^{-1} as the heating rate accelerates from 5 to 10°C/min. At this stage, the Shorea parvifolia has the highest E_a of 62kJmol^{-1} when heated at the rate of 10°C/min and lowest E_a of 27.26kJmol^{-1} at the heating rate of 5°C/min. In the third-stage of pyrolysis, the E_a of the Shorea laevis and Koompassia malaccensis seems to decrease as the heating rates increased. The Shorea laevis exhibited lowest E_a of 8.97kJmol^{-1} while the Shorea parvifolia has the highest E_a of 95.59 kJmol^{-1}, both at the heating rate of 20°C/min. The assessment of A has a secure connection to the E_a; the higher the E_a indicates higher value of A.

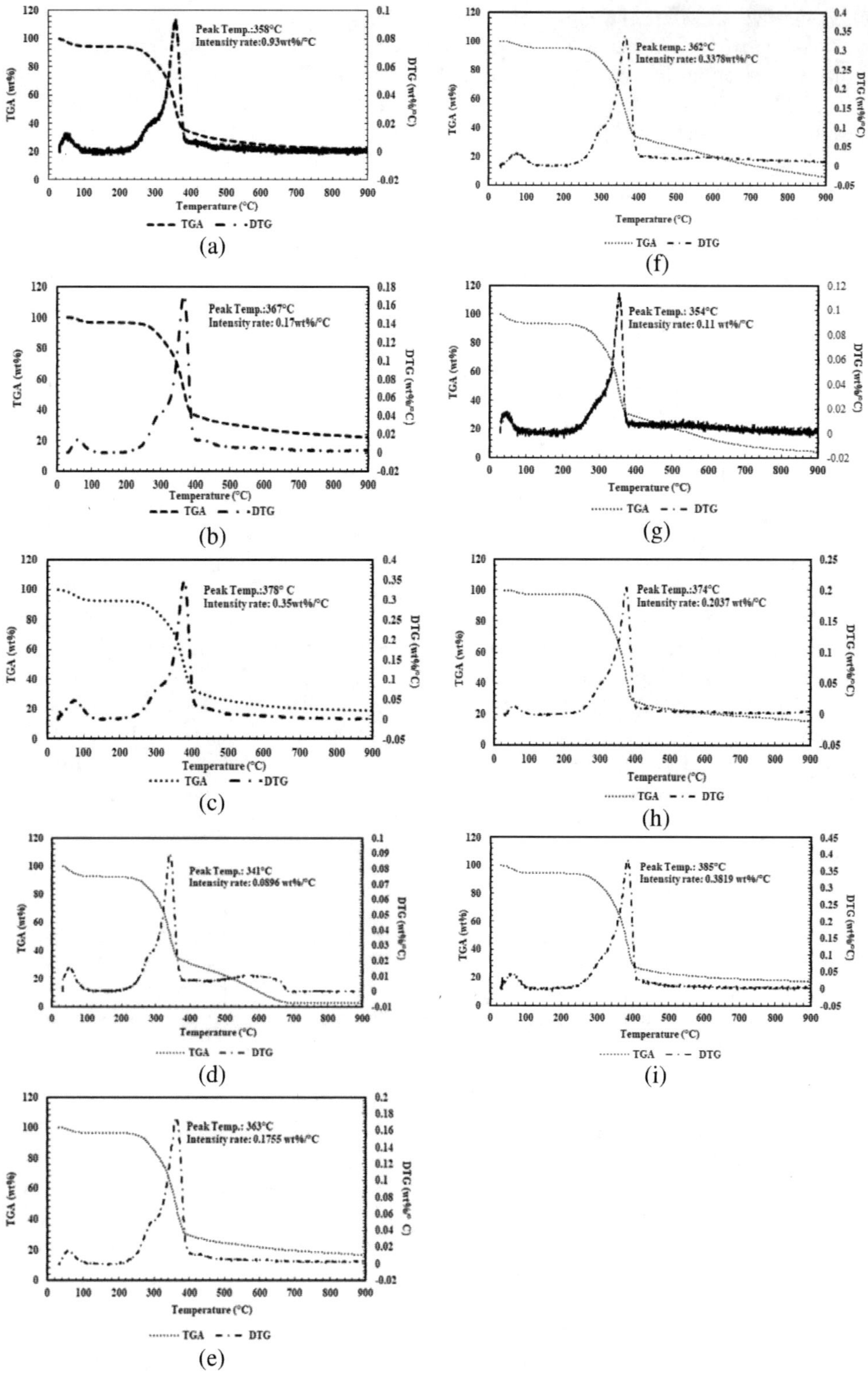

Figure 1. The mass fraction (TGA) and the time derivative of the mass fraction (DTG) as functions of temperature for Shorea laevis, Koompassia malaccensis, and Shorea parvifolia at 5 [(a), (d), (g)], 10 [(b),(e),(h)] and 20[(c), (f), (i)]°C/min heating rates.

IOP Conf. Series: Materials Science and Engineering **811** (2020) 012003 doi:10.1088/1757-899X/811/1/012003

Table 1. Thermal degradations and chemical kinetics of Malaysian wood at the heating rates of 5, 10, and 20°C/min from 30 to 900°C.

Row	Species		Shorea laevis			Koompassia malaccensis			Shorea parvifolia		
1	Classification		Heavy hardwood			Medium hardwood			Light hardwood		
2	Density	kg/m3	905			862			440		
3	Moisture content	%	14			16			14		
4	Heating rate	°C//min	5	10	20	5	10	20	5	10	20
5	Temp. at the 1st peak	°C	358	367	378	341	363	362	354	374	385
6	Temperature at the 2nd peak *	°C	-	-	-	573	-	-	-	-	-
7	Peak MLR	wt%/°C)	0.94	0.17	0.35	0.09	0.18	0.33	0.11	0.20	0.38
Second-stage reaction											
8	E_a	kJ mol^{-1}	41.28	49.11	31.41	30	45.54	49.05	27.26	63	45.06
9	A	s^{-1}	$4.89x10^{3}$	$1.85x10^{-2}$	$6.22x10^{-4}$	$4.8x10^{-4}$	$9.78x10^{-3}$	$1.62x10^{-2}$	$2.19x10^{-4}$	$2.5x10^{-1}$	$7.15x10^{-3}$
10	R^2		0.93	0.96	0.90	0.86	0.92	0.96	0.85	0.98	0.93
Third-stage reaction											
11	E_a	kJ mol^{-1}	17.22	16.55	8.97	26.68	14.86	12.87	57.51	46.95	95.59
12	A	s^{-1}	$9.2x10^{-5}$	$7.23x10^{-5}$	$4.75x10^{-6}$	$4.2x10^{-4}$	$5.89x10^{-5}$	$2.73x10^{-5}$	$1.36x10^{-1}$	$1.95x10^{-2}$	$1.13x10^{2}$
13	R^2		0.81	0.82	0.93	0.91	0.8	0.81	0.86	0.87	0.99

Notes:
* = Second peak temperature due to volatile evolution of extractive at the third-stage of pyrolysis.

4. Conclusions

The pyrolysis of Malaysian wood; (1) BalauShorea laevis (Balau); (2) Koompassia malaccensis (Kempas); and (3) Shorea parvifolia (Light red meranti) have been investigated through TGA technique in an inert atmosphere of nitrogen gas at three different heating rates. The results disclosed that the maximum decomposition temperature peaks shifted towards higher temperature when the heating rates of the wood samples accelerates. Further, the activation energies were determined by adopting the Coats and Redfern (CR) method, assuming n = 1. Generally, the density and moisture content does not influence the kinetics parameter studied in this work. Overall, the thermal decomposition in the second-stage was dominated by the hemicellulose and cellulose while the lignin was eminent in the third-stage. The higher activation energy in the second-stage indicates more cellulose is contained in the sample. From the results obtained, it can be said that in this work, the activation energy of lignin is much less than the hemicellulose and cellulose except for the Shorea parvifolia. The present study is useful to understand the complete degradation process and the pyrolysis kinetics involved in the thermal degradation of the Malaysian wood as indoor and outdoor furniture material.

Acknowledgment

The authors thank the University of Putra Malaysia and; Department of Chemical and Environmental Engineering for the project grant sponsor GP/2017/9563900 and GP-IPS/2017/9573600. A special appreciation to the Director-General of Malaysian Timber Industry Board (MTIB) and the officers for advising in material selection and sponsoring the material. As this work is the pioneering work for the timber fire properties in Malaysia, we are looking forward to further research in this field to optimize the utilization of Malaysia timber holistically.

IOP Conf. Series: Materials Science and Engineering **811** (2020) 012003 doi:10.1088/1757-899X/811/1/012003

References

[1] Safri, A. Malaysia's timber exports hit RM9.1b in first five months of 2019. *The Sun Daily*, July 25, 2019. https://www.thesundaily.my/business/malaysia-s-timber-exports-hit-rm9-1b-in-first-five-months-of-2019-CD1151848 (accessed July 25, 2019).

[2] Hostikka, S.; Matala, A. Pyrolysis Model for Predicting The Heat ReleaseRrate of Birch Wood. *Combust. Sci. Technol.* **2017**, 189, 1373–1393. DOI: 10.1080/00102202.2017.1295959.

[3] Sharma, P.; Diwan, P.K. Study of Thermal Decomposition Process and The Reaction Mechanism of The Eucalyptus Wood. *Wood Sci. Technol.* **2017,** 51, 1081–1094. DOI: 10.1007/s00226-017-0924-7.

[4] Lu, K.M.; Lee, W.J.; Chen, W.H.; Lin, T.C. Thermogravimetric Analysis and Kinetics of Co-pyrolysis of Raw/Torrefied Wood and Coal Blends. *Appl. Energy*. **2013**, 105, 57–65. DOI: 10.1016/j.apenergy.2012.12.050.

[5] Wong, W. S. Timber Structures in Malaysian Architecture and Buildings. Ph.D. Dissertation, University of Tasmania, Launceston, Australia, 1995.

[6] Coats, A.W.; Redfern, J.P. Kinetic Parameters From Thermogravimetric Data. *Nature*. **1964**, 201, 68–69. DOI: 10.1038/201068a0.

[7] Malaysia Timber Council. Commercials Malaysian Timbers; Malaysia Timber Council: Kuala Lumpur, MY, 2016.

[8] Antal, M.J.; Varhegyi, G. Cellulose Pyrolysis Kinetic: The Current State of Knowledge. *Ind. Eng. Chem. Res.* **1995**, 34, 703–717. DOI: 10.1021/ie00042a001.

[9] Gašparoviè, L.; Labovský, J.; Markoš, J. Calculation of Kinetic Parameters of The Thermal Decomposition of Wood by Distributed Activation Energy Model (DAEM). Chem. Biochem. Eng. Q. 2012, 26, 45–53.

IOP Conf. Series: Materials Science and Engineering **811** (2020) 012004 doi:10.1088/1757-899X/811/1/012004

Analysis of Circuit Design of a Rowing Electric Installation with a Distributed DC Bus of a Port Icebreaker

V V Romanovskiy[1], A S Bezhik[1]

[1]Admiral Makarov State University of Maritime and Inland Shipping, 5/7 Dvinskaya Str., Saint-Petersburg, 198035, Russia

Abstract. Currently, there is a steady trend in increasing the volume of electrical equipment used in shipboard needs. At the same time, the primary electric power system of ships has long been based on an alternating current system. However, due to the changing and increasing requirements for the quality of the generated electric energy, the marine industry is developing more and more promising directions for the development of electrical systems. Thus, the most efficient and environmentally friendly shipboard power distribution system is the direct current (DC) system, which was adopted for a single power plant installation in combination with an alternating current network. This article discusses the current topic of electric propulsion on ships. The circuit design of a rowing electric installation with a distributed DC bus of a port icebreaker is considered. The expressions used to calculate the static load and build an electromagnetic model of a permanent magnet synchronous motor are given. Computer models for calculating the load in static and dynamic modes are shown.

1. Introduction

Recently, there has been a steady trend in increasing the volume of electrical equipment used on board the vessel. As a result, there are growing demands on the quality of the generated electric energy [1]. Shipbuilding companies are faced with urgent tasks, the electric power systems installed by them must ensure the necessary quality of the generated electric energy [2], be reliable, including have an acceptable combination of the cost of its installation and operation with acceptable weight and size parameters [3]. In this case, the technical and economic indicators are determined mainly by a rational choice of the voltage value and the nature of the current [4].

Thus, the most efficient and environmentally friendly shipboard power distribution system is the direct current (DC) system, which was adopted for a single power plant installation in combination with an alternating current network. When it was created, more attention was paid to variable speed generators, the use of special energy storage devices, space saving, and increasing the plant's performance in static and dynamic modes [5].

2. Methods and Materials

In the system under consideration, almost all energy is generated by generators with diesel primary engines. The use of an electric power plant, where the power transmission, as well as the drive and traction motors are variable speed electric drives, can reduce fuel consumption on some types of ships [6]. But in some special operating conditions, such as dynamic positioning, the load changes significantly [7], for example, with the influence of sea waves on the hull of the vessel and the influence of weather. A sudden change in load represents continuous distortion in the electrical system and in prime movers. In addition, in order to maintain safe operation limits of power plants, it is necessary to keep the average load of working drives low, which increases fuel consumption and emissions of harmful substances into the environment. High-speed energy storage systems can solve these problems by effectively reducing fluctuations in the load power in the power system [8], smoothing out sudden changes in energy consumption, improving system stability and increasing the average power of energy produced with fewer running generators [9], and therefore, reducing fuel consumption and maintenance.

The concept of using a rowing electric installation (REI) with a distributed DC bus (DDCB) has found application in the port icebreaker, which is designed to provide navigation of vessels on approach channels and in port water areas with an ice cover of up to 1.5 meters and has the parameters shown in the table 1.

Table 1. Main dimensions and characteristics of the port icebreaker.

Name of the parameter	Value
Full length	89,2 m
Waterline length	77,6 m
Waterline width	19,9 m
Maximum draft	7,5 m
Deadweight at max. draft	2000 t
Main DG	3 * 4880 kW
Harbor DG	800 kW
REI	4 × 3000kW electric motor
Ice penetration	2 knots at 1.5 m in flat ice
Mooring emphasis	115 t
Speed in clear water	15 узлов
Ice class	PMPC- Icebreaker 7, or MAKO- PC3

A distinctive feature of the vessel is a propulsion system consisting of four independent azimuthal propulsion systems with an electric drive Azipod ABB Ice-1400 traction type with a capacity of about 3 MW each. The helical-steering columns (HSC) are located in pairs in the bow and stern of the vessel, which makes it possible to work as efficiently as possible in the ice with both stern and bow, maneuvering and performing special tasks in the waters of any port [10].

The main current circuit shown in figure 1 includes:

- three main synchronous generators with a capacity of 4730 kW, adjustable 750 rpm, adjustable output voltage of synchronous generators 810 V;
- three controlled rectifiers, converting 810 V AC to adjustable 890 V DC;
- four frequency inverters based on ACS800 (ABB company) convert a constant voltage into an adjustable voltage within 0 ÷ 660 V of an alternating voltage for controlling Azipod's systems;
- four HSC of the Azipod ICE 1400 3000 kW type incorporate permanent magnet motors, which greatly simplify the maintenance of the HSC;
- two transformers with a power of 1500 kVA 610V / 400V are used for power supply of a ship power station, after which, after filtration, from the higher harmonics receive 400 V 50 Hz.

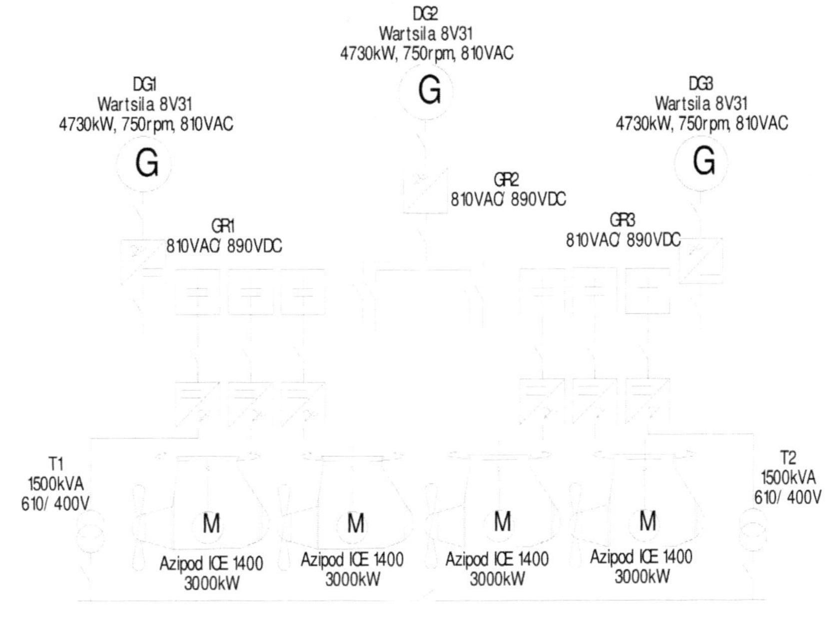

Figure 1. Diagram of the electric power system of the port icebreaker.

The main objective of frequency converters operating at a power plant is to maintain the constancy of standard voltage and frequency parameters. The quality of electricity must be ensured by a filter system. At the same time, on ships with a REI with a DDCB, the block diagram of which is shown in figure 2, special energy storage devices are installed to save fuel.

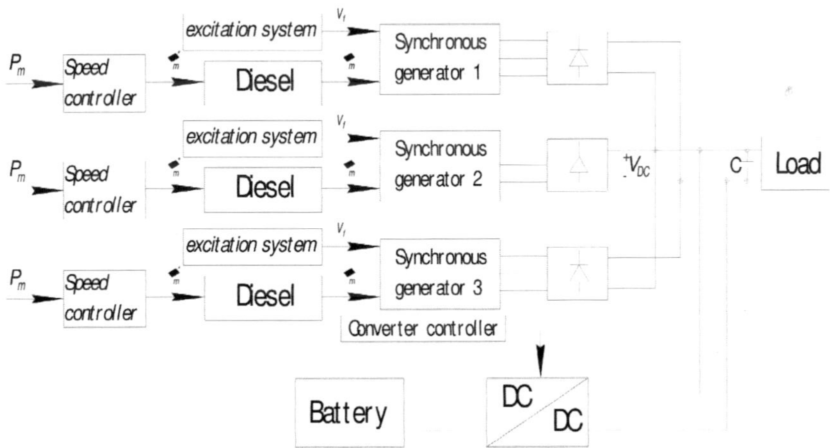

Figure 2. Block diagram of the REI with a DDCB of the port icebreaker.

The main load of the REI with a DDCB, the computer model of which was created and modeled in Matlab / Simulink using the SimPowerSystem tool kit, is shown in figure 3, is connected to the electric propulsion system, including DC / AC converters and electric motors. Such systems demonstrate a constant nature of the load as they are able to consume the same total power at different voltage levels [11]. This phenomenon is caused by an electromechanical modulator, which adjusts the switching time depending on the load.

IOP Conf. Series: Materials Science and Engineering **811** (2020) 012004 doi:10.1088/1757-899X/811/1/012004

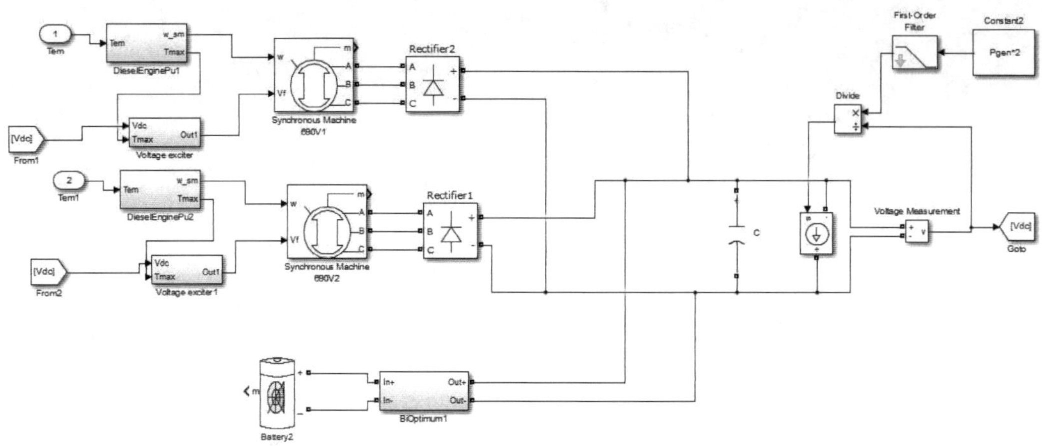

Figure 3. Computer model of the REI with a DDCB based on the Matlab/Simulink program.

In order to simulate a constant load (CL), a controlled voltage source can be used where the required current is obtained by dividing the set power by the real DC voltage, as shown in equation (1):

$$I_{load}=\frac{P_{ref}}{U} \tag{1}$$

The linearized expression of the constant load is found from equation (2), where v is a small voltage wave:

$$i=-\frac{P_{ref}}{U^2}+2\frac{P}{U} \tag{2}$$

At this operating point, the constant load can be represented by a negative resistance parallel with constant power, given in equations (3) and (4). A negative increment of resistance has a destabilizing effect on the automatic system and is known as negative unstable impedance:

$$R_{CPL}=-\frac{U^2}{P} \tag{3}$$

$$I_{CPL}=2\frac{P}{U} \tag{4}$$

In reality, with a dynamic load, the use of a permanent magnet synchronous motor (PMSM) is necessary.

The vector diagram shown in figure 4 describes the processes that occur in the engine, which are considered in a rotating coordinate system *d* and *q*, where the *d* axis is oriented relative to the rotor flow.

The motor rotor, which is a permanent magnet, forms a flux linkage $\left(\overrightarrow{\psi_f}\right)$, equal to the product of the rotor flux by the number of turns of the stator winding. The vector of this flux linkage has a direction along the rotor axis (*d*) from the positive to the negative pole and lags behind the stator current vector $\left(\overrightarrow{I_s}\right)$ by some angle φ.

When the rotor rotates, a constant flux creates an EMF vector (E) in the stator windings, directed at right angles to the flux and 90° behind it (derivative of flux linkage) [12].

IOP Conf. Series: Materials Science and Engineering **811** (2020) 012004 doi:10.1088/1757-899X/811/1/012004

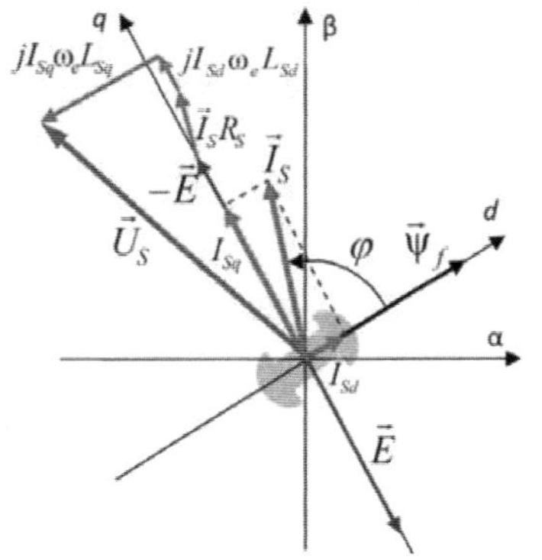

Figure 4. Vector diagram.

The EMF value is determined by equation (5):

$$E = \psi_f \omega_e \tag{5}$$

where:
ψ_f – flux linkage;
$\omega_e = Z_p \omega_R$ – field rotation speed;
ω_R – rotor speed;
Z_p – the number of pairs of motor poles.
Based on figure 3, the following vector relation (6) can be formed:

$$\overrightarrow{U_s} = -\vec{E} + \overrightarrow{I_s} R_s + j\omega_e \left(\overrightarrow{I_{sd}} L_{sd} + \overrightarrow{I_{sq}} L_{sq} \right) \tag{6}$$

where:
$\overrightarrow{U_s}$ – stator voltage vector;
$\overrightarrow{I_s}$ – the stator current vector, while I_{Sd} and I_{Sq} are its component;
L_{sd} and L_{sq} – stator inductance along the d and q axes;
R_S – stator resistance.

In accordance with figure 4, the stator voltage vector is balanced by the EMF vector and the voltage drop across the active (R_S) and reactive (L_{sd} , L_{sq})) resistances of the stator winding.

Considering that flux linkage is a product of current and inductance, and the motor stator is a three-phase inductance coil [13], in this case, excluding magnetic losses, expression (7) can be composed:

$$\begin{cases} U_{sd} = L_{sd} \dfrac{dI_{sd}}{dt} + R_s I_{sd} - \omega_e L_{sq} I_{sq} \\ U_{sq} = L_{sq} \dfrac{dI_{sq}}{dt} + E + R_s I_{sq} + \omega_e L_{sd} I_{sd} \end{cases} \tag{7}$$

Figure 4 shows the formation of the components of the stator current differential with respect to a rotating coordinate system.

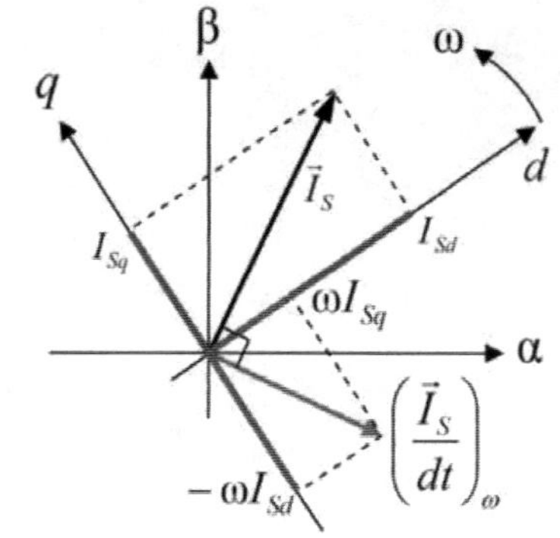

Figure 5. Vector diagram.

Taking into account figure 5, that flux linkage is the current multiplied by inductance, confirming expression 7, we can compose the following:

$$\begin{cases} f_d(\omega_e) = -\omega_e L_{sq} I_{sq} \\ f_q(\omega_e) = \omega_e L_{sd} I_{sd} \end{cases} \tag{8}$$

Given that $E = \psi_f \omega_e$, we compose expressions (7) with respect to differentials. Also, introducing the moment equation, we obtain equations (9), with which it is possible to build an electromagnetic model of the engine:

$$\begin{cases} pI_{sd} = \dfrac{1}{L_{sd}}\left(U_{sd} - R_s I_{sd} + \omega L_{sq} I_{sq}\right) \\ pI_{sq} = \dfrac{1}{L_{sq}}\left(U_{sq} - R_s I_{sq} - \omega L_{sd} I_{sd} - \psi_f \omega_e\right) \\ M_{em} = \dfrac{3Z_p}{2}(I_{sq}\psi_f + I_{sd} I_{sq}(L_{sd} - L_{sq})) \end{cases} \tag{9}$$

The model of a permanent magnet synchronous motor created in the Matlab/Simulink program is shown in figure 6.

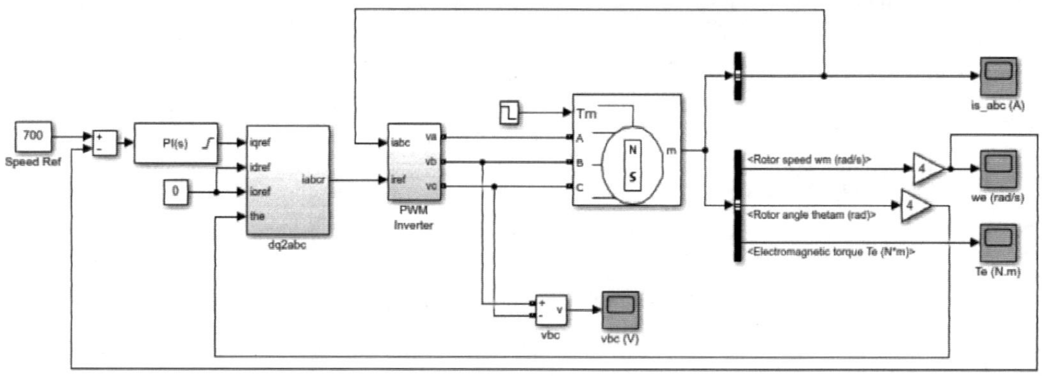

Figure 6. Model of a PMSM.

3. Conclusion

Based on the analysis of the circuit design for a RE with DDCB, we can conclude that this concept improves the efficiency of using a single electric power system, combining the advantages of using

AC and DC. Moreover, the system complies with the rules of selectivity and protection of equipment, can be used for any marine application with a capacity of 20 MW and operates at a rated voltage of 1000 V DC and higher.

The use of the above expressions allows one to mathematically determine the static load of a rowing electric installation and construct an electromagnetic model of a PMSM.

4. References

[1] McCoy T J 2015 Electric ships past, present, and future, *IEEE Electrification Magazine* **3(2)** 4-11.

[2] Bolvashenkov I, Herzog H G, Rubinraut A, Romanovskiy V 2014 Possible ways to improve the efficiency and competitiveness of modern ships with electric propulsion systems, In *2014 IEEE Vehicle Power and Propulsion Conference* (Coimbra) 1-9.

[3] Zahedi B, Norum L E, Ludvigsen K B 2014 Optimized efficiency of all-electric ships by dc hybrid power systems, *J. Power Sources* **255** 341-54.

[4] Zahedi B, Norum L E 2012 Modeling and simulation of all-electric ships with low-voltage DC hybrid power systems, *IEEE Transact. Power Electronic.* **28(10)** 4525-37.

[5] Geertsma R D, Negenborn R R, Visser K, Hopman J J 2017 Design and control of hybrid power and propulsion systems for smart ships: A review of developments, *Appl. Energy* **194** 30-54.

[6] Jin Z, Sulligoi G, Cuzner R, Meng L, Vasquez J C, Guerrero J M 2016 Next-generation shipboard dc power system: Introduction smart grid and dc microgrid technologies into maritime electrical networks, *IEEE Electrification Magazine* **4(2)** 45-57.

[7] Seenumani G, Sun J, Peng H 2011 Real-time power management of integrated power systems in all electric ships leveraging multi time scale property, *IEEE Transact. Control Syst. Technol.* **20(1)** 232-40.

[8] De Breucker S, Peeters E, Driesen J 2009 Possible applications of plug-in hybrid electric ships, *IEEE Electric Ship Technol. Symposium* (Baltimore) 310-7.

[9] Vicenzutti A, Bosich D, Giadrossi G, Sulligoi G 2015 The Role of Voltage Controls in Modern All-Electric Ships: Toward the all-electric ship, *IEEE Electrification Magazine* **3(2)** 49-65.

[10] Bolvashenkov I, Herzog H G, Rubinraut A, Romanovskiy V V 2014 Promising ships propulsion systems with electric motion and steering gondolas, *2014 Ninth Int. Conf. Ecologic. Vehicles Renew. Energies* (New York) 1-4.

[11] Reed G F, Grainger B M, Sparacino A R, Mao Z H 2012 Ship to grid: Medium-voltage DC concepts in theory and practice, *IEEE Power and Energy Magazine* **10(6)** 70-9.

[12] Lin F, Zuo S, Deng W, Wu S 2016 Modeling and analysis of electromagnetic force, vibration, and noise in permanent-magnet synchronous motor considering current harmonics, *IEEE Transact. Indus. Electronics* **63(12)** 7455-66.

[13] Kemmetmuller W, Faustner D, Kugi A 2014 Modeling of a permanent magnet synchronous machine with internal magnets using magnetic equivalent circuits, *IEEE Transact. Magnet.* **50(6)** 1-14.

IOP Conf. Series: Materials Science and Engineering **811** (2020) 012005 doi:10.1088/1757-899X/811/1/012005

Experience of Use High-Precise Position Systems for Inland Waters

V Karetnikov[1], I Pashchenko[1], V Kozlov[1], A Butsanets[1]

[1]Department of Shipping on Inland Waterways, Admiral Makarov State University of Maritime and Inland Shipping, 5/7 Dvinskaya Str., Saint-Petersburg, 198035, Russia

Abstract. Real Time Kinematic (RTK) is a differential method of object positioning using GNSS in which code corrections are supplemented by phase ones and transmitted from a base station (with precisely known coordinates) to a movable receiver in real time [1]. Having a series of advantages over classical differential methods of object positioning the most essential of which is centimeter precision of positioning, the method has found wide use in various spheres. Networks providing RTK solutions in real time on a commercial basis, and also free of charge in certain countries within the frameworks of state geodetic networks cover considerable areas with a field of high-precision corrections.

1. Introduction

Currently in water transport there is enhanced interest in robotized self-contained above-water objects that function using unmanned technologies. Many problems can be solved using remotely controlled or automated robots [2, 3]. Robots enable surveys of a reservoir and dams [4] excluding human factor and reducing personnel costs. Above-water unmanned vessels assist in intelligence operations in shallow waters under conditions of limited visibility [5]. A bathymetric survey with the use of robots allows charts of shallow and protected areas to be prepared with a considerable reduction of costs of such operations [6-8]. Not of less importance is environmental monitoring, for instance, there are already unmanned vessels with automatic water sampling [9, 10].

Thus, it can be concluded that the use of unmanned vessels in operations is a sufficiently constructive solution allowing the efficiency and safety of operations to be increased.

The development of unmanned technologies is impossible without the development of remote control [4]. If the said technology is implemented there is no crew on board the vessel, and control is performed by the operator remotely. The operator receives data to take decisions aimed at ensuring safety of navigation from various data sensors located on board the vessel in real time. Here one should bear in mind that the introduction of unmanned vessels should not have negative effects on safety of navigation.

Data sensors should be very accurate and transmit data on the state of the vessel in time. The concept of the unmanned vessel implies a high level of automation. The larger the vessel, the more sensors it has.

As the vessel is significantly affected by external forces (wind, current, the heaving of the sea), one of the most important problems is the vessel's positioning accuracy. The scientific publications discuss uses of RTK to prevent tsunamis [11], research aimed at increasing positioning accuracy [12, 13], there are no cases, however, of comparing popular (in the opinion of the employed experts) high-precision positioning systems already operating in the spheres of satellite navigation of inshore objects and flying vessels for their use in internal waterways.

2. Methods and Materials

In September 2018 the workgroup of the Scientific and Educational Center «Unmanned technologies in water transport» attempted an experiment to research whether it is possible to use RTK in practice for navigation in internal waterways.

In the course of the experiment it was planned:

- to establish the precision of determining the RTK position of a vessel proceeding on an area of inland waterways at working (safe speed);

- to determine the safety of the RTK solutions being received, as well as the actual number of failures and the re-initialization time of onboard RTK equipment, taking into account the specifics of the area of inland waterways selected for the experiment;

- to give a comparative characteristic of the data obtained through RTK on the position of the vessel with the reference data (the post-processed data were taken as the reference ones).

The experiment used the vessel (of project 3052) Viktor Shurpitsky (figure 1) the main characteristics of which are given in table 1. [14]. The vessel is a work vessel of the Neva-Ladoga region of waterways and navigation ensuring servicing and monitoring of navigation aids.

Figure 1. Vessel «Viktor Shurpitsky».

Table 1. Basic characteristics of the vessel.

Characteristic	Dimension	Value
Length overall	m	35.65
Width overall	m	5.80
Full height	m	2.60
Draft (river/sea)	m	1.51/1.48
Displacement with % load	t	157
Power of propulsion machinery	kW	2x200
Crew	persons	10
Freeboard, (river/sea)	m	1.096/1.126
Cruising capacity	days	6

On top of the ship's deckhouse two GNSS antennas and a radial system SA-2 (figure 2) were placed:

1. TW8829 Antenna GLONASS/GPS L1/L2 – an external active professional antenna to receive GPS L1/L2, GLONASS L1/L2 and SBAS (WAAS, EGNOS and MSAS) signals in the frequency band (L1 from 1,574 to 1,606 MHz and L2 1,215MHz-1,261MHz). It is part of the development kit NV08C-EVK-RTK-M
2. Geodetic antenna GPS/GLONASS L1/L2;
3. SA-2 Radial system (manufactured by the Russian company Design Bureau NAVIS)

IOP Conf. Series: Materials Science and Engineering **811** (2020) 012005 doi:10.1088/1757-899X/811/1/012005

Figure 2. Location of GNSS antenna and SA-2.

The distances between the antennas were measured for subsequent bringing their centers of radiation into one point. The measurements were 90 cm by midship line and 66 cm between the antennas.

Geodetic antenna No.2 is connected to the receiver GLONASS/GPS JAVAD MAXOR (manufactured by the company JAVAD, U.S.) and was in the course of the experiment used jointly with the receiver for continuous recording using protocol BINR [15] of the data in RINEX format of the version 2.11. These data were later adjusted by the corrections, which allowed the coordinates to be determined with the accuracy 5 mm and these data to be accepted as the reference.

3. Results

The data on the ship's position obtained with the use of antenna No.1 and the receiver NV08C-RTK-M (dual-frequency multisystem high-precision receiver GNSS to support navigation with centimeter precision using phase measurements GNSS GLONASS and GPS at the frequencies L1 and L2), for subsequent analysis and comparison were recorded using protocol NMEA [15] (figure 3). To evaluate and visualize the data obtained from the navigational receivers in real time, the software of Design Bureau Navis, «Storegis» was used (figure 4). The use of СУ-2 enabled determination of the spatial angles of the ship's orientation with the accuracy up to 0.2° per one meter of base.

Figure 3. Navigational receivers.

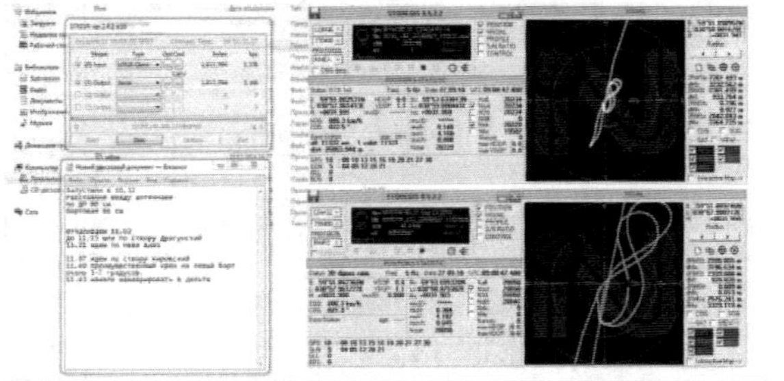

Figure 4. Software «Storegis».

The selection of the area and time for testing was one of the priority conditions for the conduct of the experiment.

First, it was necessary to choose a place with strong additive noises. Depending on the properties the most widespread additive noises can be divided into three basic classes: fluctuation, concentrated, and pulse.

Of the greatest interest for our case is the last class, for it includes a subclass of pulse noises. Such noises are a regular or chaotic alternation of pulses that in a certain measure complicate the reception of a useful signal. An almost any noise lasting less than the duration of a signal element can act as a single noise. Despite an insignificant duration of action of separate pulse noises (usually $10^{-5} - 10^{-8}$ s), they are able to reduce the fidelity of the transmitted information significantly. This occurs due to the spectrum of such noise filling almost the entire frequency band of the receiver.

Pulse noises include:

- Industrial noise;
- Atmospheric noise;
- Electrization noise.

The said noises mostly occur due to the processes occurring in the Earth's atmosphere, and industrial activities. Thus, the water area in immediate proximity to the moorage wall of a shipbuilding or other large industrial facility can be the most constructive place for testing. In areas of the said objects there are much more sources of industrial noises than in a port area. The main source of the said noises will be electric arc welding, power components, electric motors, etc.

Besides, if possible, preference should be given to the time when there is precipitation, for under such conditions atmospheric noises are highly likely to occur.

Second, in choosing the area it is necessary to pay attention to the presence of a cross current. This will allow dynamic testing ensuring the misalignment angle between the midship line and the ship's center-of-gravity path.

The time period from 26 September 2018 to 28 September 2018 was selected as the expected testing time. In the said period a rainy weather with double-layer cloud should be observed in the expected testing area, which is confirmed by the information presented in forecast charts with frontal analysis (figure 5). Such conditions may be considered as highly favorable for atmospheric noises to arise.

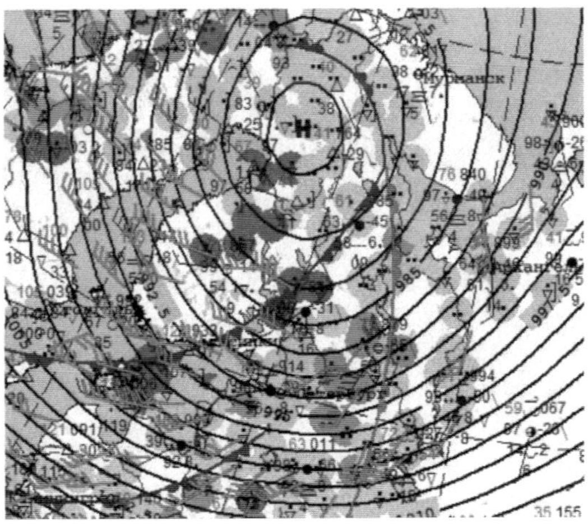

Figure 5. Forecast charts.

The suggested selection of place and time of the testing seems to be highly constructive, for it will allow simultaneous dynamic and static measurements to be taken in the complex jamming and situational environment, which is especially important taking into account the basic requirement for supporting safety of navigation.

A section of inland waterways of the Neva-Ladoga area of waterways and navigation 1,316–1,327 km was selected for conducting the experiment. The presence of shipbuilding yards (a potential source of industrial noises), ship traffic density (meeting ships and ships of opportunity as sources of cross-noise), a complex fairway and developed infrastructure have made this area the most interesting one for the conduct of this experiment.

4. Discussion

The complete coverage of the area with a cellular network enabled transmission of RTK corrections through Internet.

A base station (BS) located on the roof of the office building of Design Bureau Navis in Saint Petersburg (figure 6) the coordinates of which are precisely known through accumulation over long time and averaging of navigational information on its location, is connected to Internet. Through special software «NTRIP Caster» [15] a point of access was installed to transmit RTK corrections to the receiver in RTCM SC-104 format. The RTK receiver located on board the ship was connected to the BS point of access through the ship's network of WiFi-router (in which Internet was accessed through a cellular modem).

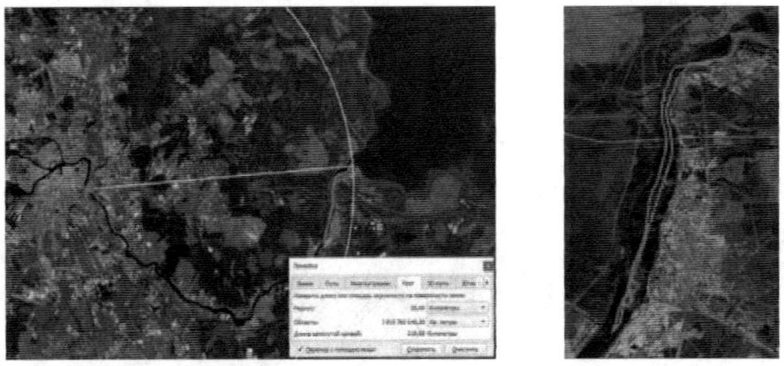

Figure 6. BS location and ship's route.

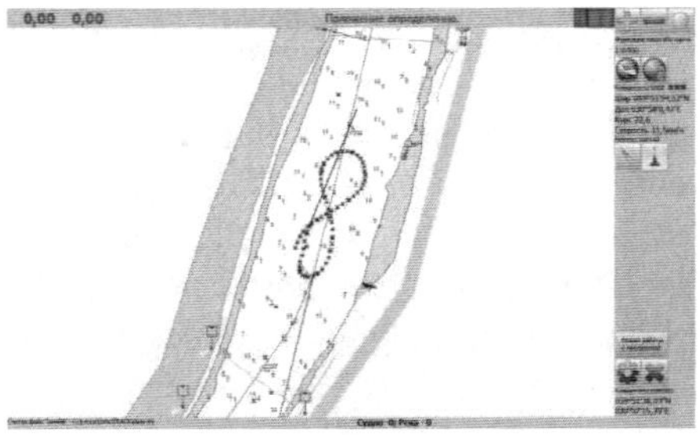

Figure 7. Eight maneuver on the ship's ECDIS.

The initialization time of the onboard RTK receiver on the «NTRIP Caster» server of BS, the commencement of reception of RTK solutions and transition from floating solutions (RTK Float) to the fixed ones (RTK Fix) [16] was about 3-4 min. during the ship's moorage. The age of the RTK correction is 1-2 sec.

Having recorded within 10 min. the navigational information for post-processing and the fixed solution, the RTK ship set to motion. Having performed a passage on the river Neva from 1,316 km to 1,327 km and back, having performed two Eight maneuvers (figure 7) and having safely passed in the closest proximity to the «Nevsky Shipbuilding Yard», the ship returned and got moored to the wharf.

5. Conclusion

In the course of the ship's movement on the experiment's route, short-time (2-3 sec.) losses of the fixed RTK solution were observed.

The positioning accuracy in RTK varied from 3-10 cm. There were no observed initialization failures of the onboard RTK receiver on the BS server.

According to the post-processed data and RTK solution graphs were constructed (figure 8), where the blue graph is navigational data accepted as reference, and the green graph is RTK solutions.

The performed experiment shows that, having solved a number of issues related to the specifics of operation of water transport:

- effects of pitching and rolling on the orientation of the GNSS antenna in space, inclusion into the antenna complex of a ship's lean angle and differential sensors
- selection of the optimal way of transmitting RTK corrections to the onboard receivers (in absence of the coverage with a cellular network of an area of inland waterways, the use of satellite communications, etc.)
- to pay special attention to the electromagnetic protection of navigational information transmission channels for RTK under the effects of cross-noise, industrial noises on them [17]. Especially on areas of inland waterways passing near large cities and industrial areas
- creation of a network of base stations, as well as their control center in order to support a reliable field of high-precision RTK corrections

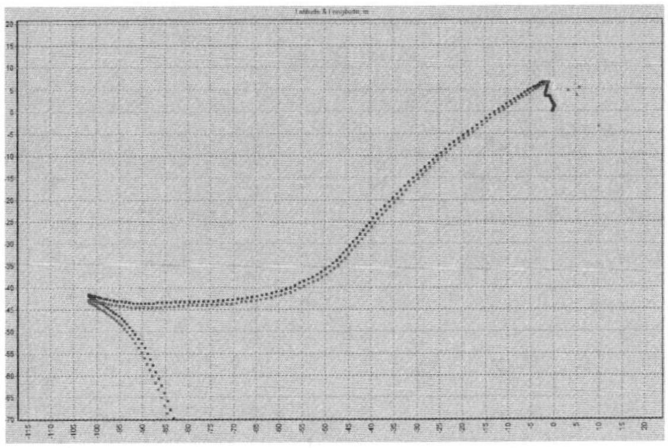

Figure 8. Comparison of reference data and RTK solutions.

Using the already available developed infrastructure of inland waterways, it is possible to set up a new information and communication network of transmitting high-precision corrections of centimeter precision capable of supporting unmanned ship navigation in inland waterways of the Russian Federation.

References

[1] Rizos C 2009 Network RTK research and implementation-a geodetic perspecti, *Position.* **1(02)** 144-50.

[2] Wróbel K, Montewka J, Kujala P 2018 System-theoretic approach to safety of remotely-controlled merchant vessel, *Ocean Eng.* **152** 334-45.

[3] Kondo M, Shoji R, Miyake K, Furuya T, Ohshima K, Shimizu E, Nakagawa M 2018 Monitor System for Remotely Small Vessel Navigating, *Int. Conf. Human Interface Manage. Info.* (Springer, Cham) 419-28.

[4] Hue N H, Huong D C, Thanh N H 2019 A Technical Solution to Improve the Existing River Dikes for both Flood Prevention and Transportation Purposes, *IOP Conf. Series: Mater. Sci. Eng.* **507(1)** 012012.

[5] Metcalfe B, Thomas B, Treloar A, Rymansaib Z, Hunter A, Wilson P 2017 A compact, low-cost unmanned surface vehicle for shallow inshore applications, *2017 Intel. Syst. Conf. (IntelliSys)* (IEEE) pp 961-8.

[6] Kitts C, Mahacek P, Adamek T, Rasal K, Howard V, Li S, Hulme S 2012 Field operation of a robotic small waterplane area twin hull boat for shallow-water bathymetric characterization, *J. field Robot.* **29(6)** 924-38.

[7] Jin J, Zhang J, Shao F, Lyu Z, Wang D 2018 A novel ocean bathymetry technology based on an unmanned surface vehicle, *Acta Oceanologica Sinica* **37(9)** 99-106.

[8] Karetnikov V, Ratner E, Ageeva A 2017 Introduction of the automated hydrographie survey systems for creation of electronic nautical charts of internal waterways in Southern Africa, *2017 IEEE AFRICON* (IEEE) pp 1479-84.

[9] Shuo J, Yonghui Z, Wen R, Kebin T 2017 The Unmanned Autonomous Cruise Ship for Water Quality Monitoring and Sampling, *2017 Int. Conf. Comput. Syst. Electronic. Control (ICCSEC)* (IEEE) pp 700-3.

[10] Chen Y J, Liu P X, Jiang B H, Zhang S X, Feng F 2018 Study on the application of water quality monitoring technology based on unmanned ship, *Desalination water treat.* **121** 311-5.

[11] Ohta Y, Kobayashi T, Tsushima H, Miura S, Hino R, Takasu T, Sato T 2012 Quasi real-time fault model estimation for near-field tsunami forecasting based on RTK-GPS analysis: Application to the 2011 Tohoku-Oki earthquake (Mw 9.0), *J. Geophysic. Res. Solid Earth* **117(B2)**.

[12] Parkins A 2011 Increasing GNSS RTK availability with a new single-epoch batch partial ambiguity resolution algorithm, *GPS solutions* **15(4)** 391-402.

IOP Conf. Series: Materials Science and Engineering **811** (2020) 012005 doi:10.1088/1757-899X/811/1/012005

[13] Heinrich M, Sperl A, Mittmann U, Henkel P 2018 Reliable Multi-GNSS Real-Time Kinematic Positioning, *2018 Int. Symposium ELMAR* (IEEE) pp 103-8.

[14] Weber G, Dettmering D, Gebhard H, Kalafus R 2005 Networked transport of RTCM via internet protocol (Ntrip)-IP-streaming for real-time GNSS applications, *ION GNSS 18th Int. Technic. Meet. Satellite Division* pp. 13-6.

[15] Lenz E 2004 Networked transport of RTCM via internet protocol (NTRIP)–application and benefit in modern surveying systems, *FIG Working Week* pp. 22-7.

[16] Paziewski J 2016 Study on desirable ionospheric corrections accuracy for network-RTK positioning and its impact on time-to-fix and probability of successful single-epoch ambiguity resolution, *Adv. Space Res.* **57(4)** 1098-111.

[17] Vishnevsky Yu G, Sikarev A A 2006 *Signal destruction fields and electromagnetic protection of information channels in automatic ship traffic control systems* (SPb: Sudostroyeniye) p 356.

IOP Conf. Series: Materials Science and Engineering **811** (2020) 012006 doi:10.1088/1757-899X/811/1/012006

Eulerian-Eulerian Approach to CFD Simulation of Two-Phase Bubble Column

Nur Khairunnisa Abd Halim[1], Siti Aslina Hussain[1,3], Mus'ab Abd. Razak[1], Mohd Amirul Syafiq Mohd Yunos[2]

[1]Department of Chemical and Environmental Engineering, Faculty of Engineering, Universiti Putra Malaysia, 43400 UPM Serdang, Selangor Darul Ehsan, Malaysia
[2]Plant Assessment Technology, Industrial Technology Division, Malaysian Nuclear Agency, Bangi 43000 Kajang, Selangor Darul Ehsan, Malaysia

aslina@upm.edu.my

Abstract. Calculation work is carried out in the present project to study the effect of different parameters on hydrodynamics of gas-liquid flow by using CFD. The steady state and transient, three dimensional simulations are performed of the whole bubble column for various inlet gas and sparger designs. A standard k-ε is used to describe turbulence occurring in the continuous fluid. Results show he asymmetric recirculation stronger with an increase in the number of holes giving better mixing. The gas hold-up increased with superficial gas velocity but independent of the sparger design at high velocity. Comparison of simulation results with the previous experimental work data has provided a successful validation of the model.

1. Introduction

Bubble column reactors are frequently opted for multiphase flow system due to its easiness in mechanical part, low maintenance and cost operation, and have good heat and mass transfer characteristics [1]. The limitation of bubble column reactors is of their non-uniformity in back-mixing, bubble size distribution and gas hold-up have large diversities. Although bubble column reactors are widely used in industrial gas-liquid processes [2], detailed understanding is still lacking due to their complex flow encountered inside bubble column despite of its simple construction.

A range of works have been published to capture the dynamic characteristic of gas-liquid flow through experimental [3, 5, 6] and numerical [4, 7, 8]. Several number of frameworks for the description of flow dynamics in bubble column reactor have been considered particularly extensively, are gas distributor [2-4, 9]; superficial gas velocity [8]; gas hold-up distribution [2, 10]; operating and design variables [3]; as well as flow pattern [8, 11]. In most previous publications on computational studies, the model is evaluated based on data collected with a single sparger with a pair of air-water system.

The brief review of previous works points out the need to study the sparger configuration and its influence of gas flow rate on dynamics of gas-liquid flows. In order to compliment the simulation results, two different design of spargers were applied on the same domain model to observe the effect of sparger design on gas hold-up distribution along axial direction. Comparison of results of both spargers will develop the relationship between model design and gas velocity, and their effects on flow dynamics of vertical bubble column.

IOP Conf. Series: Materials Science and Engineering **811** (2020) 012006 doi:10.1088/1757-899X/811/1/012006

2. CFD Model

A 3D-computational vertical domain of rectangular shaped bubble column with dimensions of 0.2 m width, 0.2 m depth and 1.0 m height are used in this numerical modelling. The schematic diagram of actual design of bubble column reactor is shown in figure 1. Two different design of sparger is modelled to provide adequate mixing behaviour in reactor for scale-up purposes. The plate designs are frequently used by current chemical and petrochemical plants in their bubble column reactors. The specifications of spargers are given in table 1. The superficial gas velocity of 0.0125 m/s, 0.0501 m/s and 0.0627 m/s are applied for individual spargers.

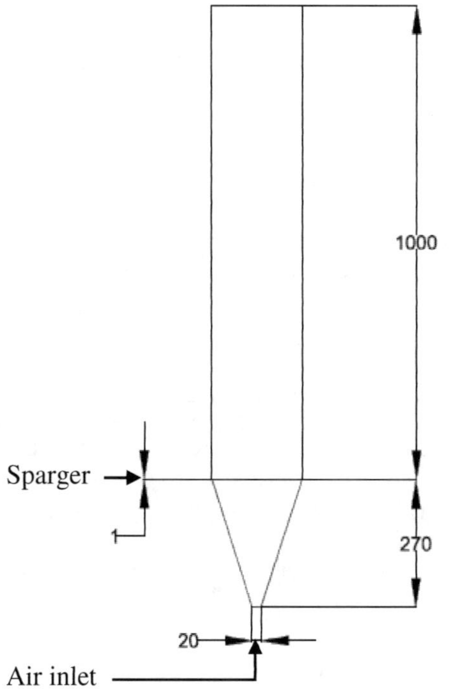

All dimensions are in millimetres (mm)

Figure 1. Schematic layout of bubble column.

Table 1. Details of sparger design.

Sparger design	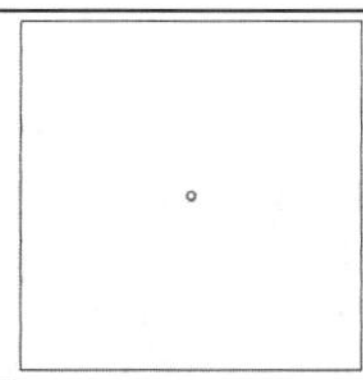	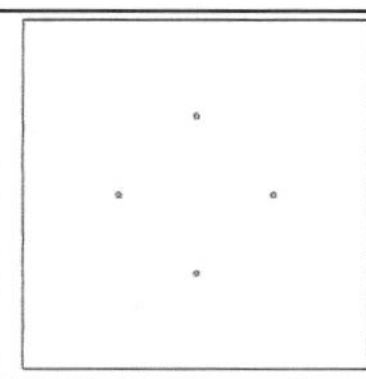
Sparger code	A	B
No. of holes	1	4
Holes diameter (mm)	5.0	2.6

Commercially available CFD code, ANSYS™ 19.0 software is used to numerically model the hydrodynamics of the bubble column reactor. The code technique bases on the finite volume methods with Euler-Euler two-fluid models as proposed by Sato and Sekoguchi [12, 13]. The simulations are carried out with air and water taken as dispersed and continuous respectively. The model equations for the two-phase turbulence flow were solved in each cell of domain. No slip boundary conditions were used at all the impermeable cells. Normal liquid velocity was set to zero. Turbulence were modelled using k-ε equations [13, 14]. Grace correlation was used for interphase momentum transfer and Lopez de Bertonado for turbulence dispersion, with turbulent dispersion coefficient value of 0.3 [14, 12]. Calculations were performed in time dependent mode with first order implicit formulation. The Schiller and Naumann model (default in CFX) was used for interphase drag coefficient. The timestep of 0.1 s was used constantly throughout the simulation. Convergence criteria for continuity, momentum, turbulence, kinetic energy and rate of dissipation were set as 1×10^{-6} for each timestep.

3. Results and Discussions

3.1. Effect of sparger design on flow pattern and gas hold-up profile

There are mainly two configurations of sparger plate used in this study which are as listed in table 1. To study the effect of the distributor configuration, simulations were performed at constant gas velocity. Figure 2 shows snapshots of contour plots of axial liquid velocity for both spargers at timestep of 0.1 s with constant bubble size diameter of 4 mm for superficial gas velocity of 0.0125 m/s. The bubble size diameter was specified based on experimental measurements using referral bubble column design and specification by Yunos et al. [6]. By comparing of the snapshots, it shows that the flow pattern evolution with different sparger was quite dissimilar. sparger B has the smaller diameter of holes and higher number of holes compared to sparger A. The contour plot shows the fluid flows upward towards dispersed area where the bubbles expanded and dispersed up to the surface of the liquid. Therefore, the asymmetry in the recirculation velocity pattern of fluid increases as the number of holes decreases which is in accordance with McClure et al. [9].

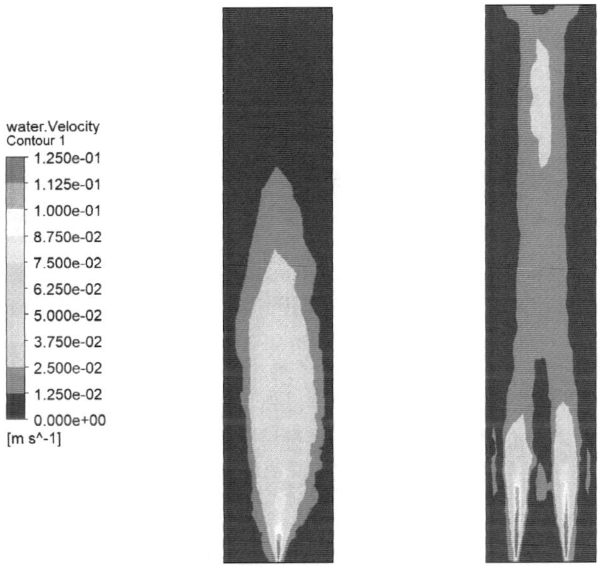

Figure 2. Plot contour of axial liquid velocity at vertical mid-plane of both sparger A and B at constant superficial gas velocity of 0.0125 m/s.

In contrast to liquid flow pattern, the gas hold-up is rather interrelated in complex way with the sparger hole size and numbers. However, one would expect that an increase in number of holes will increase the value of hold-up. This complex issue can be attributed into different modelling and physical phenomena. Firstly, from the experimental work by McClure et al. [9], it is explained that the difference in gas hold-up among different sparger becomes small at higher superficial gas velocity. For this present work, the superficial gas velocity is used toward the highest value could obtained of 0.0627 m/s, thus it is expected the variation of gas hold-up with respect to sparger design would be small. Secondly, gas hold-up is primarily function of the superficial gas velocity. Shah et al. [15] said that the gas hold-up increases linearly with the increasing superficial gas velocity. In these simulations, a constant value of superficial gas velocity is used therefore, the gas hold-up may remain independent of sparger design. Lastly, the gas hold-up is directly affected by the bubble size distribution in the bubble column. In this case, the bubble size is set constant at 4 mm to simplify numerical calculations. This proves another reason for the invariance of the gas hold-up distribution is required. However, according to Sanyal et al. [10] this model can be safely relied upon if the volume averaged steady state gas hold-up is of interest.

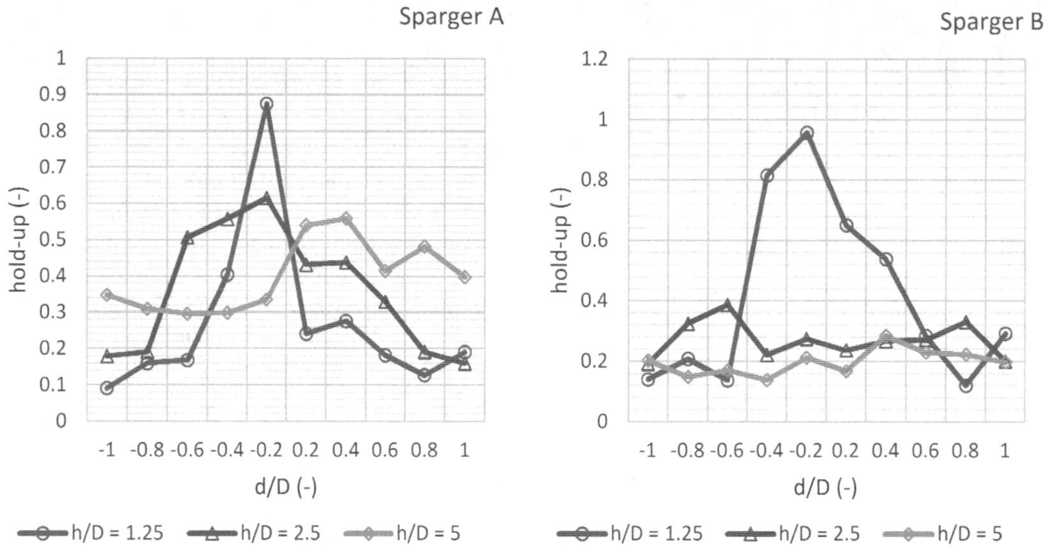

Figure 3. Axial hold-up velocity over column width at different heights for sparger A and B with constant superficial gas velocity of 0.0501 m/s.

Axial hold-up velocity accordance to width of column are plotted in figure 3 for both sparger at superficial gas velocity of 0.0501 m/s within three different heights. From figure 3, sparger B shows wider distribution of gas bubbles in the column compared to sparger A. Thus, for this case we have calculated the average holdup along vertical central line of column width for each height. As observed in the graphs plotted, we can predict on how the hydrodynamics of fluid occurs in the bubble column. The axial hold-up for sparger A starts lower and reached the highest values of 0.875 at h/D of 1.25. However, the velocity drops again after that specified width of column. On the other hand, sparger B shows highest value of axial hold-up at 0.975 near to the sparger plate and decreasing slowly with respect to width of column. This is because it assumed that the liquid enters the column uniformly over the entire cross section. As the liquid moves up the column, it develops a parabolic profile. The velocity of the liquid varies with time and location in the column which is consistent with existence literature values [8].

3.2. Effect of superficial gas velocity on flow pattern and local hold-up profile

To study the effect of varies values of superficial gas velocity, the simulations were performed using sparger with highest number of holes, which is sparger B to all superficial gas velocity values. Figure 4 shows liquid velocity vectors for the above simulation at timestep of 0.1 s for sparger B. Relatively axisymmetric flow pattern disappears for higher superficial gas velocity (>0.05 m/s) [9]. Thus, bubble column flow pattern should be treated with caution before create any qualitative conclusion, particularly when it involves extrapolations of process variables such as the superficial gas velocity.

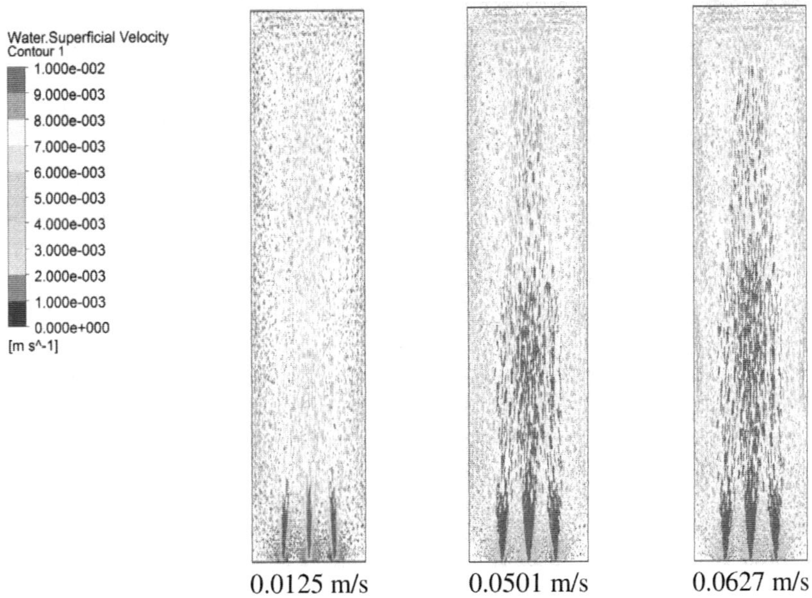

Figure 4. Effect of superficial gas velocity on liquid velocity vectors for sparger B.

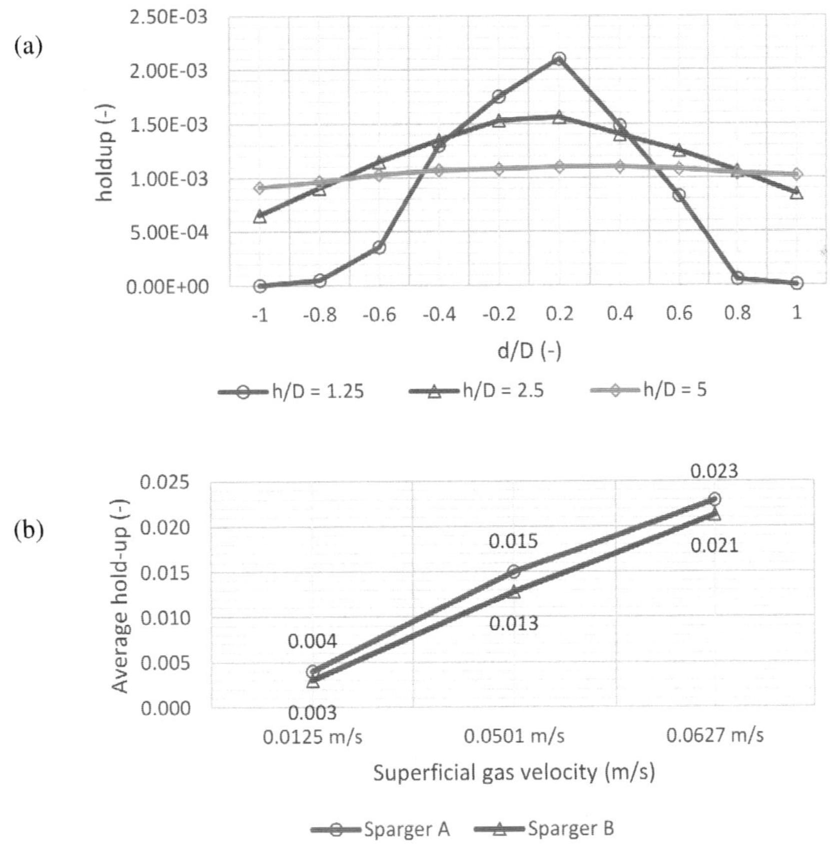

Figure 5. Distribution of average gas hold-up (a) at different heights along column width with sparger A and (b) comparison between sparger A and B at h/D value of 1.25.

IOP Conf. Series: Materials Science and Engineering **811** (2020) 012006 doi:10.1088/1757-899X/811/1/012006

Figure 5 shows the graphical plots of the average values of gas hold-up as a function of superficial gas velocity at total time of 5 s. From the graph plotted in figure 5(b), it shows that the values of gas hold-up increase as superficial gas velocity increases due to the increment in turbulence of the liquid phase. However, the values of average hold-up of sparger B is lower compared to values of sparger A shows that the gas hold-up is not interrelated with sparger design particularly at high superficial velocities. The hold-up starts to differs at lower superficial velocities where the sparger with smaller hole diameter has much higher hold-up. Such results are similarly reported by other authors [3, 4, 9] Therefore, we could say that the gas hold-up is directly proportional to the superficial gas velocity which is accordance with previous literature by Wagh et al. [2].

3.3. Model validation

To achieve validation of the model used, axial velocity data of experimental and numerical results were plotted both sparger A and B at constant superficial velocity of 0.0501 m/s shown in figure 6. The numerical data were taken from earlier study by Yunos et al. [6] obtained by using high speed camera technique. From the graph plotted, it shows that the values plotted in sparger A is higher compared to data plotted in sparger B showing that the axial velocity is ignoring the distributor effect at high superficial velocities. Since this is an incipent studies, there is not much validation has been done to compare with the experimental data.

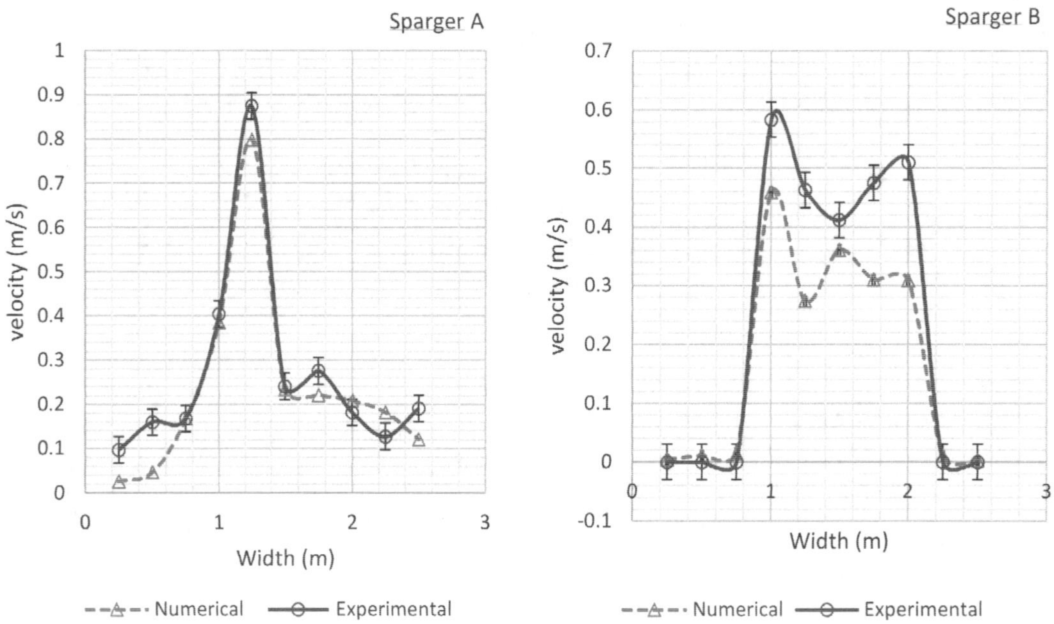

Figure 6. Comparison of numerical and experimental data on axial velocity values for both sparger A and B.

4. Conclusion

The overall gas hold-up and its distribution along axial direction of column is discussed. From the analysis, it shows that there is significance impact of sparger design and superficial gas velocity on dynamics of bubble column. The transition velocity from bubbly to churn-turbulent flow increases with increasing superficial gas velocity value. It was also found that the orifice diameter of sparger affect column performance the most at low superficial velocities (<0.05 m/s). Higher overall hold-up occurred at sparger with the smaller orifice diameter. However, the sparger holes does not affect much at higher superficial velocity. Therefore, we could say that the gas hold-up increased with superficial gas velocity but is independent on sparger design at high superficial velocity (>0.05 m/s).

Acknowledgment
The authors gratefully acknowledged Universiti Putra Malaysia for awarding the funds to conduct this work under project number GP-IPS/2017/9573900. The authors are also cordially acknowledged to the Department of Chemical and Environmental Engineering, Faculty of Engineering, Universiti Putra Malaysia and Industrial Technology Division, Malaysian Nuclear Agency for support and supervision.

References

[1] Rzehak R, Krauß M, Kováts P, Zähringer K 2017 Fluid dynamics in a bubble column: New experiments and simulations, *Int. J. Multiph. Flow* **89** 299-312.

[2] Wagh S M, Ansari M E A, Kene P T 2014 Axial and radial gas holdup in bubble column reactors, *Bulletin Korean Chem. Soc.* **35(6)** 1703-5.

[3] Buwa V V, Ranade V V 2002 Dynamics of gas-liquid flow in a rectangular bubble column: Experiments and single/multi-group CFD simulations, *Chem. Eng. Sci.* **57** 4715-36.

[4] Akhtar M A, Tadé M O, Pareek V K 2006 Two-fluid Eulerian simulation of bubble column reactors with distributors, *J. Chem. Eng. Japan* **39** 831-41.

[5] Ziegenhein T, Lucas D, Rzehak R, Krepper E 2013 Closure relations for CFD simulation of bubble columns 1-12.

[6] Yunos M A S M, Halim N K A, Hussain S A 2017 Investigations of bubble size, gas hold-up, and bubble rise velocity in quadrilateral bubble column using high-speed camera, *Am. J. Eng. Technol. Soc.* 1-18.

[7] Fletcher D F, McClure D D, Kavanagh J M, Barton G W 2017 CFD simulation of industrial bubble columns: Numerical challenges and model validation sucesses, *Appl. Math. Model.* **44** 25-42.

[8] Saleh S N, Mohammed A A, Al-Jubory F K, Barghi S 2018 CFD asessement of uniform bubbly flow in a bubble column, *J. Pet. Sci. Eng.* **161** 96-107.

[9] McClure D D, Wang C, Kavanagh J M, Fletcher D F, Barton G W 2016 Experimental investigation into the impact of sparger design on bubble columns at high superficial velocities, *Chem. Eng. Res. Des.* **106** 205-13.

[10] Sanyal J, Vásquez S, Roy S, Dudukovic M P 1999 Numerical simulation of gas–liquid dynamics in cylindrical bubble column reactors, *Chem. Eng. Sci.* **54** 5071-83.

[11] Li G, Yang X, Dai G 2009 CFD simulation of effects of the configurations of gas distributors on gas-liquid flow and mixing in bubble column, *Chem. Eng. Sci.* **64** 5104-16.

[12] Sato Y, Sekoguchi K 1975 Liquid velocity distribution in two-phase bubbly flow, *Int. J. Multiphase Flow* **2** 79-95.

[13] Lopez de Bertodano 1991 Turbulent bubble flow in a triangular duct PhD Thesis Rensselaer, *Polytech. Inst.* Troy New York.

[14] Pourtousi M, Zeinali M, Ganesan P, Sahu J N 2015 Prediction of multiphase flow pattern inside a 3D bubble column reactor using a combination of CFD and ANFIS, *Rsc Adv.* **5(104)** 85652-72.

[15] Shah Y T, Kelkar B G, Godbole S P, Deckwer W D 1982 Design parameters estimations for bubble column reactors, *AIChE J.* **28** 353-79.

IOP Conf. Series: Materials Science and Engineering **811** (2020) 012007 doi:10.1088/1757-899X/811/1/012007

Unidirectional Heat Transfer into a Rotating Cylinder Containing R-134a Refrigerant

David Park[1], Sunwoo Kim[1]

[1]Department of Mechanical Engineering, University of Alaska Fairbanks, Fairbanks, AK, 99775-5905, USA

Abstract. An analysis was conducted on the feasibility of designing a unidirectional heat transfer cylindrical wall suitable for microgravity conditions. The heat transfer wall is unidirectional in the sense that heat is only allowed to flow in one direction, which is made possible by the principles of phase-change heat transfer (PCHT). The wall is to consist of two concentric cylindrical walls and is to direct heat from the outer wall to the inner wall radially. A mathematical model is presented for the proposed unidirectional heat transfer wall. Preexisting heat transfer correlations were utilized to calculate the heat transfer rate of the unidirectional heat transfer wall with R-134a as the working fluid. The present study demonstrates that the unidirectional heat transfer wall can switch from a heat conductor to a thermal insulator by adjusting the loading of the working fluid within the walls of the proposed device.

1. Introduction

One of the challenges of designing a thermal control system for space missions is the presence of extreme temperature fluctuations in space. For this, phase-change heat transfer (PCHT) that are highly coveted in engineering due to the high heat transfer rates can be a key tool. However, in the application of space missions, separation of the liquid and vapor phases does not occur due to the absence of gravity. In lieu of the gravitational force, the present study suggests that a centrifugal force is utilized for the separation of phases. The design of the unidirectional heat transfer walls in figure 1 consists of a double-walled cylindrical tube, in which fluid is contained within the wall. The heat transfer wall is designed to enclose an environment to be maintained at a lower temperature than the temperature of the surroundings. When the device remains still, the vapor covers the inside surface of the outer wall, leading to insulating the wall. When the device is rotated axially, the centrifugal force separates liquid and vapor phases of the working fluid allowing heat transfer from outside to inside.

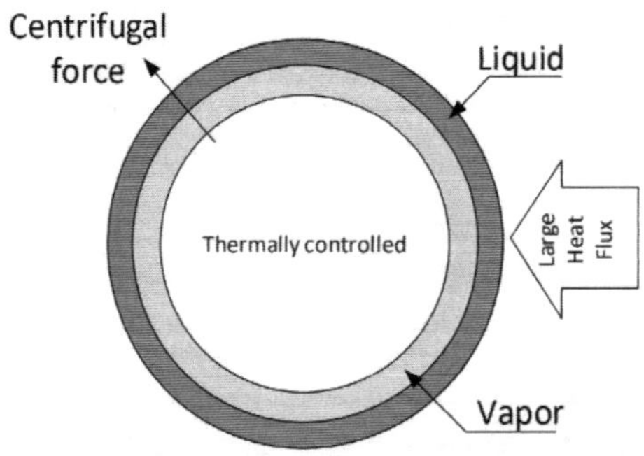

Figure 1. Cross-sectional view of the unidirectional heat transfer wall.

In the preliminary stages of research, efforts were made to find literature that modeled systems similar to that of the proposed unidirectional heat transfer wall. A topic of particular interest was the modeling of axially rotating heat pipes. Rotating heat pipes are similar in that they also utilize both PCHT and rotational motion to achieve effective heat transfer. Rotating heat pipes are PCHT devices that are commonly implemented on rotating machinery, such as high-speed drills, motors, and compressors [1-3]. The principles of rotating heat pipes provided conceptual insight into the design of the unidirectional heat transfer wall. However, their mathematical models were not suitable for the purposes of the present research. Therefore, the present research focused on formulating a suitable model for the unidirectional heat transfer wall—a model that serves as a proof of concept for the feasibility of the heat transfer wall's design. The calculation of the heat transfer coefficient provides insight into the range of adjustability in performance of the unidirectional heat transfer wall.

2. Model Development

The inner and outer walls of the device are assumed to spin at the same angular velocity in our model. No-slip condition for both the inner and outer walls allows the working fluid to be stationary relative to the walls and consequently leads to pool boiling of the working fluid. Figure 2 shows the orientation of the Cartesian coordinate system. The liquid velocity in the x-direction is assumed to be negligible. The working fluid is R-134a. The heat transfer analysis of the present study is limited to the space between the inner and outer walls of the system.

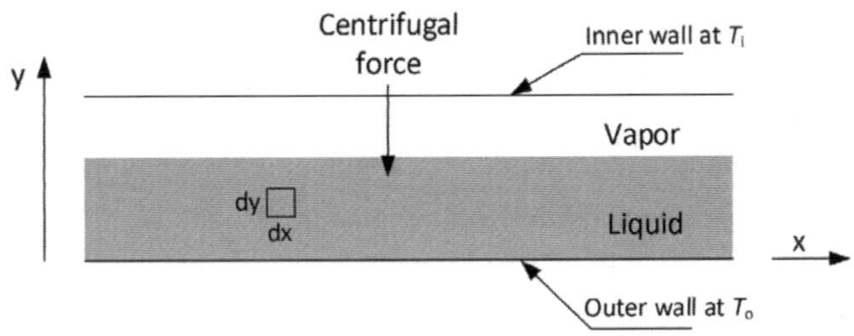

Figure 2. Diagram of the coordinate system of the unidirectional heat transfer wall.

As liquid velocity is not of concern in modeling the system, simple existing Nusselt correlations for boiling and condensation can be used in calculating the heat transfer coefficient of the unidirectional heat transfer wall. In respects to analysis of evaporation of the liquid film, no forced convection is assumed in the x-direction – heat transfer in the liquid film is assumed to be of natural convection alone. Marto [4] reported on a correlation formulated by Korner for natural convection in nucleate pool boiling:

$$Nu_e = 0.133Ra^{0.375} \tag{1}$$

where *Ra* is the Rayleigh number based on the centrifugal acceleration *a* and the liquid film thickness δ. This Nusselt correlation allows for the calculation of the Nusselt number for the process of evaporation for the heat transfer wall. This Nusselt number is subsequently used to calculate the evaporation heat transfer coefficient, h_e.

Gerstmann et al. [5] suggested an analysis of the thickness of the condensate film for horizontal downward facing horizontal plates. Such an analysis provides a order of magnitude estimate to the relative size of the condensate droplets. The film thickness was determined to be relatively small, which gives rationale to conclude that the effects of the returning condensate drops on the liquid film are negligible.

As the system is dependent on the processes of both evaporation and condensation, Nusselt correlations for both processes must be obtained in order to calculate the overall heat transfer coefficient of the wall. In the study of Nusselt correlations for condensation, most models stem from the analytical model of condensation of a laminar film on a vertical surface, as presented by Nusselt [6]:

$$Nu_c = 0.943 \left[\frac{L_c^3 h'_{fg} a(\rho_l - \rho_g)}{k_l \nu_l (T_{sat} - T_i)} \right]^{\frac{1}{4}} \tag{2}$$

where h'_{fg} is a modified expression given for the latent heat of vaporization, L_c is the characterist length, a is the centrifugal acceleration, ρ_l and ρ_v are the density of liquid and vapor, respectively, k_l is the liquid thermal conductivity, ν_l is the liquid kinematic viscosity, and T_{sat} and T_i are the saturation and inner wall temperature:

$$h'_{fg} = h_{fg}(1 + 0.68Ja) \tag{3}$$

where h_{fg} is the latent heat of vaporization and Ja is the Jacob number. This condensation can be written in a general form, which is as follows:

$$Nu_c = c \left[\frac{L_c^3 h'_{fg} a(\rho_l - \rho_g)}{k_l \nu_l (T_{sat} - T_i)} \right]^{n} \tag{4}$$

where the values of c and n account for the different geometric configurations of condensation. The values of c and n can be obtained analytically; however, for more complex configurations, these values are often found empirically. In modeling the unidirectional heat transfer wall, the geometric configuration for the process of condensation was assumed to be a horizontal downward facing plate. Gerstmann et al. [5] reported on correlations for laminar film condensation on the underside of horizontal and inclined surfaces. The paper provides the following correlation for laminar film condensation on horizontal downward facing plates:

$$Nu_c = \begin{cases} 0.69Ra^{0.200}, 10^6 < Ra < 10^8 \\ 0.81Ra^{0.193}, 10^8 < Ra < 10^{10} \end{cases} \tag{5}$$

where the characteristic length used in the analysis is defined as the following:

$$L_c = \left(\frac{\sigma}{a(\rho_g - \rho_v)} \right)^{\frac{1}{2}} \tag{6}$$

The Nusselt number allows for the calculation of the condensation heat transfer coefficient, h_c. A notable observation of the characteristic length used in this analysis is that the characteristic length is independent of any physical dimensions of the system. Rather, the characteristic length is a function of the acceleration, respective densities, and surface tension σ of the fluid.

Once the heat transfer coefficients for both evaporation and condensation are obtained, the heat flux can be calculated for the system. The overall heat flux of the boiling/condensation model is expressed as the product of the overall heat transfer coefficient and the outer and inner wall temperature difference:

$$q" = U(T_o - T_i) \tag{7}$$

U is defined as the combined heat transfer coefficient of the system, where h_e and h_c are the heat transfer coefficients for evaporation and condensation, respectively:

$$U = \left(\frac{1}{h_e} + \frac{1}{h_c} \right)^{-1} \tag{8}$$

3. Adjustable Heat Transfer Rate

The heat transfer characteristics and performance were investigated using the present model. The inner and outer walls rotate at a constant angular velocity; however, the inner and outer wall accelerations are proportional to their respective diameters. Note that the acceleration of the outer wall is greater than that of the inner wall, as the larger radius of the outer wall produces a larger tangential acceleration. Taking this into account, the inner wall acceleration was used for all evaporation calculations, while the outer wall acceleration was used for all condensation calculations. The acceleration on the outer wall were set to be $a_o = 1000$ m/s^2. A wall temperature differential, ΔT was considered in determining the temperatures of the inner and outer walls. For instance, a ΔT value of 5 °C corresponds to constant wall temperatures assumed at values of 30 °C and 20 °C for the outer and inner walls, respectively (given a saturation temperature of 25 °C).

Table 1 provides the thermophysical properties of R-134a, and the dimensions of the heat transfer wall used in this study. All properties were assumed to be at saturation conditions, for both evaporation and condensation. For this particular analysis, properties of R-134a at a saturation temperature of 25 °C were used. Properties were obtained from Cengel et al. [7].

Table 1. Dimensions of heat transfer wall and thermophysical properties of saturated R-134a.

Parameters	Notation	Value	Units
outer wall diameter	D_o	0.100	m
inner wall diameter	D_i	0.090	m
saturation temperature	T_{sat}	25	°C
thermal conductivity (liquid)	k_l	8.33 x 10^{-2}	W m^{-1} K^{-1}
volume expansion coefficient	β	3.24 x 10^{-3}	K^{-1}
density (liquid)	ρ_l	1207	kg m^{-3}
density (vapor)	ρ_g	32.34	kg m^{-3}
enthalpy of vaporization	h_{fg}	177,800	J kg^{-1}
dynamic viscosity (liquid)	μ_l	2.012 x 10^{-4}	kg m^{-1} s^{-1}
kinematic viscosity (liquid)	ν_l	1.667 x 10^{-7}	m^2 s^{-1}
surface tension	σ	8.08 x 10^{-3}	N m^{-1}
specific heat (liquid)	$c_{p,l}$	1427	J kg^{-1} K^{-1}
thermal diffusivity (liquid)	α	4.836 x 10^{-8}	m^2 s^{-1}

Figures 3 and 4 summarize the results of the heat flux and heat transfer coefficient calculations. q'' and U values are given for various ΔT values, as a function of the thickness δ of the liquid layer. The graphs show that the heat transfer characteristics of the system can be adjusted significantly by varying ΔT.

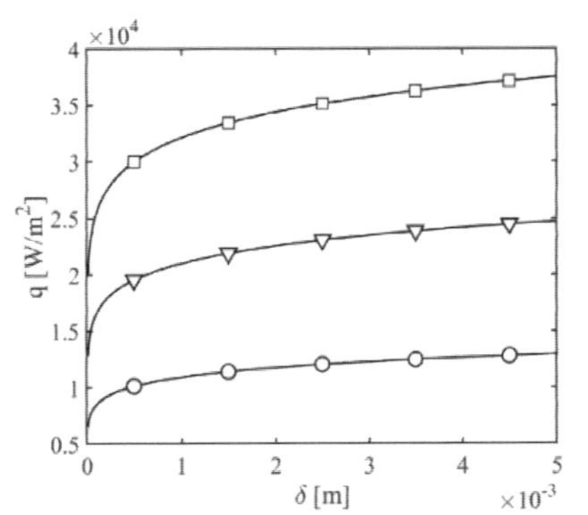

Figure 3. q'' as a function of δ, given ΔT: ○, 3 °C; ▽, 5 °C; □, 7 °C.

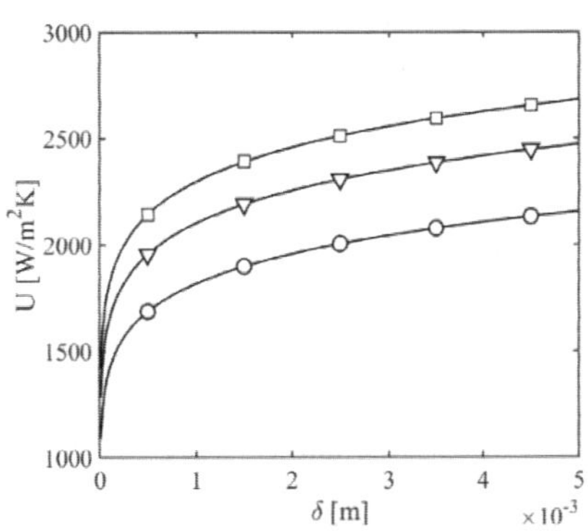

Figure 4. U as a function of δ, given ΔT: ○, 3 °C; ▽, 5 °C; □, 7 °C.

The graphs show that for the given temperature differentials, the heat flux was calculated to be on the order of 10,000 W/m^2. This validates the hypothesis that the boiling/condensation system will produce a high heat flux for the heat transfer wall. One thing we must note is that as the no-slip condition for the liquid and vapor were applied to the inner and outer walls of the system, we assumed zero relative velocity of the liquid and of the vapor in the tangential direction. When the outer wall only spins while the inner wall is stationary, we will expect to achieve higher heat transfer performances, as the tangential velocities of the liquid and vapor will not be negligible, which increases heat transfer by mode of forced convection.

In isolating just one individual curve from the graph, we can see that varying the liquid film thickness δ has a significant effect on the associated heat transfer coefficient and heat flux of the heat transfer wall. This means that indeed, the thermal resistance of the system can be adjusted by modification of fluid loading. The thermal resistance can also be adjusted with the modification of rotational speed and wall temperature differential, which is demonstrated by the graphs.

4. Conclusions

By calculating the heat flux of the unidirectional heat transfer wall for various conditions, we were able to validate that the device can operate as both a heat conductor and a thermal insulator. Many assumptions had to be made in developing a model for the heat transfer wall. A summary of future studies of the heat transfer wall is as follows:

- In respects to fluid mechanics, the liquid and vapor were assumed to be stationary in the tangential direction. This was assumed under the no slip boundary condition applied to the working fluid within the system. However, this may not be the case for a real physical system. Further studies will be made to account for the associated velocities of the working fluid, with subsequent investigation into how the fluid mechanics of the system affects the convective heat transfer of the system.
- The effects of returning condensate drop on the liquid evaporation film were neglected in our model. Although the relative size of drops is predicted to be small, the velocity of the returning condensate drops is expected to enhance convective heat transfer performance to some degree. The study of returning condensate drops will be thoroughly studied by experimental means.
- Further study will be conducted on the thickness of the condensate liquid film, and how this film thickness varies with fluid loading. This is necessary to be able to quantify the amount of working fluid needed in the system to produce the associated evaporation and condensation liquid film thicknesses.

Acknowledgments

This material is based in part upon work supported by the Alaska NASA EPSCoR Program (NNX13AB28A).

References

[1] Song F, Ewing D, Ching C Y 2003 Fluid flow and heat transfer model for high-speed rotating heat pipes *Int. J. Heat Mass Transfer* **46** 4393-401.
[2] Bertossi R, Guilhem N, Ayel V, Romestant C, Bertin Y 2012 Modeling of heat and mass transfer in the liquid film of rotating heat pipes *Int. J. Therm. Sci.* **52** 40-9.
[3] Song F, Ewing D, Ching C Y 2008 Heat transfer in the evaporator section of moderate-speed rotating heat pipes *Int. J. Heat Mass Transfer* **51** 1542-50.
[4] Marto P J 1984 Rotating heat pipes *Heat Mass Transfer Rotating Machin.* 609-32.
[5] Gerstmann J, Griffith P 1966 Laminar film condensation on the underside of horizontal and inclined surfaces *Int. J. Heat Mass Transfer* **10** 567-80.
[6] Nusselt W 1916 Die Oberfllchen-kondensation der Wasser-dampfes *Zeitschrift des Vereines Deutscher Ingenieure* **60** 541-69.
[7] Cengel Y A, Ghajar A J 2015 *Heat and Mass Transfer* (New York: McGraw-Hill) 553-4.

A Method of Selection the Textile Materials for Virtual Reconstruction

S C Zhang[1], V E Kuzmichev[1, 2]

[1]Ivanovo State Polytechnic University, Department of Garment Design, 21 Sheremetev Av., Ivanovo, 153000, Russia
[2]Wuhan Textile University, Fashion Faculty, No.1 Sunshine Avenue, Jiangxia District, Wuhan, 430200, Hubei Province, China

wkd37@list.ru

Abstract. A reconstruction of historical costume can be applied to the movies, theatres, and online museums. Additionally, in the contemporary fashion and industry, the historical European costume has been one of the most essential inspiration for new design. This article developed a method of selection the analogues of textile materials which have been used in European men's suits from 1850s to the 1920s based on published books and real historical male coat. This investigation will be useful to reconstruct historical men costume in virtual reality with high-precision.

1. Introduction

Heritage comes in many ways – in tangible forms such as sites, buildings, landscapes, or as intangibles, like memories, emotions, values and customs [1]. During runways of latest fashion weeks in Paris, Milan, New York and London, the relevant elements and features of historical costumes have been appeared in many notable brands' design [2]. Especially the turn of 20th century, men's everyday suits were similar to contemporary styles, emphasized quality tailoring, practicality and well-dressed [3]. Therefore, the better understanding of historical costumes can provide an inspiration for contemporary fashion designers. To date, however, full appreciation of historical costume still affected by lots of restrictions, such as costume heritage has a relatively short life, since costumes normally consist of old organic fibres which are vulnerable to the environment and quickly degrade, compared to other relics [4]. For this purpose, many science researchers are working on reconstructing historical costumes by 3D technology which not only on the realistic simulation of fabric mechanical properties but also virtual garment on the synthetic avatar and some other aspects.

In order to accurately reconstruct historical costume and improve construction efficiency, the primary factor is to find textile materials that are consistent with historical prototype. In 3D CAD system, not only the surface appearance of the fabric, but also its physical and chemical properties such as buckling ratio, buckling stiffness, density etc. are optional. An overview of academic historic textile and costume collections which presents the guidelines for the management of such collections has been published in the monograph of Welters LM [5]. Kang et al. demonstrated the way about how to create the design and the fabric image in Photoshop CS5, and express surface detail and adjusted the capabilities of different fabrics in DC Suite [6]. Cybulska M. done the exploration about the properties that archaeological textiles had in the past and developed the way of virtual reconstructions of textile [7, 8]. The research team, led by N. Magnenat-Thalmann developed a method of 3D modeling and simulating for different styles of historical costume according to the characteristics of textile materials [9].

However, it is still unable to find the prototypes of historical textile materials of European men's suit during 1850s to 1920s from current scientific research results and the fabric market. And it is also inaccuracy to test textiles from the real historical men's suit directly, because it degraded by many factors. In order to selection the perfect substitute for historical textile materials, this research is going through the following process. Firstly, the structure and layer of each garment of men's suit should be analyzed to determine which textile material need to de study. Secondly, the information of textile material based on books and historical coat prototype will be established and searching for the contemporary substitute. Thirdly, selection the similar textile materials in CLO 3D to prepare for virtual reconstruction. This approach will be helpful not only for traditional museums and material cultural heritage inheritors, but also for the 3D clothing technology and new digital media.

2. Textile Materials for Men'S Suit

From 1850s to 1920s, men's suit usually consists from shirt, drawers, trousers, waistcoat and coat, Theses five garments shown in figure 1 (a) [10, 11]. Figure 1 (b) shows the location of each garment which were formed the men's suit during this period [3].

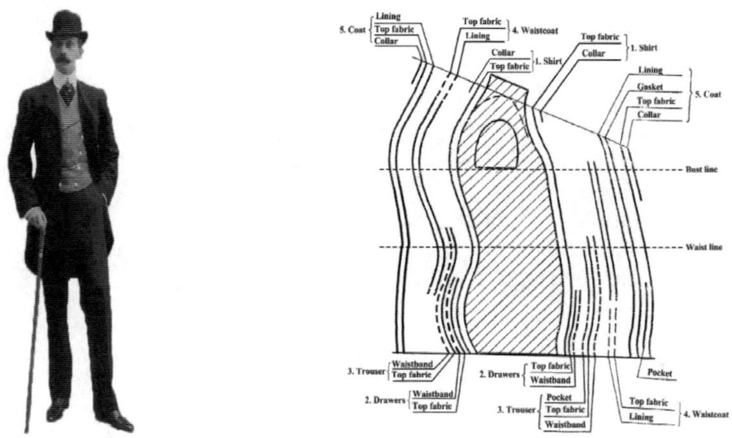

(a) (b)

Figure 1. Men's suit: (a) prototype (1895-1915) [12], (b) cutaway view.

As we can see from figure 1 (b), men's suit was made from many kinds of fabrics at the turn of 20th century, such as cotton, silk, wool, etc., that can let each garment have different functions. According to our research on the layers of men's suit, there are totally nine kinds of textile fabrics which we need to study as shown in table 1.

Table 1. Different historical textile materials of the men's suit.

No.	Garment	Area	No.	Garment	Area
1 & 2	Shirt, drawers	Top fabric	6	Coat	Top fabric, lining on forepart
3	Trouser	Top fabric	7		Pad on forepart
4	Waistcoat	Front	8		Lining on Side, skirt and tail
5		Back, lining	9		Lining on sleeve

3. Methods

For each textile material of men' s suit, the information we can find was limited, so we need to adopt different approaches for different fabrics. Here are two methods for us to speculate the contemporary fabrics which are similar with historical fabrics:

- Finding properties from historical books such as fiber content, type and weight (density).
- Simple testing and measuring of real historical men's coat such as fiber, thickness, weave and fabric count.

4. Results and Discussion

4.1. Results from the 1st method

According to Grace (1923) [13], Dyer (1923) [14], Waugh (1964) [11], Cole and Deihl (2015) [10], the fibers and the type of common fabrics can be find. It should be noted that the thickness of the wool will be changed depending on a season. Weight of fabric per yard has been published on Emanuel (1920) [15], but according to the International System of Units, we need to convert the weight of fabric from oz. per yard to g. per m2. Based on these three basic properties, we could found the similar contemporary textile material from fabric market. Table 2 shows the results of parts of historical fabrics and the similar modern fabrics. The No. of each fabric are same with table 1.

Table 2. Information about historical textile materials for the men's suit and its contemporary substitutes – Part 1.

	Historical			Contemporary
No.	Fiber	Type	Weight (density, g/m^2)	Details(density, g/m^2)
1 & 2	Cotton	Madras, percale, oxford, white cambric	176.48	Cotton, percale, 175
3	Wool	Usually lighter in weight than for the coat, high quality wool, jersey-weaves, merinos, doeskins, cassimere, kerseymere, ratiné.	337.85	Wool, worsted, 150 ~ 337.84
4	Wool	Plain, checks, tweed or cotton pique.	119.8	Wool, worsted plaid, 122.58 ~ 200
5	Silk	Silk satin or brocade, and cotton in different weaves and finishes	82.09	Silk, brocade, 83.33
6	Wool	Superfine wool, worsteds, cassimere, tweeds, vicunas.	506.79	Wool, worsted melton, 300 ~ 506.67
7	Linen	Canvas, cotton and linen mixtures.	---	---
8	Cotton	---	---	---
9	Cotton	---	---	---

4.2. Results from 2nd method

Three fabrics No. 7- No. 9 couldn't be ascertaining by using 1st method so we used 2nd method to find fiber, thickness, weave and fabric count.

Real historical men's coat which was made in 1900s from Garment Design Department (Ivanovo State Polytechnic University) collection as shown in figure 2 (a) has been chosen for measuring. Figure 2 (b) presents the cloth thickness gauge for measuring the fabric thickness. This product has an error of ± 0.01mm. Figure 2 (c) presents the 60-fold fabric analysing glass for obseving the weave and fabric count. Cotton and wool fabrics are expressed in terms of metric density, that is, the number of warp or weft yarns within a width of 10 cm. The two-digit latitude and longitude density of the fabric is indicated by the symbol "X". For example, 236×220 means that the fabric has a warp density of 236 threads/10 cm and a latitudinal density of 220 threads /10 cm [16].

(a) (b) (c)

Figure 2. Object and tools for measuring.

Table 3 presents the results of No.7~9 historical fabrics. Based on these properties, we could found the scope of the similar contemporary fabric.

Table 3. Information about historical textile materials for the men's suit and its contemporary substitutes – Part 2.

	Historical				Contemporary
No.	Fiber	Thickness (mm)	Weave	Fabric count	Details
7	Linen	0.34	Plain	130×110	55% linen, 45% cotton, plain, 0.35mm, 162.5 g/m^2
8	Cotton	0.34	Twill	320×290	Cotton, twill, 0.35mm, 189.19 g/m^2
9	Cotton	0.21	Plain	320×260	Cotton, plain, 0.21mm, 100 g/m^2

4.3. Selection of textile material in CLO 3D for virtual reconstruction

We compared the historical textile materials information with contemporary fabrics in virtual try-on software, especially the type, composition, weight and thickness. Finally, identify the properties of digital contemporary fabrics which are similar to historical fabric prototypes. The selected digital textile materials are shown in table 4 which from library of CLO 3D.

Table 4. Properties of digital textile materials in CLO 3D.

No.	Type	Classification	Weight (density, g/m^2)	Thickness (mm)	Stretch-weft stiffness (g/s^2)	Stretch-warp stiffness (g/s^2)	Bending-weft stiffness (g/mm^2 /s^2 /grad)	Bending-warp stiffness (g/mm^2 /s^2 /grad)
1&2	Cotton	Cotton Oxford	165	0.37	217,880	440,584	1445	1289
3	Wool	Super 120s Wool	157.5	0.27	165,263	182,216	976	976
4	Wool	Super 120s Wool	157.5	0.27	165,263	182,216	976	976
5	Silk	Silk Charmeuse	81.1	0.19	9672	48.714	156	375
6	Wool	Coatweight Twill	345	0.84	368,706	489,034	1300	1300
7	Linen	Linen	173.39	0.36	215,456	799,917	850	1500
8	Cotton	Cotton Gabardine	189	0.35	1,700,000	1,700,000	2200	4500
9	Cotton	50s Cotton Poplin	105.03	0.21	280,769	356,091	938	1289

5. Conclusion

The method of selection textile materials of historical men's suit based on written materials and real historical male coat prototypes have been developed and the information of both real and digital textile materials have been established. Based on these results, we are reconstructing real costume now. At the same time, combined with the digital twin with historical male body and method of historical clothing shaping during making and finishing, the more accurate simulations will be produced.

Acknowledgments

This research is the part of the project No. RFMEFI61619X0113 "Simulation (modeling) of historical costume digital replica by using reverse engineering technology" supported by Ministry of Science and High Education of Russian Federation and French Partenariat Hubert Curien Kolmogorov 2019.

References

[1] Tomas N, Kristina T 2018 *Cultural Heritage Preservation: the Past, the Present and the Future* (Sweden: Halmstad University Press) p 9.

[2] Zhang S C, Kuzmichev V E 2018 Calculation of the body measurements after analyzing the historical pattern block, *IOP Conf. Series: Mater. Sci. Eng.* **459** 2019 012087.

[3] Zhang S C, Kuzmichev V E 2019 Retrospective analysis of European men's suit from 1850s to 1920s for virtual reconstruction, *2019 3rd Int. Conf. Adv. Edu. Manage. Sci. (AEMS 2019).*

[4] Myungsook A 2007 Study on the Conservation and Restoration of Excavated Costume, *J. Korean Soc. Costume* **57(3)** 14-22.

[5] Welters L M, Margaret O 2011 *Historic Textile and Costume Collections in the Academic Setting* ITAA Monograph # 11 International Textile and Apparel Association.

[6] Kang Y, Wu S, Ko Y A, Kim A, Kim N, Kim J C, Cheeyong K, Ko H S 2014 A Study on Men's Wear Production Using Digital Clothing Program, *J. Korea Multimedia Soc.* **17(2)** 266-75.

[7] Cybulska M, Maik J 2007 Archaeological textiles - a need of new methods of analysis and reconstruction, *FIBRES & TEXTILES in Eastern Europe* **18(3)** 100-5.

[8] Cybulska M 2012 To see the unseen - Computer graphics in visualisation and reconstruction of archaeological and historical textiles, *In Mukai N ed Computer Graphics* InTech Rijeka pp 213-28.

[9] Magnenat T N, Volino P 2005 From early draping to haute couture models 20 years of research, *Visual Comput.* **21** 506-19.

[10] Cole D J, Deihl N 2015 *The History of Modern Fashion From 1850* (London: Laurence King Publishing Ltd).

[11] Waugh N 1964 *The Cut of Men's Clothes 1600-1900* (New York: Theatre Art Books).

[12] Information on https://www.pinterest.com/pin/122441683592660108/

[13] Grace G D 1923 *Fabrics* (Chicago: J. B. Lippincott Company).

[14] Dyer E 1923 *Textile fabrics* (Boston: Houghton Mifflin Company).

[15] Emanuel A P 1920 *Hand books of the textile industry vol 4 Fabrics Analysis* (Philadelphia: Textile Publish Company).

[16] Yao M 2015 *Textile Materials* (Beijing: Chine Textile Press).

Silica Aerogel Thermal Insulation Coating as Commodity Usage

Noppakun Sanpo[1], Jaturong Jitputti[1], Peeranut Prakulpawong[2], Chadapon Srikamut[2], Supan Yodyingyong[3], Darapond Triampo[2,]*

[1]SCG Chemical Co., Ltd., Siam Cement Group (SCG), 271 Sukhumvit Road, Muang District, Rayong, 21150, Thailand
[2]Department of Chemistry and Center for Innovation in Chemistry, Faculty of Science, Mahidol University, Phuttamonthon Sai 4 Road, Salaya, Nakhon Pathom, 73170, Thailand
[3]Institute for Innovative Learning, Mahidol University, Phuttamonthon Sai 4 Road, Salaya, Nakhon Pathom, 73170, Thailand

darapond.tri@mahidol.edu

Abstract. Silica aerogel-based thermal insulation coating (SA-coating) was prepared from a commercial acrylic binder. The purpose of this investigation is to determine the effectiveness of SA-coating with the application in energy-efficient home design for temperature insulation purposes. The weather acceleration test and thermal insulation property of SA-coating were investigated and compared to the original commercial binder. The weather acceleration test of SA-coating showed equivalent weathering stability and robustness compared to the original binder. The thermal insulation property was performed from an in-lab setup, called temperature difference (TD) measurement. In a closed chamber, without air circulation, the surface temperature with SA-coating was reduced by as much as 26 degrees from 90°C to ~64°C. More so, if TD measurement was performed in a ventilated area, the temperature can be reduced from 50°C to 36°C (room temperature was 31°C). The thermal conductivity of the coating at different temperatures was also measured. The water contact angle measurements and the scanning electron micrographs showed that SA-coating can be made hydrophilic to hydrophobic by simple abrasion.

1. Introduction

Thermal insulation materials for energy-efficient construction design have been widely discussed and commercialized for applications to lower energy consumption [1-3]. Materials such as, glass fiber, glass wool, or mineral wool, insulation blankets installed under the roof of buildings for cooling purposes [4]. Heat reflective thermal paints and cool-roof shields are a few examples of commercialized products available in the market. The key to these thermal insulation systems is to fill them with porous core material or material with low thermal conductivity (20-80 mW/(m K)) [5-8].

Silica aerogel (SA) is a promising material for insulation applications because of its very low thermal conductivity [9]. The thermal conductivity of silica aerogels is lower than that of still air. Besides its low thermal conductivity, the silica aerogel is extremely lightweight, hydrophobic, and has good fire and acoustic resistances [9-11]. These characteristics make silica aerogel a premier additive material for energy efficiency and protection in building materials and industrial coating.

IOP Conf. Series: Materials Science and Engineering **811** (2020) 012009 doi:10.1088/1757-899X/811/1/012009

In this paper, varying volume ratios of binder to SA were prepared. Temperature difference, thermal conductivity, weathering test, surface microstructure, and hydrophobicity of the coating were investigated to determine the optimum volume of SA needed in an acrylic binder. The objective of this research was to optimize the addition of SA in improving the thermal insulation performance of commercial acrylic binders for application as energy-saving roof-coatings for buildings.

2. Materials and Experiments

2.1. Materials and sample preparation
Commercial acrylic binder was purchased from HomeMart, Thailand. It is a liquid acrylic, waterproof product for roofs and facades with very highly flexible film. Silica aerogel (SA) was obtained from Thillium, Thailand. Tween-20 was purchased from Sigma-Aldrich, USA.

Tween-20 was first dispersed in 100 mL of water in a blender, and then 9.0 g of SA was added to the mixture. The purpose of dispersing SA in the Tween is to make SA more dispersible in acrylic binder. Tween-20 was varied from 0.0 – 0.4 grams. Tween and SA mixture were then blended with the acrylic binder. A series of binder to SA in the volume ratios of 1:1, 1:2, 1:3, 1:4, and 1:5 were prepared. Tween, SA, and binder mixtures were then cast on an aluminum sheet and left for drying at 50°C for further characterization. The casted coatings were of 1-mm and 2-mm thick.

2.2. Temperature difference (TD) measurements
The thermal insulation property was performed from an in-lab setup, called temperature difference (TD) measurement. As shown in figure 1, the temperature of the hotplate serves as the constant heat source, T_source, to be compared to the surface with SA-coating, T_coat. T_source was set to 50, 60, 70, 80, and 90°C.

Figure 1. Setup for temperature difference (TD) measurement.
Data logger attached with two thermocouples records T_source, the surface of the hotplate; and T_coat, the surface of the coating, simultaneously.

2.3. Measurement of thermal conductivity
The thermal conductivity of the coating was measured by using the C-Therm TCi, C-Therm Technologies Ltd. The instrument employs the modified transient plane source for thermal conductivity measurement. The test was conducted at room temperature (31°C), 50, 60, 70, 80, and 90°C. An accuracy check was performed on the instrument before running any tests as standard reference material and confirmed the instrument was performing well within the stated accuracy specification of 5%.

2.4. Weather acceleration test

The weather acceleration test was performed following ASTM D154. The condition runs for 4 cycles. Each cycle runs for 7 days. One cycle estimates 1.5 years equivalent. A cycle consisted of an alternating exposure of condensation at 50°C for 4 hours and exposure of UVB 313 nm at 60°C for 4 hours.

2.5. Water contact angle measurements

The water contact angle measurements were performed using a KRÜSS model G23 apparatus (KRÜSS GmbH, Germany) with a CCD camera (20x objective lens). The contact angle photographs were taken by the ManyCam software. The water drop volume used was 10 μL. The results were measured from the photographs with the ImageJ angle analysis tool. The water contact angles were measured before and after abrasion of the coated surface.

2.6. Scanning electron microscopy (SEM)

A Hitachi SU3500 scanning electron microscope operating at 5 kV was used to capture surface image of the coatings. No sputtering was required because the instrument was operated under low vacuum.

3. Results and Discussion

For the recent years, there has been extreme temperature occurrence all around the world, extreme heat (40°C or higher) has now, seems to be a common phenomenon in some parts of the world. While in other areas, extreme coldness has also been somewhat a familiar occurrence. We cope with these extreme heat with air-conditionings and extreme coldness with heaters. Both of which consume much energy in houses and buildings. For this research work, a simple thermal insulation coating was made by mixing commercial SA with commercial acrylic binder. Table 1 summarized the T_coat measured for 1-mm and 2-mm thickness coating. Volume ratios of binder to SA were varied, 1:1, 1:2, 1:3, 1:4, and 1:5. TD were measured with the heat source (T_source) at 60°C, 70°C, 80°C, and 90°C. As expected the thicker the coating, more TD could be achieved.

Table 1. Temperature difference measurements of (a) 1- and (b) 2-mm thickness coating with 0.4 g TWEEN.

(a) Coating thickness 1 mm

Ratio of Binder to Silica aerogel	T_coat (°C)			
	Set T_source 60°C	Set T_source 70°C	Set T_source 80°C	Set T_source 90°C
1:0	59.7 ± 0.5	71.7 ± 1.0	81.7 ± 1.0	91.3 ± 0.8
1:1	57.0 ± 1.0	65.3 ± 0.6	71.7 ± 0.6	74.3 ± 0.6
1:2	56.3 ± 0.6	61.7 ± 1.2	69.7 ± 0.6	72.3 ± 0.6
1:3	53.3 ± 1.2	57.3 ± 0.6	67.7 ± 0.6	70.3 ± 0.6
1:4	51.7 ± 0.6	55.3 ± 0.6	66.3 ± 0.6	68.7 ± 0.6
1:5	50.0 ± 1.0	53.7 ± 0.6	64.7 ± 0.6	67.0 ± 1.0

(b) Coating thickness 2 mm

Ratio of Binder to Silica aerogel	T_coat (°C)				
	Set T_source 50°C (ventilated area)	Set T_source 60°C	Set T_source 70°C	Set T_source 80°C	Set T_source 90°C
1:0	51.3 ± 0.5	59.3 ± 0.6	70.7 ± 1.0	79.7±0.5	89.3±0.6
1:1	46.7 ± 0.8	54.0 ± 1.0	58.0 ± 1.0	68.3±0.6	70.7±0.6
1:2	42.5 ± 0.9	52.3 ± 0.6	55.7 ± 0.6	65.3±0.6	68.7±0.6
1:3	38.8 ± 1.1	50.0 ± 1.0	52.7 ± 0.6	62.0±1.0	66.7±0.6
1:4	38.2 ± 0.9	49.0 ± 1.0	52.3 ± 0.6	60.3±1.2	64.3±0.6
1:5	36.1 ± 0.9	48.0 ± 1.0	51.3 ± 0.6	58.7±0.6	63.7±1.2

Moreover, when TD was measured with T_source at 50°C in a ventilated area, T_coat on the surface of SA-coating was only ~35°C, slightly warmer than room temperature by 4°C. This is significant because it best simulates the real usage of the coating, i.e., in a ventilated environment. T_coat could be much lower if the environment had a cooling system, like air-conditioning. We believe that this SA-coating can be used in homes to reduce temperature fluctuation and help minimize energy consumption in homes, office buildings, and industrial constructions.

Thermal conductivity of SA-coating increased with increasing temperature, table 2. It is unclear why for the same coating at higher temperature, though, having higher thermal conductivity; it was able to obtain a larger temperature difference.

Table 2. Thermal conductivity of SA-coating measured at different temperatures.

Ratio of Binder to Silica aerogel	Thermal conductivity ($mW\ m^{-1}\ K^{-1}$) @ 31°C	@ 50°C	@ 60°C	@ 70°C	@ 80°C	@ 90°C
1:0 (pure binder)	201 ± 34	n/a	n/a	n/a	n/a	n/a
1:1	129 ± 6	133 ± 6	161 ± 9	180 ± 14	185 ± 4	193 ± 8
1:4	66 ± 2	68 ± 4	72 ± 3	81 ± 4	84 ± 3	88 ± 8

*n/a, the thermal conductivity exceeds the limit of the instrument

The effect of increasing SA in the composite largely decreased thermal conductivity. Comparing the ratios 1:1 and 1:4 with T_source of 70°C, 2-mm thick; though, the thermal conductivity was 180 ± 14 $mW\ m^{-1}\ K^{-1}$ to 81 ± 4 $mW\ m^{-1}\ K^{-1}$, respectively. But T_coat of 1:1 was 58.0 ± 1.0 °C compared to 52.3 ± 0.6 °C that is only ~6°C difference. While when the thickness of the coating was increased from 1 mm to 2 mm, the temperature difference measurement at 70°C of 1:1 ratio could also be reduced by ~6°C (65.3 ± 0.6 °C as to 58.0 ± 1.0 °C, respectively). We concluded here that the huge difference in the value of the thermal conductivity does not necessarily be reflected in the temperature difference measurement. The thickness of the coating, however, has more contributions to the ability of a coating to reduce temperature in this case.

Weathering test (following ASTM D154) is a common standard used to test weathering stability and robustness of coatings. The test was designed to expose the coating to a simulated accelerated outdoor conditions with alternating UV exposures and humidity exposures. Chalky failure is one criteria in the standard used to determine whether the coating passes or fails. A standard cloth of testing is used to rub on the coating after each cycle; if chalk like material or debris is detected on the cloth, then the coating failed to pass the test. Chalky failure was not observed for all ratios of SA-coating of at least 4 cycles (or an equivalent of 6 years). SA-coating showed equivalent weathering stability and robustness compare to that of the original binder.

Water contact angle measurements characterized the hydrophilicity or hydrophobicity of a surface. Table 3 tabulated the contact angle of all ratios. Pure binder (1:0) had contact angle of 60 degrees and after abraded it increased to 63 degrees. This was not a significant increase. During the drying process, the more hydrophilic part of the binder was air-dried on the surface, but after abrasion the inner less hydrophilic surface of the binder was exposed.

Table 3. Water contact angle on SA coating of different ratios of Binder to Silica Aerogel before and after abrasion.

Binder : Silica Aerogel	Contact angle before abrasion	Contact angle after abrasion
1:0 (pure binder)	60	63
1:1	83	101
1:2	92	124
1:3	97	141
1:4	101	141
1:5	103	144

In 1:1 ratio, the contact angle was increased to 83 degrees before abrasion and 101 degrees after abrasion. This was because the original binder is slightly hydrophilic in nature, but SA is hydrophobic. With the exposure of SA to the surface after abrasion, the surface became hydrophobic. For volume ratios of binder:SA, with SA greater than 2 parts, the coating became hydrophobic because more SA could be found on the surface of the coating. Scanning electron micrographs of SA coating, figure 2, had irregular openings with SA protruding from the coating.

Figure 2. Scanning electron micrographs of SA coating of at (a) 50x and (b) 1,000x magnification.

4. Conclusions

In this work, a simple thermal insulation coating was made by mixing commercial silica aerogel with commercial acrylic binder. Weathering test of silica aerogel coating showed equivalent weathering stability and robustness compare to that of the original binder. Following ASTM D154, all ratios of silica aerogel coating passed at least 4 cycles (or approximately 6 years equivalent). The water contact angle measurements showed that silica aerogel coating could be made hydrophilic or hydrophobic depending of the amount of silica aerogel added. The volume ratio of binder to silica aerogel, with silica aerogel greater than 2 parts will give a hydrophobic surface. Scanning electron micrographs show silica aerogel protruding from coating.

In a closed chamber, the temperature difference measurement of silica aerogel coating can reduce the surface temperature by an average of 25%. But thermal insulation can be greatly appreciated in a ventilated environment, temperature can be reduced from a heat source of 50°C to 35°C (slightly warmer than room temperature by 4°C). Tween of 0.4 g gives best processing in blending silica aerogel into the binder.

We believe that this silica aerogel coating can be used to reduce temperature fluctuations in homes and help minimize energy consumption in homes, office buildings, and industrial constructions.

Acknowledgements

The financial support from the Center of Excellence for Innovation in Chemistry (PERCH-CIC), Ministry of Higher Education, Science, Research, and Innovation is thankfully acknowledged. Furthermore, this work was also partially supported by the Faculty of Science, Mahidol University; Thailand Research Fund (grant no. IRG5980007); NSTDA-EGAT joint program (grant no. FDA-CO-2561-7467-TH); Faculty of Graduate Studies, Mahidol University; Institute for Innovative Learning, Mahidol University; and the Center of Nanoimaging, Mahidol University.

References

[1] Cuce E, Cuce P M, Wood C J, Riffat S B 2014 Toward aerogel based thermal superinsulation in buildings: a comprehensive review, *Renew. Sustain. Energy Rev.* **34** 273-99.

[2] Jelle B P 2011 Traditional, state-of-the-art and future thermal building insulation materials and solutions–Properties, requirements and possibilities, *Energy Build.* **43** 2549-63.

[3] Al-Homoud M S 2005 Performance characteristics and practical applications of common building thermal insulation materials, *Build. Environ.* **40** 353-66.

[4] Lyons A 2010 Materials for Architects and Builders, fourth ed., Elsevier Ltd., United Kingdom.

[5] Jelle B P, Gao T, Sandberg L I C, Ng S, Tilset B G, Grandcolas M, Gustavsen A 2015 Development of nano insulation materials for building constructions, *Proc. of 5th Int. Symp. Nanotechnology in Construction (NICOM5)* 429-34.

[6] Gao T, Jelle B P, Sandberg L I C, Gustavsen A 2013 Monodisperse hollow silica nanospheres for nano insulation materials: synthesis, characterization, and life cycle assessment, *ACS Appl. Mater. Interfaces* **5** 761-7.

[7] Sandberg L I C, Gao T, Jelle B P, Gustavsen A 2013 Synthesis of hollow silica nanospheres by sacrificial polystyrene templates for thermal insulation applications, *Adv. Mater. Sci. Eng.* **2013** 1-6.

[8] Jelle B P, Gustavsen A, Baetens R 2010 The path to the high performance thermal building insulation materials and solutions of tomorrow, *J. Build. Phys.* **34** 99-123.

[9] Hüsing N, Schubert U 1998 Aerogels—Airy Materials: Chemistry, Structure, and Properties, *Angew. Chem. Int. Ed.* **37** 22-45.

[10] Gao T, Jelle B P, Gustavsen A, Jacobsen S 2014 Aerogel-incorporated concrete: an experimental study, *Constr. Build. Mater.* **52** 130-6.

[11] Baetens R, Jelle B P, Gustavsen A 2011 Aerogel insulation for building applications: a state-of-the-art review, *Energy Build.* **43** 761-9.

[12] Kiil S 2015 Quantitative analysis of silica aerogel-based thermal insulation coatings, *Prog. Org. Coat.* **89** 26-34.

IOP Conf. Series: Materials Science and Engineering **811** (2020) 012010 doi:10.1088/1757-899X/811/1/012010

Copper-Zinc Oxide Synergistic Approach as Low-Emissivity Material for Energy-Saving Windows

Sireethorn Termkleebbuppa[1], Supan Yodyingyong[2], Jeerapond Leelawattanachai[3], Wannapong Triampo[4], Noppakun Sanpo[5], Jaturong Jitputti[5], Darapond Triampo[1]

[1]Department of Chemistry and Center for Innovation in Chemistry, Faculty of Science, Mahidol University, Phuttamonthon Sai 4 Road, Salaya, Nakhon Pathom, 73170, Thailand
[2]Institute for Innovative Learning, Mahidol University, Phuttamonthon Sai 4 Road, Salaya, Nakhon Pathom, 73170, Thailand
[3]National Science and Technology Development Agency (NSTDA), 111 Thailand Science Park, Phahonyothin Road, Khlong Nueng, Khlong Luang, Pathum Thani 12120, Thailand
[4]Department of Physics, Faculty of Science, Mahidol University, Phuttamonthon Sai 4 Road, Phuttamonthon, Salaya, Nakhon Pathom, 73170, Thailand
[5]SCG Chemical Co., Ltd., Siam Cement Group (SCG), 271 Sukhumvit Road, Muang District, Rayong, 21150, Thailand

darapond.tri@mahidol.edu

Abstract. In luxurious glass buildings and constructions, heat gain (or heat loss) mostly comes through the windows. Infrared (IR) radiation is an important factor that causes the entry of heat into the buildings. To save energy consumption of air conditioners, low-emissivity (Low-E) glass coating applications are focused. In the energy-saving field, transparent conductive oxide (TCOs) are used as coatings to minimized IR entry. In this work, copper-doped zinc oxide (CZO) is synthesized for thermal reflective material. Cu(II) ions are doped to ZnO by a sol-gel method to obtain CZO nanoparticles and is coated on a glass substrate. The morphology of CZO is investigated with scanning electron microscopy (SEM); phase crystallinity is determined by X-ray diffraction (XRD); and UV-Vis-NIR spectroscopy is used to characterize UV/IR-shielding and also the optical transparency. IR cameras and in-lab thermal insulation setup are used to test the heat insulation properties. The result shows that 15% copper-doped zinc oxide has the best insulation from IR rays with the lowest temperature in the interior (T3) of 50.6°C while the outer temperature (T2) was at 85.5°C the lowering of the temperature by 34.9°C (cooler by 59%). CZO synthesized from the sol-gel method has promising properties for Low-E glass coating applications.

1. Introduction
It is undeniable that glass buildings look attractive and luxurious. However, glass in building constructions is a poor thermal insulator and it serves as an opening entry to solar radiation. Solar energy distribution comprises of 5% from UV (295 - 400 nm), 50% from visible (400 - 700 nm) and 45% from IR region (700 - 2500 nm). The majority of the IR rays from 700 – 1100 nm radiates heat into the

atmosphere [1]. Blocking of the NIR radiation is an important factor to study in fabricating energy-saving window.

Transparent conductive oxides (TCO) are promising candidate as energy-saving window because it can reflect (or sometimes absorb) NIR while allowing visible light to be transmitted. Examples of TCO in various applications are indium tin oxide (ITO) used in liquid crystal displays (LCDs) [2], fluorine-doped tin oxide (FTO) in solar cells [3], zirconium-doped indium oxide (ZrIO) in organic light-emitting diodes (OLEDs) [4], indium-doped cadmium oxide (ICO) in optoelectronic devices [5] and more. Zinc oxide (ZnO) by itself absorbs UV and NIR, but is not transparent to visible light. Doping impurity to ZnO, such as, B-doped ZnO, Ti-doped ZnO [6], Al-doped ZnO [7], Sn-doped ZnO [8] are among the many dopings of ZnO that can improve transmission of visible light. Copper-doped ZnO (CZO) is to be consider in this research.

CZO is regarded as a p-type semiconductor. Previous literatures have determined that 5% doping of Cu is the maximum limit to dope into ZnO [9-12]. If the concentration of Cu exceeds 5%, CuO is formed and would result in a p-n junction in the nanoscale, of ZnO as n-type and CuO as p-type [13]. The synergistic effect of ZnO-CuO has been utilized in photocatalytic application [14].

To obtain CZO nanoparticle, there are many methods from previous research, such as, by sol-gel [15], co-precipitation [16], vapor transport method [17], solvothermal synthesis [18], and RF magnetron sputtering [19]. CZO nanoparticles were used in many applications, for example, photoluminescence, transparent electrode, and electroluminescence material.

The purpose of this work is to fabricate CZO by sol-gel method as a thermal reflective material for energy-saving window coating applications. Different mole percentages of Cu will be loaded to ZnO. Effect of percentage of dopant will be described in this work. Different amount of dopant was varied from 0 to 25 mol% Cu. Calcination condition was fixed at 550°C for 3 hours. Morphology of CZO particles was determined by SEM; and crystallinity was verified by XRD characterization. IR camera was used to study IR reflective property by detecting the temperature on the surface of the coated glass. An in-lab thermal insulation set-up was used to examine the heat insulation property of CZO nanoparticles.

2. Experimental

2.1. Materials

Zinc acetate dihydrate ($Zn(CH_3COO)_2{\cdot}2H_2O$), monoethanolamine (MEA, $H_2N(CH_2CH_2OH)$, antimony doped tin oxide (ATO) and copper (II) acetate monohydrate ($Cu(CH_3COO)_2{\cdot}H_2O$) were purchased from Sigma-Aldrich. 2-propanol ($CH_3C_2H_5OH$) was purchased from Fluka. All chemicals were used without further purification.

2.2. Sol-Gel Preperation of Cu-ZnO

CZO nanoparticles were synthesized following previous research by Jongnavakit *et. al.* [20] and Saidani *et.al.* [21]. First, 1 mL of MEA was added to 50 mL isopropanol and stirred for 5 minutes at 65°C. Then $Zn(CH_3COO)_2{\cdot}2H_2O$ was added to the isopropanol prepared and stirred at 65ºC for another 1 hour to make 0.3M zinc-acetate solution (ZnAcetate-solution). Similarly, 0.15M copper-acetate solution (CuAcetate-solution) was prepared from $Cu(CH_3COO)_2{\cdot}H_2O$. The CuAcetate-solution was then added drop-wised to ZnAcetate-solution at 0, 1, 5, 10, 15, 20, and 25 mol% Cu. The two solutions were stirred together for another 1 hour at 65°C. The mixtures of ZnAcetate and CuAcetate solutions were of no precipitates. Then the solutions were calcined at 550°C for 3 hours. The final product of CZO nanoparticles were obtained with grey colour.

2.3. Cu-ZnO Characterization

The crystallinity of the nanoparticles was studied with X-ray diffractometer (XRD, D8 Advance Bruker) using Cu Kα radiation operated at step size 0.0200 degrees from 20° to 80° with 1.000 second time per step. The scanning electron micrographs of undoped and Cu-doped ZnO were taken by SU3500 Hitachi at different magnifications from 10 Kx to 100 Kx. All samples of 0 - 25% CZO were prepared with the same condition. The UV/IR-shielding and optical transparency of the nanoparticle were characterized by UV-Vis-NIR spectrometer (model Carry 5000, Agilent Technologies).

2.4. Thermal insulation properties and application of Cu-ZnO

To study the emissivity of CZO nanoparticles, 100 mg of CZO powder of different mol% Cu were re-dispersed in isopropanol and coated on a glass slide. Infrared lamp (150 watts) was turned-on and shined onto the glass slide for 120 seconds. Infrared (IR) thermal photographs of the coated glass slide were taken every 30 seconds while the IR-lamp was turned-on and continued to take the thermal images after the lamp was turned-off for another 240 seconds. The infrared thermal camera (Fluke TIS20) was used to determine the temperature on the coated glass surface. The set-up is shown in figure 1. Commercial ATO and bare glass are used as the reference standard for comparison.

To study the thermal insulation property application, an in-lab set-up to mimic a house with a window that is exposed to sunlight was used, set-up shown in figure 2. The coated glass slide was placed on the foam box with a cut out opening with the size of the glass slide. IR lamp (150 watts) shined onto the sample for 60 minutes. The data-logger measured the temperature at T1 (5.5 cm from the light source) with an average temperature of 68.6°C. The data-logger at T2 presents the temperature close to the outer glass surface of each sample. T3 and T4 measure the temperature inside the box where T3 is close to the glass surface and T4 is further away representing the common areas where we would normally stay in a room. All 4 areas (T1 – T4) were measured at every 1 minute and detected by a data-logger (Testo175T3). The set-up is modified from previous work [22].

Figure 1. Equipment set-up used to determine the emissivity property of CZO nanoparticles.

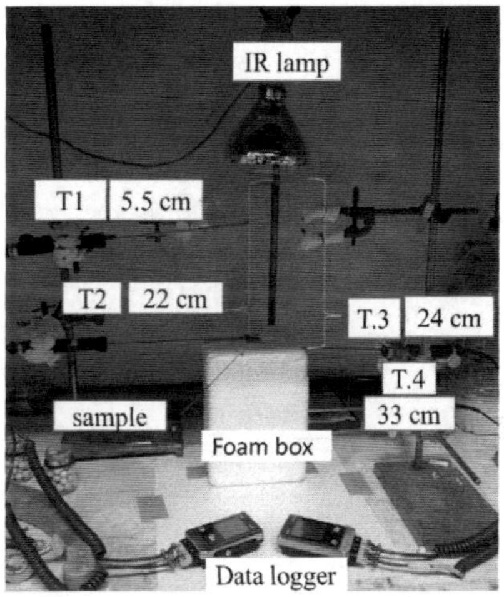

Figure 2. In-lab thermal insulation equipment used to study the thermal insulation property.

3. Results and Discussion

XRD patterns, figure 3(a), gave many diffraction peaks which implied that ZnO grew in random orientation. Three main peaks of ZnO figure 3(b) corresponded to (100), (002), and (101) indicate that undoped ZnO and Cu-doped ZnO had hexagonal wurtzite structure [23]. With 1 – 5 mol% Cu, the spectrum in figure 3(b) shifted to a higher angle and new absorption peaks were not found because Cu ions were completely substituted in the ZnO lattice [20]. On the other hand, with 10% Cu new absorption peaks appeared at (111) and (200) referring to CuO phase [24, 25]. With 10 – 25 mol% Cu, the more Cu was added, an increase of CuO phase was found due to the excess of Cu ions. This is because the maximum solubility limit of Cu ions in ZnO is 5 mol% [9-12]. When Cu is added in large quantities, some Cu ions will substitute the Zn lattice and some excess Cu filled the vacancy of ZnO lattice resulting in larger particle size and higher intensity of CuO phase [26]. The peak intensity of ZnO decreased with increase mol% of Cu [27]. The intensity of CuO phase in XRD pattern of 15%, 20%, and 25% CZO is much higher than of 10% CZO because more Cu ions cannot substitute into the ZnO lattice, and CuO phase were formed instead.

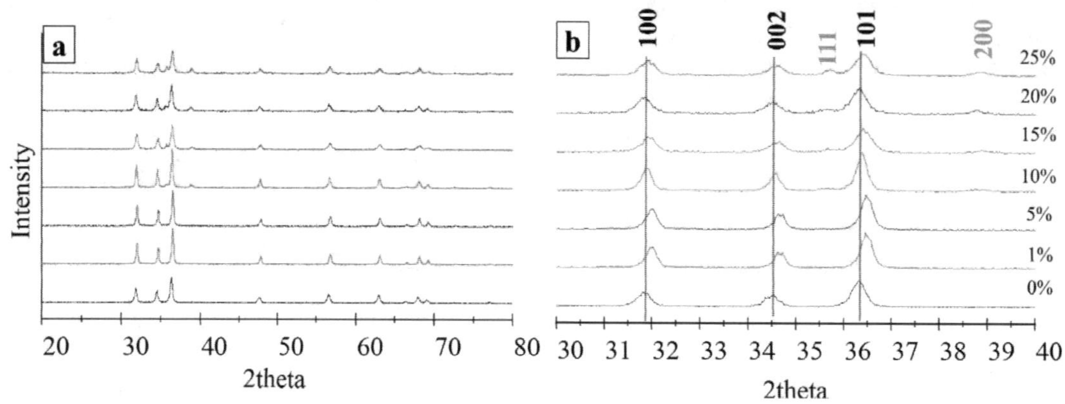

Figure 3. XRD spectra of CZO nanoparticle with different mol% of Cu (a) scanning from 20° - 80° and (b) a zoom-in of (a) from 30° – 40° to show the three dominant peaks of ZnO.

From the SEM micrographs in figure 4 when %Cu is less than 10% CZO, samples had the same morphology, this indicate that %Cu had no effect on the morphology of CZO nanoparticles. But when %Cu was greater than 10%, CZO particles showed more spherical shape aggregation as a result of an excess of Cu ions which filled in the vacancy of ZnO lattice. The excess of Cu caused a lattice alteration in 15 - 25% CZO which showed larger particle size than 0 - 10% CZO [16, 27].

Figure 5 plotted the %reflection of CZO in UV-Vis-NIR region from 200 – 2400 nm. In general, there are two types of IR-shielding material. One is NIR reflective material or low-emissivity coating. Another is NIR absorptive material used in solar-thermal energy conversion. UV-Vis-NIR spectra indicate that CZO has ability to reflect wavelengths in the NIR region while commercial ATO has outstanding IR absorption property. Comparing among 1 – 10% CZO in the range of 700 - 1100 nm found that 1% CZO has ability to reflect in NIR region higher than 5% and 10% CZO. For longer wavelength than 1100 nm, 5% and 10%Cu have very similar reflection values, and both 5% and 10% CZO have a better reflection ability than 1% CZO. However, excess mol% of Cu result in less reflection in visible region. For lowest %reflection in visible region, 10% CZO would have a better visible transparency. In 15 – 25% CZO, they have higher %reflection in both visible and NIR region compare to 1 - 10% CZO this is because of the larger particle size found in 15 – 25% CZO. From these results, 15% CZO is the most reasonable to use for NIR reflective material because the reflection of NIR region is the highest and reflects visible light only about 36%.

Figure 6 plotted the temperatures on the coated glass slide taken from the thermal IR camera while the IR lamp was turned-on (from 0 – 120 seconds) and turned-off (from 120 – 360 seconds). Consider for 0 – 10% CZO, it was found that the more mol% of Cu, the higher the temperature on glass surface. However, with excess of %Cu (15 – 25% Cu), the surface temperature reduced from that of 10% CZO. Nevertheless 15% CZO had the best IR insulation property which will be described next as to why it is so. Series of thermal photographs of 15% CZO taken every 30 seconds obtained from the IR camera were shown in figure 7. Similar thermal insulation experiments like in figure 7 were conducted for all samples. Data from figure 7 were used to plot figure 6.

Results from the in-lab set-up to mimic a house with a glass window, set-up shown in figure 2, the average temperatures at T2, T3 and T4 were plotted in a bar graph in figure 8. Considering 0 – 10% CZO, it indicated that the more mol% Cu, the higher temperature at T2. Hence 10% CZO displays the highest temperature at T2 (88.8°C) and the lowest temperatures at T3 (51.8°C) and T4 (41.1°C). This implied that 10% CZO was suitable to block the transmission of IR rays than 0 – 5% CZO. The higher temperature at T2 means a better reflectivity to the IR rays which is proper for Low-E glass application since IR rays cannot pass through the coated glass. Although the surface of coated glass of commercial

ATO showed the highest temperature (94.9°C), the temperature inside the box (T3 and T4) was still higher than of 10% CZO. This is due to the NIR absorption property ATO. ATO absorbs the IR radiation on the outer surface (T2) and the accumulated heat is transferred into the box. Conversely, 10% CZO reflected IR radiation causing the temperature at T2 to increase, however the heat is not cumulated in the coated glass and is not transferred to the interior of the box resulting in a lower T3 compare to ATO.

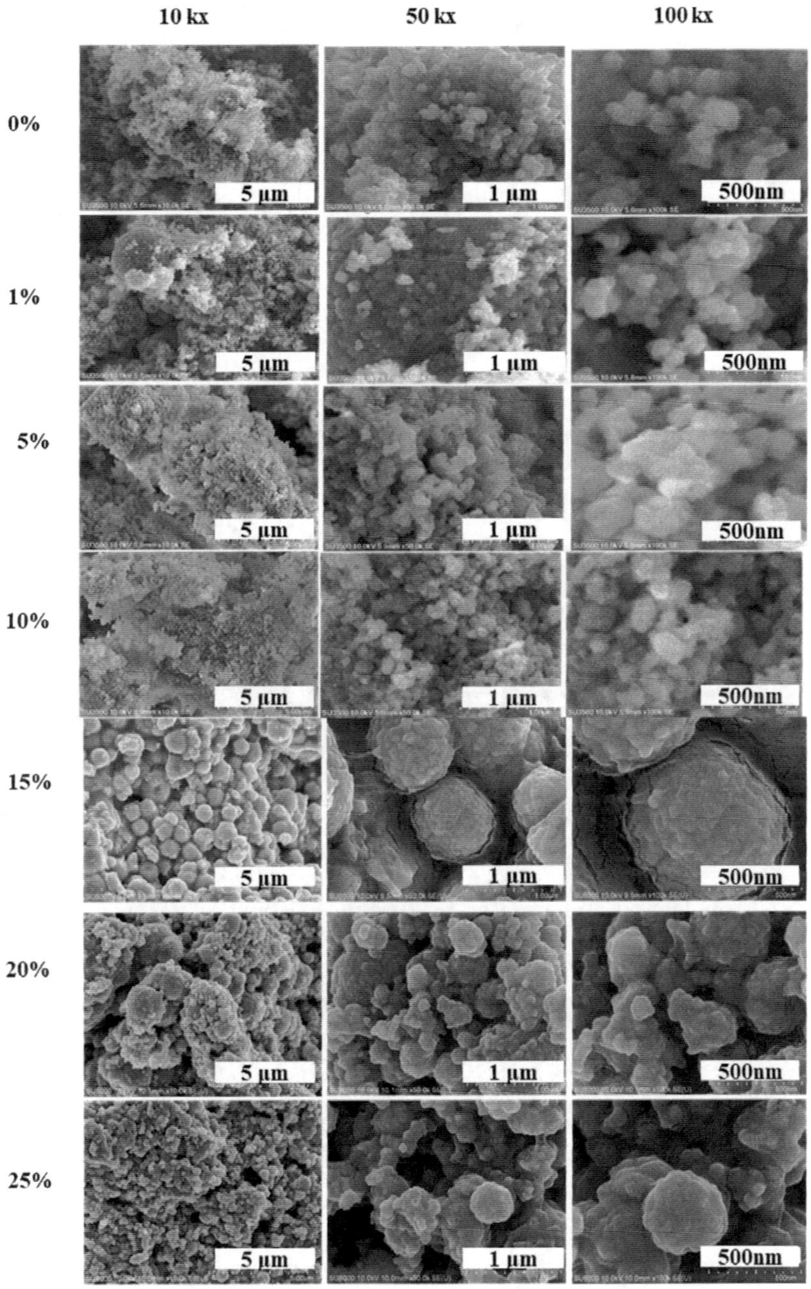

Figure 4. SEM micrographs of undoped and doped ZnO with 0 – 25 mol% Cu at different magnifications.

For 15 – 25% CZO, with n-ZnO and p-CuO (a nano p-n junction), 15% CZO was most suitable as heat insulation with large uniform shape and particle size, and the high %reflection of the NIR region, and the lowest temperature for T3 and T4. It is to be noted that the temperature difference (figure 8) for the different samples were rather small, this is because our coated glass slide covered only ~5% of the total exposure area. A more significant difference is expected with larger area of the coated glass or windows.

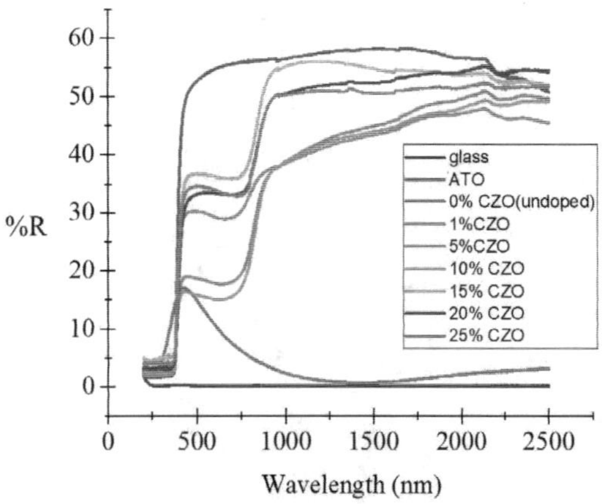

Figure 5. UV-Vis-NIR spectra in reflection mode.

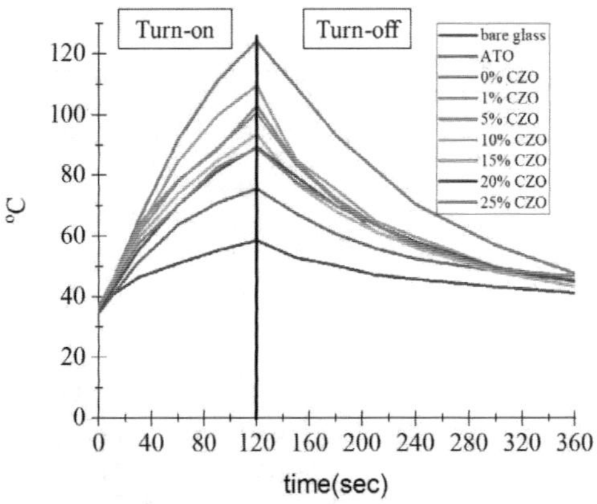

Figure 6. Plot of the temperatures on the coated glass slide when the IR lamp was turned-on and turned-off.

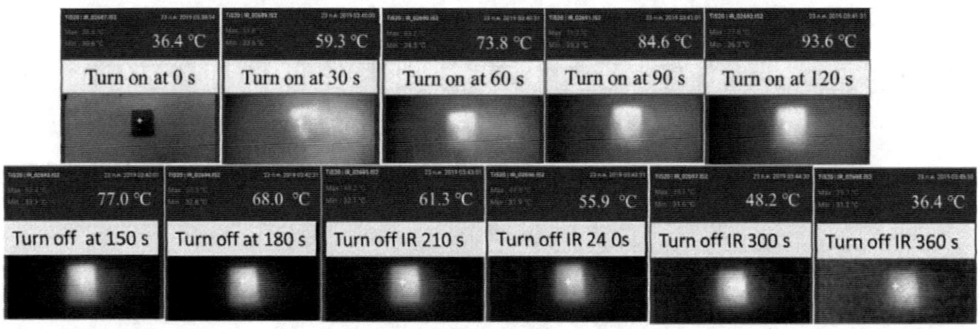

Figure 7. Thermal photographs of 15% CZO obtained from IR camera at different times.

4. Conclusions

Copper-doped zinc oxide (CZO) was successfully prepared from $Zn(CH_3COO)_2$ $2H_2O$ and $Cu(CH_3COO)_2$ $2H_2O$ with MEA and isopropanol as a stabilizer and solvent, respectively. To obtain the CZO powder the solution was calcined at 550°C for 3 hours. CZO powder was confirmed by XRD spectra and found that 1 – 5 mol% Cu does not change the ZnO structure. All of samples still have a high degree of crystallinity with a hexagonal wurtzite structure. When mol% of Cu was higher than 5%, the excess Cu ions formed CuO as shown in the XRD spectra. Because the solubility limit of Cu ions in ZnO structure was 5%, the morphology of CZO was the same as undoped ZnO. However, with excess Cu ions, 15 – 25 mol% CZO, the nanoparticles exhibit larger particle size.

The emissivity of each sample were examined by the IR thermal camera. For 0 – 10% CZO, increasing mol% of Cu can promote the IR reflective property of CZO. High IR reflective characteristic of CZO lead to high temperature on outer glass surface (T2). From the emissivity experiment, it proved that 10% CZO was the best sample as an insulation material. The emissivity results were consistent with the thermal insulation experiment because the lowest temperature at T3 and T4 were also from 10% CZO.

On the other hand, for 15 – 25% CZO synergistic effect of n-ZnO and p-CuO, 15% CZO yielded the most appropriate mol% Cu with the lowest temperature inside the insulation box (T3). Notwithstanding this synergistic effect of 15% CZO had more influence than the heat reflection of 10% CZO. The 15% CZO is a better heat insulator than 10% CZO. UV-Vis-NIR spectra indicated that 15% CZO had the highest % reflection of IR region and reflects lesser of visible spectrum which is a bonus for the material in transparent coating application. Finally, the addition of Cu impurity to ZnO can improve ZnO property, i.e. reducing % reflection in visible region and promote IR insulation property.

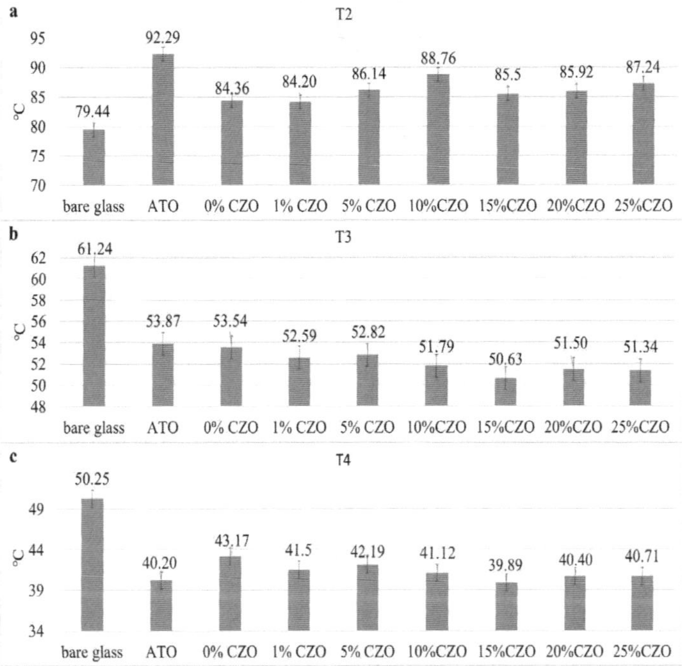

Figure 8. Bar graphs of the average temperature at (a) T2 (b) T3 and (c) T4 while the IR lamp was turned-on with the average source temperature of ~70°C.

Acknowledgments
The financial support from the Center of Excellence for Innovation in Chemistry (PERCH-CIC), Ministry of Higher Education, Science, Research, and Innovation; the Faculty of Science, Mahidol University; the Thailand Research Fund (grant no. IRG5980007); the NSTDA-EGAT joint program (grant no. FDA-CO-2561-7467-TH); the Faculty of Graduate Studies, Mahidol University; the Institute for Innovative Learning, Mahidol University; and the Center of Nanoimaging, Mahidol University are thankfully acknowledged.

References

[1] Bendiganavale A K, Malshe V C 2008 Infrared Reflective Inorganic Pigments, *Chem. Engr.* **1** 67-79.
[2] Minami T 2008 Present status of transparent conducting oxide thin-film development for Indium-Tin-Oxide (ITO) substitutes, *Thin Solid Film* **516** 5822-8.
[3] Ikhmayies S, Ahmad-Bitar R 2008 Effect of Processing on the Electrical Properties of Spray-Deposited SnO_2:F Thin Films, *Am .J .Appl .Sci* .**6** 672-7.
[4] Kim H, Horwitz J S, Kushto G P, Qadri S B, Kafafi Z H, Chrisey D B 2001 Transparent conducting Zr-doped In_2O_3 thin films for organic light-emitting diodes, *Appl. Phys. Lett* .**78** 1050-2.

[5] Gupta R K, Ghosh K, Patel R, Mishra S R, Kahol P K 2008 Structural, optical and electrical properties of In doped CdO thin films for optoelectronic applications, *Mater .Lett* .**62** 337-5.

[6] Soumya S, Mohamed A P, Mohan K, Ananthakumar S 2015 Enhanced near-infrared reflectance and functional characteristics of Al-doped ZnO nano-pigments embedded PMMA coatings, *Sol. Energ. Mat. Sol.* C **143** 335-46.

[7] Shahzad M B, Qi Y, Lu H, Wang X 2013 A study on the Al doping behavior with sol aging time and its effect on structural and optical properties of sol–gel prepared ZnO thin films, *Thin Solid Films* **534** 242-8.

[8] Soumya S, Sheemol V N, Amba P, Mohamed A P, Ananthakumar S 2018 Sn and Ag doped ZnO quantum dots with PMMA by in situ polymerization for UV/IR protective, photochromic multifunctional hybrid coatings, *Sol. Energ. Mat. Sol.* C **174** 554-65.

[9] Fu M, Li Y, Wu S, Lu P, Dong F 2011 Sol-gel preparation and enhanced photo catalytic performance of Cu-doped ZnO nanoparticles, *Appl. Surf. Sci.* **258** 1587-91.

[10] Singhal S, Kaur J, Namgyal T, Sharma R 2012 Cu-doped ZnO nanoparticles: Synthesis, structural and electrical properties, *Physica* B **407** 1223-6.

[11] Elilarassi R, Chandrasekaran G 2013 Structural, optical and electron paramagnetic resonance studies on Cu-doped ZnO nanoparticles synthesized using a novel auto-combustion method, *Frontiers Mater. Sci.* **7** 196-201.

[12] Agarwal D C, Singh U B, Gupta S, Singhal R, Kulriya P K, Singh F, Tripathi A, Singh J, Joshi U S, Avasthi D K 2019 Enhanced room temperature ferromagnetism and green photoluminescence in Cu doped ZnO thin film synthesised by neutral beam sputtering, *Scientific Report* **9** 6675.

[13] Qin C, Wang Y, Gong Y, Zhang Z, Cao J 2019 CuO-ZnO hetero-junctions decorated graphitic carbon nitride hybrid nanocomposite: Hydrothermal synthesis and ethanol gas sensing application, *J. Alloys Compd.* **770** 972-80.

[14] Benxia L, Yanfen W 2010 Facile synthesis and photocatalytic activity of ZnO_CuO nanocomposite, *Superlattice Microst.* **47** 615-23.

[15] Wang D, Zhou J, Liu G 2009 The microstructure and photoluminescence of Cu-doped ZnO nano-crystal thin films prepared by sol–gel method, *J. Alloys Compd.* **487** 545-9.

[16] Muhammad S, Inam U, Khan M I, Jamshid K, Yaqoob K M, Tauseef Q M 2018 Structural and optical properties of pure and copper doped zinc oxide nanoparticles, *Results in Physics* **9** 1301-9.

[17] Mohana R, Krishnamoorthy K, Kim S J 2012 Enhanced photocatalytic activity of Cu-doped ZnO nanorods, *Solid State Commun* .**152** 375-80.

[18] Wu C, Shen L,Yu H, Zhang Y C, Huang Q 2012 Solvothermal synthesis of Cu-doped ZnO nanowires with visible light-driven photocatalytic activity, *Mater .Lett* .**74** 236-8.

[19] Girija K G, Somasundaram K, Topkar A, Vatsa R K 2016 Highly selective H2S gas sensor based on Cu-doped ZnO nanocrystalline films deposited by RF magnetron sputtering of powder target, *J .Alloys Compd* **684** 15-20.

[20] Jongnavakit P, Amornpitoksuk P, Suwanboon S, Ndiege N 2012 Preparation and photocatalytic activity of Cu-doped ZnO thin films prepared by the sol–gel method, *Appl. Surf. Sci.* **258** 8192-8.

[21] Saidani T, Zaabat M, Aida M S, Benaboud A, Benzitouni S, Boudine A 2014 Influence of annealing temperature on the structural, morphological and optical properties of Cu doped ZnO thin films deposited by the sol–gel method, *Superlattice Microst.* **75** 47-53.

[22] Long J, Jiang C, Zhu J, Song Q, Hu J 2019 Controlled TiO_2 coating on hollow glass microspheres and their reflective thermal insulation properties, *Particuology*.

[23] Yuan H, Xua M, Huang Q Z 2014 Effects of pH of the precursor sol on structural and optical properties of Cu-doped ZnO thin films, *J. Alloys Compd.* **616** 401-7.

[24] Joshi K, Rawat M, Gautam S K, Singh R G, Ramola R C, Singh F 2016 Band gap widening and narrowing in Cu-doped ZnO thin films, *J. Alloys Compd.* **680** 252-8.

[25] Zhoul S M, Zhang X H, Meng X M,Zou1 K, Fan X, Wu S K, Lee S T 2004 The fabrication and optical properties of highly crystalline ultra-long Cu-doped ZnO nanowires, *Institute of Physics Publishing Nanotechnol.* **15** 1152-5.

[26] Marselie J, Fauzia V, Sugihartono I 2017 The effect of Cu dopant on morphological, structural and optical properties of ZnO nanorods grown on indium tin oxide substrate, *IOP J. Physics: Conf. Ser.* **817** 012014.

[27] Chen L C, Hsieh C A, Zhang X 2014 Electrical Properties of CZO Films Prepared by Ultrasonic Spray Pyrolysis, *Mater.* **7** 7304-13.

IOP Conf. Series: Materials Science and Engineering **811** (2020) 012011 doi:10.1088/1757-899X/811/1/012011

EDM Process by Using Copper Electrode with INCONEL 625 Material

N. Yuvaraj[1, a], R. Arshath Raja[1, b], Prakash Palanivel[3, c], Dr. N.V. Kousik[4, d]

[1]Research Associate, ICT Academy, India
[2]Information Security Analyst, Skill TeQ, Adambakkam, Chennai, Tamilnadu, India
[3]Associate Professor, Galgotias University, Greater Noida, GB District, UP, India

[a]yraj1989@gmail.com, [b]arshathraja.ru@gmail.com, [c]prakash.palanivel@skillteq.com, [d]nvkousik@galgotiasuniversity.edu.in

Abstract. Electrical discharge method (EDM), based on the thermo electrical energy of the workpiece and the mechanical electrical machining, are non-traditional machining methods. Current development of the nano mixed dielectric discharge device has the latest trends in tough and soft materials finishing. This research assesses the impact on nano-particle mixed electric discharge extraction of the inconel 625 product by means of a copper electrode of the material removal frequency (MRR), device wear level (TWR), and superficial surface roughness. In contrast to dielectric fluid are the output parameters such as current, period and frequency, and nano-substances of titanium carbide. During this phase, a sequential, separate discharges between the electrode and the workpiece contribute to the product removal electro thermally. The method quality largely depends on the Electrode content, the production of the electrode workpiece product. Appropriate electrode choice may reduce machining costs. Die-Sinker EDM with the copper electrode is therefore able to provide for improvement of output parameters and reduction of production costs. In case of surface roughness, a copper electrode has improved efficiency and is better suited for nano powdered dielectric medium.

1. Introduction

EDM is a proven processing method for the manufacturing of geometrically complicated and/or hard material parts, which by traditional machining methods are extremely difficult to operate.

The company's production method for complex shapes and narrow apertures with high accuracy is one of the most popular and widely accepted methods. This process is typically used for the profile joining of diamond wheels in plastic, the development of micro nozzles, boiling of composites and the creation of moulds and dies in hardened steel. Such rough and fragile machining products create an excessive wear and cost of equipment. For many years, the mechanical properties of tool steels have been thoroughly studied.

For EDM the device or job component are broken down into dielectric fluid and is divided up by a small distance. The release energy is produced on the surface of the workpiece at the spark point with very high temperatures. The sample has a temperature increase of up to 40 000 K, which results in the melting and vaporization of a minute portion of the workpiece. The top layer of the workpiece resolidifies and cools very quickly afterwards.

The fundamental cycle of EDM is very straightforward. Between an electrode and a workpiece is an electric spark formed. The light indicates that the energy streams. It induces intense heat and temperatures hitting between 8000 and 12000 ° C, nearly burning everything. This is an electronic flame. The fire is managed and positioned with great care so that it impacts only the product layer. The heat treatment under the surface is not normally influenced by the EDM system.

In this process, the material removal is occurred electro thermally by a series of successive discrete discharges between electrode and the work piece. The performance of the process, to a large extent, depends on the Electrode material, Work piece material manufacturing method of the electrodes. A suitable selection of electrode can reduce the cost of machining. So Die- Sinker EDM using copper electrode will give for optimizing Performance parameters and reducing cost of manufacturing, and it is that a copper electrode will give better performance in certain characteristics and is more suitable with nano powder mixed dielectric medium in case of surface roughness.

2. Literature Survey

A study of state of the art technology for advanced materials processing using Die Sinking EDM, WEDM, Micro-EDM, Dry EDM and RDE-EDM was performed by Anand Pandey et al. [1].

A. A. Khan et al. [2] also examined the use of aluminum or mild steel with copper and blade electrodes in the measurement of electrode damage along the cross-section of an electrode contrasted with that throughout its duration throughout EDM.

M. R. Shabgard et al. [3] have researched the most effective, practical chromium-based product welded steel study in the FW4. The use of selling product raises mold life by about 150% and the regular maintenance duration up to 200% relative with moulds made from normal metal.

El Taweel et al. [4] researched the relationship of system parameters with a tool content, like the composite of Al-Cu-Si-Tic, created from powder metallurgy (P/M) manufacturing, in electrical discharge of CK45 iron.

The study of surfaces roughness work and MRR for brass-machined and copper electrical-discharge electrode machined device metal (EDG) methods was performed by Helmi et al. [5].

The key process of making tungsten carbide (WC-Co) moulds and dies is Pichai Janmanee et al. [6] conducted a study of the mechanical electrical discharge machining (EDM). The EDM method usually creates microcracks on the workpiece layer.

Prasad Bari et al. [7] have analyzed the right dielectric fluid for a given work piece and instrument content to improve MRR and minimize TWR.

Mohd Amri Lajis et al. [8] have studied about an analysis on Experiments were performed using a Charmilles Electrical Discharge Machine, Series-Roboform. An electrode was used with a flat graphite ring with a diameter of 9 mm.

In an experiment on CNC regulated die sinking form EDM, Rajesh Choudhary et al. [9] analyzed the study. The EN31 metal is constructed of 25 mm x 25 mm x 20 mm length.

3. Problem Defnition

The challenge found by the previous study is that many machining parameters have an effect on surface ruggedness and it was difficult to achieve an optimal surface performance by these parameters. We are able to achieve optimal surface quality through the use of the various parameters such as pulse current, pulse time, pulsation pause time, voltage, dielectric liquid and the electrical material. With rising current and voltage, the surface finish is degrading.

A higher voltage or current gives the work surface a stronger spark and makes it more profoundly crater. It renders the ground redder. The existing effect on the ground finish on other electrodes is greater. Brass electrodes absorb more fluid than other electrodes which build up in the space between job and electrode. As a consequence, the job surface becomes gritty by further uncoordinated sparks.

Therefore, during EDM activities, this work can produce more information and experience.

1. To learn more about the method of electric discharge.
2. Analysis of ground rigidity with an electrical discharge device of Inconel 625.
3. In order to determine the right parameter array in EDM for the extraction frequency and wear level of devices.

4. Material Selection

Inconel 625 is a strong nickel-based nickel alloy with a high strength, resilient temperature (HSTR). It is commonly used in aerospace systems as gas turbines, rocket engines, rockets, pumps and machinery. Due to its poor thermal stability, low strength, high stiffness, high work endurance, high carbide particle density and the strong tendency to weld to the device for the edge design, Inconel-625 is challenging to operate. Due to the wide range of applications in different fields the behavioural features of Inconel-625 are better known with EDM.

Figure 1. Inconel 625 material.

The table.1 shows the Chemical Composition in inconel 625 Material

Table 1. Chemical composition in inconel 625.

Chemical Composition	Percentage
Nickel	58.23%
Chromium	20.0-23.0 %
Iron	5.0%
Molybdenum	8.0-10.0%
Niobium (plus Tantalum)	3.15-4.15%
Carbon	0.10 %
Manganese	0.50 %
Silicon	0.50 %
Phosphorus	0.015 %
Sulfur	0.015 %
Aluminum	0.40 %
Titanium	0.40 %
Cobalt	1.0%

4.1. Tool Material – Copper Electrode

The electrical choice depends on the necessary reliability parameters (MRR, machining stability and surface roughness) and the output constraints of the electrode. The first to establish the discharges is to choose a good electrical conductor. To order to ensure geometrical integrity of an electrode, this substance must have a low melting and vaporizing temperature as well as moderate thermal diffusiveness.

This is the first EDM electrode device to choose from. Casting and machining may generate it. Copper electrodes are produced by chemical graving and electrical formation with very complex characteristics. Copper is the most appropriate EDM machine tool product.

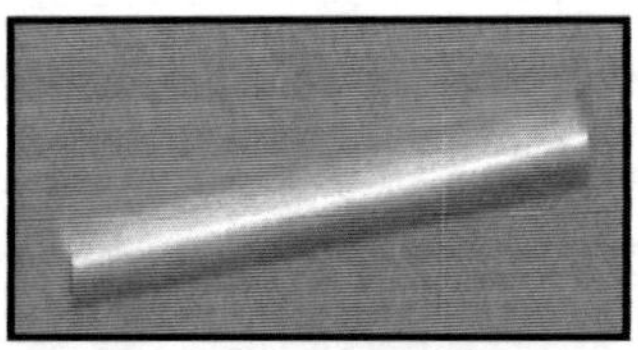

Figure 2. Copper electrode.

IOP Conf. Series: Materials Science and Engineering **811** (2020) 012011 doi:10.1088/1757-899X/811/1/012011

The table 2 shows the chemical composition in Copper electrode

Table 2. Chemical composition in Copper electrode.

Chemical composition	Percentage
Copper	56.700%
Aluminum	0.0025%
Tin	0.020%
Phosphorous	0.020%
Lead	3.000%
Iron	0.100%
Zinc	39.850%
Nickel	0.0770%

4.2. Titanium Carbide Nano Powder

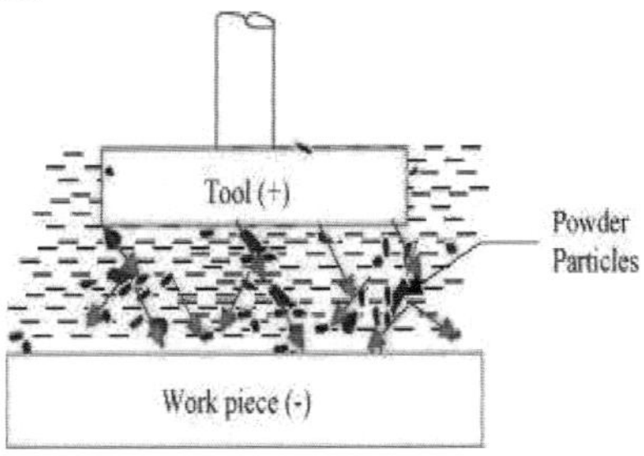

Figure 3. Nano powder mixed dielectric fluids.

Ceramics used in nano powders are more ductile than coarse grained ceramics and can be sintered at low temperatures at elevated temperatures. The strength for nano-sized iron and copper powders is about 4-6 times higher when bulk materials are dislocated.

In the conduction of ink and polymers, nano-sized copper and silver are used. It raises the storage capacity for hard disks on the device, while growing the physical dimensions of magnetic mono domains, which reduces oppressive compared to large particles.

Figure 4. Titanium carbide nano powder.

5. Design of Experiment

The experimental input parameters considered for the present study includes the following:

- Current
- Pulse ON-Time
- Flushing pressure

Here, table 3 shows the results of input parameter in terms current ON time with Nano Powder and without nano powder

Table 3. Experimental Data with Nano Powder and without nano powder.

Run Order	Input Parameter (with)			Input Parameter (without)		
	Current Amps	Current Amps	ON-Time μsec	F/Pr. Kg/sq.cm	ON-Time μsec	F/Pr. Kg/sq.cm
1	1	1	1	1	1	1
2	1	1	2	2	2	2
3	1	1	3	3	3	3
4	1	1	4	4	4	4
5	2	2	1	2	1	2
6	2	2	2	1	2	1
7	2	2	3	4	3	4
8	2	2	4	3	4	3
9	3	3	1	3	1	3
10	3	3	2	4	2	4
11	3	3	3	1	3	1
12	3	3	4	2	4	2
13	4	4	1	4	1	4
14	4	4	2	3	2	3
15	4	4	3	2	3	2
16	4	4	4	1	4	1

6. Conclusion

The study concludes that the copper electrode should enhance its efficiency, such as evaluation of the extraction speed, use of tools and roughness of the substrate. When the nano-particle is used, the electrodes absorb a significantly larger amount of working product than other electrodes collecting within the area between the workpiece and the electrode if applied to another component. The research would focus on Model of Experiment (DOE) and evaluation on product extraction speed, device level wear and surface roughness. There are several scopes that are to be conducted with some tests.

Reference

[1] Pandey A, Singh S 2010 Current research trends in variants of Electrical Discharge Machining: A review, *Int. J. Eng. Sci. Technol.* **2(6)** 2172-91.

[2] Khan A A 2008 Electrode wear and material removal rate during EDM of aluminum and mild steel using copper and brass electrodes, *Int. J. Adv. Manufact. Technol.* **39(5-6)** 482-7.

[3] Shabgard M R, Shotorbani R M 2009 Mathematical modeling of machining parameters in electrical discharge machining of FW4 welded steel, *World Academy Sci. Eng. Technol.* **52(63)** 403-9.

[4] El-Taweel T A 2009 Multi-response optimization of EDM with Al–Cu–Si–TiC P/M composite electrode, *Int. J. Adv. Manufact. Technol.* **44(1-2)** 100-13.

[5] Wagiman A 2009 Investigation of surface roughness and material removal rate (MRR) on tool steel using brass and copper electrode for electrical discharge grinding (EDG) process, *Int. J. Integrated Eng.* **1(1)** 31-9.

[6] Janmanee P, Muttamara A 2011 Optimization of electrical discharge machining of composite 90WC-10Co base on Taguchi approach, *Europ. J. Scientific Res.* **64(3)** 426-36.

[7] Bari P, Rode S, Duduke A, Shinde P, Srivastav V 2012 Dielectric Fluid in Electro Discharge Machining, *Proceed. NCNTE, Third Biennial National Conf. Nascent Technologies.*

[8] Lajis M A, Radzi H C D M, Amin A K M N 2009 The implementation of Taguchi method on EDM process of tungsten carbide, *Europ. J. Scientific Res.* **26(4)** 609-17.

[9] Choudhary R, Kumar H, Garg R K 2010 Analysis and evaluation of heat affected zones in electric discharge machining of EN-31 die steel, *Indian J. Eng. Mater. Sci.* **17** 91-8.

IOP Conf. Series: Materials Science and Engineering **811** (2020) 012012 doi:10.1088/1757-899X/811/1/012012

Damage Mode of Long Tube Structure at Different Water Depths under Strong Impact

Xiaoyan Teng[1], Yuxin Sun[1], Zhikai Wang[2*], Dongyan Shi[1], Xuanxin Tian[1]

[1]College of Mechanical and Electrical Engineering, Harbin Engineering University, China
[2]College of Ship Engineering, Harbin Engineering University, China

Abstract. In order to study the damage mode of the structure under strong impact, the finite element analysis software ABAQUS acoustic-coupling method is used to calculate and analyze the strong impact of the long tube structure under different water depths and different impact factors. The study obtained damage patterns influenced by impact factors and water depth conditions. It provides a reference for the calculation and impact design of underwater structures under strong impact.

1. Introduction

The non-contact underwater explosions of ships and submarine structures usually have severe structural vibrations and large plastic deformation, resulting in extensive impact damage and damage to the overall structure of various important hull equipment [1]. In recent years, the structure of non-contact underwater explosions has attracted more and more attention. Underwater explosion and explosion load on the structure of the impact damage process mechanism is more complex, so the experiment is the most direct and effective research approach [2]. It is difficult to carry out the explosion test of large structures and requires huge funds. Therefore, numerical methods to study the failure mechanism of underwater explosions and structural impact resistance have become effective alternatives. Fox [3] combined the DYNA3D and USA programs to analyze the nonlinear dynamic response under the lateral impact of underwater explosion. Miller [4] used DYNA3D/DAA to analyze the dynamic response of spherical and cylindrical shells under the action of underwater explosion shock waves. Yu Xiaofei [5] used MSC. DYTRAN to study the antiknock performance of reinforced cylindrical shell under underwater explosion. Geers [6] proposed a semi-empirical calculation formula for underwater explosion load that can be used underwater, and can simulate shock waves and bubble pulsation loads better. ABAQUS uses sound field and structural coupling technology, suitable for underwater explosion analysis in the far field. In this paper, ABAQUS is used to analyze the far-field explosion of the long tube structure, and the damage mode of the structure under strong impact is summarized.

2. Calculation Model

The calculation model uses a single-layer shell long tube structure. The main dimension of the calculation model is 7.0m in diameter, the total length of the model is 50m, and the thickness of the shell is 30mm.

Figure 1. Long Tube model.

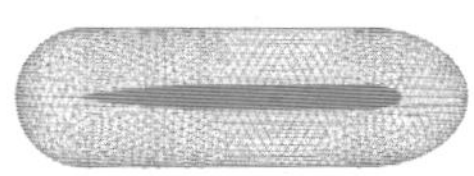

Figure 2. Model assembly drawing.

The use of finite element analysis of the transmission of underwater explosion shock waves requires careful selection of the cell size, and the thickness of the cell grid is closely related to the wavelength of the analysis. Over-grid causes low computational accuracy and even leads to non-convergence of simulation results. The denser the meshing, the higher the calculation accuracy, and the closer the solution is to the real situation, but the more computing resources are used, the more time is consumed, which is not conducive to improving the calculation efficiency.

The ABAQUS calculation of the stability limit Δt_{stable} needs to be calculated based on the cell length L^e and the material wave velocity C_d:

$$\Delta t_{stable} = L^e / C_d \tag{1}$$

The size of the mesh is related to the load frequency component of the structure. Since it is difficult to analyze the frequency of the load in the time domain, it is necessary to determine the frequency component of the load by spectral analysis of the impact load. It is necessary to divide a certain number of meshes within a stress wave wavelength to ensure that the waves propagate in the flow field and structure without distortion. The mesh size estimation formula is:

$$l_{\max} = \lambda_c / n \tag{2}$$

$$\lambda_c = \theta \cdot c \tag{3}$$

$$\theta = 8.4\times10^{-5} W^{1/3} \left(W^{1/3} / R\right)^{-0.23} \tag{4}$$

where λ_c: the wavelength of the stress wave in the structure, n:represent the number of grids required in the wavelength of a stress wave, c:the wave velocity of the stress wave, 5000 m/s in the steel. ; W: dose; R:burst distance θ: decay constant.

According to the above formula, the grid size should be 300mm, so that dividing the grid is beneficial to improve the accuracy of the calculation as much as possible without occupying too much computing resources.

Using the Cowper-Symonds model to describe the strain rate effect of a material

$$\sigma_y = [1+(\dot{\varepsilon}/c)^{\frac{1}{p}}]\sigma_0 \tag{5}$$

where, σ_y for the dynamic yield stress, σ_0 is the static yield stress, $\dot{\varepsilon}$ is the strain rate, and C and P are the strain rate parameters.

The material related parameters are shown in table 1:

Table 1. Model parameters of steel material.

ρ/kg ·m^{-3}	E/Pa	μ	σ_0/Mpa	C/s^{-1}	P
7800	2.1E+11	0.3	590	40.4	5

where ρ is density, E is Young's modulus, μ is Poisson's rati.

3. Numerical Calculation and Analysis

3.1. impact factor

Using an impact factor based on the plane wave assumption to describe the underwater explosion impact environment, the form of the impact factor is:

$$C = \sqrt{W} / R \tag{6}$$

where W is the drug pack equivalent, R is the shortest between the drug pack and the structural shell distance.

The effect of impact factor on the damage mode at 300m water depth is studied. The explosion contact point is set at 1/2 of the length of the structure. The working conditions are set as shown in table 2.

Table 2. Chang impact factor working condition setting.

Working condition number	water depth/m	impact factor/$kg^{1/2} \cdot m^{-1}$
1		3
2	300	2.5
3		2

The damage effect of the structure under the condition of 300m water depth, impact factor is $3kg^{1/2} \cdot m^{-1}$ is shown in figure 3 and figure 4. According to the displacement cloud map and PEEQ cloud map, the structure is resistant. The overall instability of the shell is presented. The surface of the pressure-resistant shell presents a plurality of in-situ concave half-wave failure modes, completely losing the bearing capacity, and the equivalent plastic strain on the surface of the pressure-resistant shell reaches or exceeds 0.3. The pressure-resistant shell will break, and it can be considered that the structure has suffered an irreversible overall instability/break under this condition.

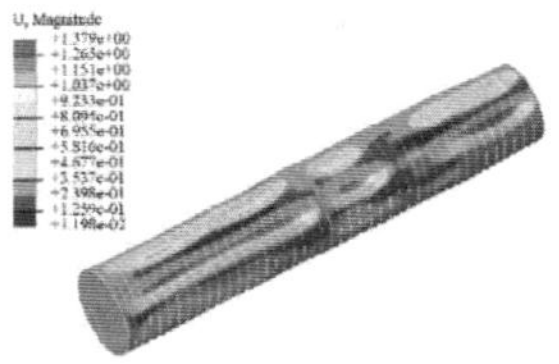

Figure 3. Local structure displacement cloud.

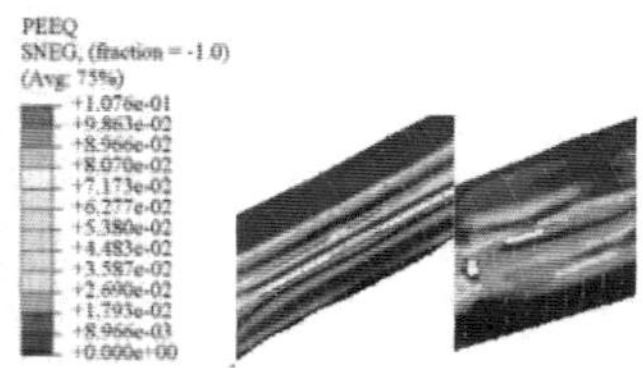

Figure 4. Local structure PEEQ and fracture.

The structural instability failure process under this condition is analyzed in detail, as shown in the following figure. Figure 5(b) is the displacement cloud diagram of the pressure-resistant shell of the explosion-proof surface at the time of 0.01s. The shock wave is applied to the pressure-resistant shell for a short time, and the impact surface of the pressure-resistant shell begins to produce obvious displacement. At 0.03s, the structure meets the explosion surface. Larger concave deformation has begun to appear. At 0.04s, the concave deformation of the structure continues to be continued. At 0.05s, the pressure-resistant shell reaches the maximum unstable shape.

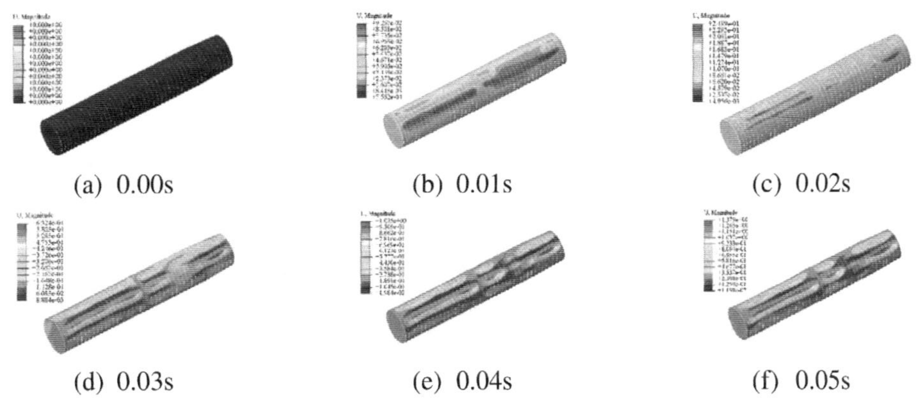

(a) 0.00s (b) 0.01s (c) 0.02s
(d) 0.03s (e) 0.04s (f) 0.05s

Figure 5. Damage process.

The damage effect of the structure under the condition of 300m water depth, impact factor is 2.5kg$^{1/2}$·m^{-1} is shown in figure 6 and figure 7. The surface of the pressure-resistant shell of the section presents a single-inward concave half-wave type of instability failure.

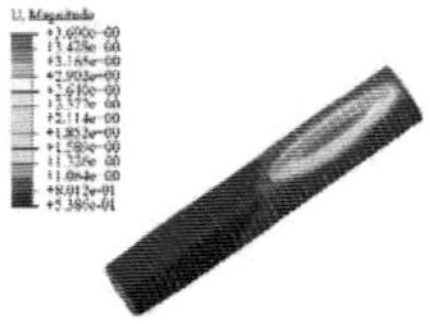

Figure 6. Local structure displacement cloud.

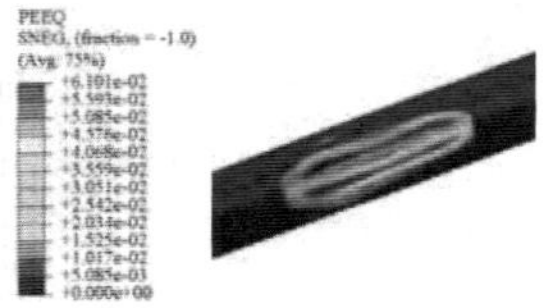

Figure 7. Local structure PEEQ.

The damage effect of the structure under the condition of 300m water depth, impact factor is 2kg$^{1/2}$·m^{-1} is shown in figure 8 and figure 9. There is no instability da-mage in the structural pressure-resistant shell, and a slight plastic strain zone appears on the impact surface of the pressure-resistant shell. Non-strength parts with a small internal thickness of the structure produce a certain plastic strain.

Figure 8. Local structure displacement cloud.

Figure 9. Local structure PEEQ.

3.2. Water depth effect

Exploring the influence of water depth on the damage mode when the impact factor is fixed, the working conditions are set as shown in table 3:

Table 3. Changing water depth working condition setting.

Working condition number	water depth/m	impact factor/kg$^{1/2}$·m^{-1}
1	600	0
2	400	2.5
3	200	

The damage effect of the structure under the condition of 600m water depth is shown in figure 10. Due to the large water depth, the hydrostatic pressure is extremely high, resulting in structural in-stability of the local compartment. Even without additional impact loads, the struure fails and loses the bear-ing capacity for hydrostatic pressure loads.

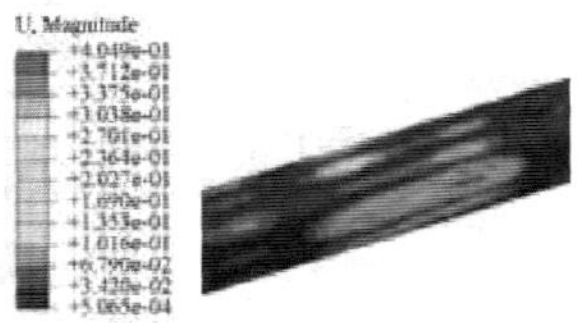

Figure 10. Local structure displacement cloud.

The damage effect of the structure under the condition of 400m water depth, impact factor is 2.5kg$^{1/2}$·m^{-1} is shown in figure 11 and figure 12. The surface of the pressure-resistant shell of the section presents a single-inward concave half-wave type of instability failure. However, the degree of failure is greater than the 300m water depth under the same impact factor, and the damage is more serious.

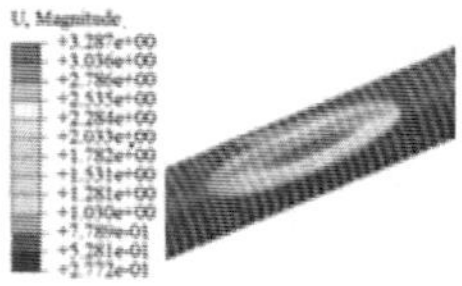

Figure 11. Local structure displacement cloud.

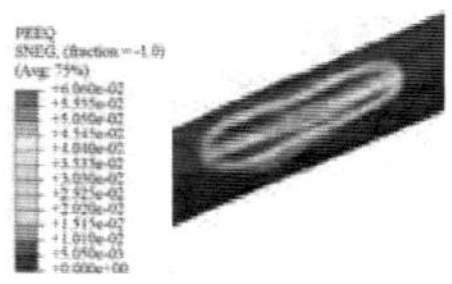

Figure 12. Local structure PEEQ.

The damage effect of the structure under the condition of 400m water depth, impact factor is $2.5kg^{1/2}\cdot m^{-1}$ is shown in figure 13 and figure 14. There is no instability damage in the structure, but a small amount of plastic strain occurs in the non-strength members inside the structure.

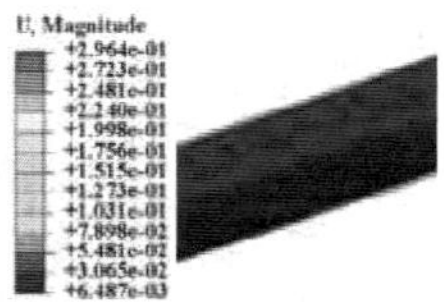

Figure 13. Local structure displacement cloud.

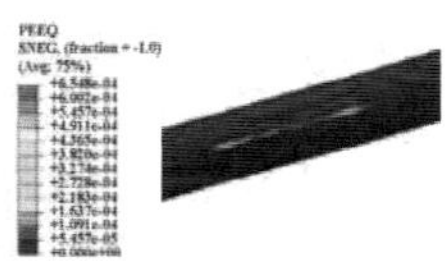

Figure 14. Local structure PEEQ.

3.3. summary

The damage effect of the structure under the impact of strong shock waves under different water depth conditions can be summarized into the following 4 categories:

Damage Mode 1: Exceeding the structural limit depth, the shell of structure is partially des-tabilized under the hydrostatic pressure load.

Damage mode 2: The shock wave is transmitted from the water to the pressure-resistant shell, and the stress wave in the pressure-resistant shell is transmitted into the non-strength member with a small thickness. This stage can be regarded as the stress wave propagating in the variable-section material structure, according to the stress wave. Propagation characteristics, transmission waves larger than the incident wave stress value act on non-strength members with small load carrying capacity, resulting in stress wave plastic damage of non-strength members, resulting in internal profile damage. Moreover, as the depth continues to increase or the burst distance continues to decrease, the degree of damage will increase, and after reaching the critical condition, damage mode 3 will be entered.

Damage mode 3: After the shock wave acts on the impact surface of the pressure-resistant shell, high-stress areas and even certain plastic damage appear on the impact surface of the pressure-resistant shell. Then, local pits appear, and the depression develops along the lateral and longitudinal directions, forming an axial failure mode of the pressure-resistant shell. As the depth continues to increase or the explosion distance continues to decrease, the damage degree will increase, and after reaching the critical condition, damage mode 4 will be entered.

Damage Mode 4: Under the action of shock wave, the whole structure is unstable, the internal support of the structure is greatly deformed, and the structure produces huge plastic strain until the shell is broken, forming a failure mode of overall buckling.

4. Conclusion

In conclusion, the simulation results show that :(1) There are four kinds of damage modes under the strong impact of the long tube structure under water, including a) structural failure under hydrostatic pressure, b) damage to the profile, c) large plastic deformation to form a depression d) the overall structure is unstable, resulting in breakage or breakage. (2) When the water depth is constant, as the impact factor increases, the degree of damage becomes larger. After reaching the critical condition, it enters a more serious damage mode. (3) When the impact factor is constant, as the water depth increases, the hydrostatic pressure becomes larger, resulting in a greater degree of structural damage. In the case of large water depth, no additional strong impact is required, and structural failure occurs under hydrostatic pressure.

Acknowledgments
This paper is funed by the International Exchange Program of Harbin Engineering University for innovation-oriented Talents Cultivation; Supported by National Natural Science Foundation of China (11602069,51779056); Supported by the Heilongjiang Natural Science Fund(E2017026).

References

[1] Pu Jinyun. Survivability of wavship[M]. Beijing: Haichao Press, 2001.
[2] Cole R H. Underwater explosion [M]. Princeton: Princeton University press, 1948.
[3] Fox P K. The dynamic response of cylindrical shells subjected to underwater explosions[D]. Naval Post Graduate School, USA, 1992
[4] Miller R D. On the explicit modeling of fluid regions using DYNA/DAA[A]. Proceedings of the 64th Shock and Vibration Symposium[C]. 1993, II: 435-445.
[5] Yu X F, Liu S G, Zhang T, Xiao H L. Response Analysis of Cylindrical Shell Subjected to Underwater Explosion Shock Waves[J]. Journal of vibration and shock, 2006, 25(5): 106-115.
[6] Geers T L. An integrated wave- effects models for an underwater explosion bubble[J]. The Journal of the Acoustical Society of America, 2002, 111(4): 1584-601.
[7] Yao X L, Qu Z J, et al. Analysis on characteristics of blast loading and stiffened plate damage due to internal blast in ship[J]. Chinese Journal of Ship Research, 2018, 13(3): 140-148.

Numerical Simulation of Explosive Welding of Marine Aluminum Steel Transition Joint Based on AUTODYN

Xuanxin Tian[1], Zhikai Wang [2*], Xiaoyan Teng [1], Dongyan Shi [1], Yuxin Sun [1]

[1]College of Mechanical and Electrical Engineering, Harbin Engineering University, China
[2]College of Ship Engineering, Harbin Engineering University, China

Abstract. The finite element analysis software AUTODYN is used to simulate the processing process of marine Aluminum-steel transitional explosion joints, and the jet, spallation and waveform distribution in the process of explosive welding are obtained. Comparing the numerical simulation results with the experimental phenomena, the correctness of the simulation results is verified. The effects of the impact velocity of the cladding, the collision angle, etc. on the welding results were further analysed. It provides a reference for studying the formation of jet phenomena in explosive welding and the selection of explosion parameters during actual welding.

Keywords: Aluminum-Steel composite plate; explosion welding; numerical simulation; jet

1. Introduction

Aluminum has been widely used as the main material for ship superstructure construction in recent years because of its low density, high strength, good plasticity and toughness, good thermal conductivity and electrical conductivity, and strong corrosion resistance. The use of the Aluminum alloy structure can reduce the weight of the ship, increase the longitudinal strength, and improve stability. The connection between the traditional Aluminum alloy superstructure and the steel ship main body is riveting, which has been gradually replaced by explosive welding in recent years.

Explosive welding is a technique in which explosives are used as an energy source[1], and the huge energy generated by the explosion of explosives is used to push the composite plate to move, and the composite plate and the substrate are strongly collided to by plastic deformation. Compared with the traditional riveting and bolting, the process of explosive welding is good, and the hardness and fatigue strength of the joint surface are higher. And it is more widely used and can achieve a wide range of dissimilar metals. Marine explosive composite joints are generally Aluminum-Steel joints or Aluminum-Titanium-Steel composite joints. For multi-layer explosive composite joints, one-time welding or multiple welding can be used. The joint performance of multiple welding is better. In this paper, the Aluminum-Steel welding process is studied, and several conclusions are proposed to provide reference for the actual welding parameters selection.

2. Basic theory of SPH

In this paper, the AUTODYYN module in the finite element software ANSYS was used to simulate the welding process by using Smooth Partical Hydrodynamic (SPH). Using this method, the jet phenomenon at the joint interface and the waveform interface morphology can be well observed, and the pressure and plastic deformation at the joint are analysed[2].

SPH is a Lagrange-based meshless calculation method. Its advantage is that it can overcome the shortcomings of the Euler method that is difficult to track structural deformation and can not identify

IOP Conf. Series: Materials Science and Engineering **811** (2020) 012013 doi:10.1088/1757-899X/811/1/012013

the interface of the material in the simulation deformation, and there is no Lagrange mesh deformation problem. The SPH method treats continuous objects as interacting particles, and motion passes between the particles. To represent the interaction between particles, the concept of a "smooth kernel" function is introduced.

$$f(x) = \int_\Omega f(x')\delta(x - x')dx' \quad (1)$$

Where f is a function of the three-dimensional spatial position vector x.

By replacing the Dirac function $\delta(x - x')$with the smooth function $W(x - x')$, the expression can be written as:

$$< f(x) >= \int_\Omega f(x')W(x - x')dx' \quad (2)$$

The smooth kernel function W should satisfy the following conditions:

(1) Regularization conditions (normalization conditions) .

$$\int_\Omega W(x - x', h)dx' = 1 \quad (3)$$

(2) The Dirac function property when the smooth length tends to zero:

$$\lim_{h\to 0} W(x - x', h) = \delta(x - x') \quad (4)$$

(3) Tight support conditions:

$W(x - x') = 0$, when $|x - x'| > \kappa h$, κ is a constant related to the smooth function at point x

3. Numerical simulation parameters and models

3.1 Model material and parameters

In order to observe the occurrence of jet phenomenon better, the density of the substrate should be less than that of the composite plate because the substrate material is the main source of jet phenomenon [3]. Therefore, Al5083H116 plate (10mm×2mm) is selected as the substrate material, and the double plate is SS304 steel plate (11mm×1mm),its length slightly larger than the substrate to ensure complete bonding, the interval is set to 1mm. The chemical composition of these two materials is shown in Table 1 and Table 2.

Table 1. Chemical composition of Al5083H116 (mass fraction, %)

Si	Fe	Cu	Mn	Mg	Cr	Zn	Ti
0.4	0.4	0.1	0.4-1.0	4.0-4.9	0.05-0.25	0.25	0.15

Table 2. Chemical composition of SS 304 (mass fraction, %)

Mn	Si	C	Cr	Ni	S	P
2.0	1.0	0.08	0.05	8.0-10.0	0.03	0.045

3.2 Collision angle and impact velocity of the composite plate

The impact velocity of the composite plate and the collision angle are all important factors affecting the welding effect[4]. If the collision angle is too small, the substrate and the composite plate cannot be well combined, and no jet phenomenon will occur. And if he collision angle is too large, it will cause over-melting. Therefore, after consulting the relevant literature, we determine the collision angles are 10°, 15° and 20°.

The formation of jets is the most basic condition for explosive welding, because in the process of explosive welding, the free jet will clean the interface to be combined, which is convenient for better combination. Therefore, the collision speed between the superstrate and the substrate must be greater than the critical speed to produce a jet phenomenon. Many scholars have done a lot of research on the lower limit of welding, and also put forward the corresponding theory. We use the formula proposed by Deribas et al to determine the minimum collision speed[5].

$$V_{pmin} = K\sqrt{\frac{H_v}{\rho}} \quad (5)$$

Where: H_v—Vickers hardness of the material; K—constant, and the value is 0.6-1.2.

The Vickers hardness of AL5083H116 material is 96 MPa. Taking a K value of 0.6, the minimum collision speed V_{pmin} obtained is 360m/s.

Therefore, in order to investigate the impact of the impact angle and the impact velocity of the composite plate on the welding results, the two influencing factors were combined and nine operating conditions as shown in Table 3 were set. The collision angles are 10°, 15° and 20°, and the impact speeds are 600 m/s, 800 m/s and 600 m/s.

Table 3. Explosive welding numerical simulation working condition setting

Working condition number	collision angles (°)	impact speeds (m/s)
1	10	600
2	10	800
3	10	1000
4	15	600
5	15	800
6	15	1000
7	20	600
8	20	800
9	20	1000

Set the particle size to 0.02mm and the established numerical model is shown in Figure 1.

Figure 1. Numerical model of explosive welding.

4. Numerical simulation results analysis

4.1 Analysis of welding simulation phenomena

By analysing the results under different working conditions, the main phenomenon in the explosion welding process are the waveform of the joint surface and the generation of the jet phenomenon. As shown in Figure 2(a), the main source of material for the jet is Aluminum, because the material density and hardness of the Aluminum sheet is less than that of the composite.

(a)Combined surface topography (b)Enlarged view of the combined surface

Figure 2. Explosive welding numerical simulation combined surface shape diagram.

Zoom in on the waveform of the joint surface, as shown in Figure 2(b). In the early collision area, the waveform and amplitude are not obvious, and with the transition of the position, the waveform of the joint surface is gradually stable and exhibits a periodic trend, which is consistent with the actual experiment.

Comparing the welding profile with different impact velocities at the same impact angle, as shown in Figure 3, it can be seen that the impact velocity has a very significant effect on the welding effect. At a speed of 400 m/s, the jet phenomenon is not obvious and the cross-section thickness at the joint surface is small. However, when the speed is 800 m/s, there is a phenomenon that the impact force is too large and cracks occur in the intermediate region substrate (layer cracking phenomenon).The spallation is related to the reflection phenomenon of stress waves. During the explosion welding process, when the pressure pulse is reflected into the tensile pulse on the free surface of the plate, it will likely cause a relatively high tensile stress somewhere near the free surface. Most engineering materials are able to withstand relatively strong compressive stress waves without damage, but cannot withstand tensile stress waves of the same strength. However, due to the weak strength of the welded joint during the explosion welding process, the interface of the just welded joint is easily pulled apart under the tensile stress, resulting in failure of the welding of the composite plate within a certain range, and the superposed plate due to the tensile stress. Figure 4(c) shows the spallation phenomenon that occurs during welding.

(a)Impact speed :400m/s (b)Impact speed :600m/s (c) Impact speed :800m/s

Figure 3. The collision angle is 10 degrees ,the combined surface topography at different impact velocities.

(a) Collision angle:10 degrees (b) Collision angle:15 degrees (c) Collision angle:20 degrees

Figure 4. The impact velocity is 600m/s, the combined surface topography at different collision angles.

4.2 Welding simulation numerical analysis

4.2.1Joint surface pressure analysis. The pressure cloud diagram and the pressure-time curve at the joint face is shown in Figure 5(a). It can be clearly seen that the maximum pressure is the position of the current collision point and moves with time. In order to accurately analyse the pressure of the contact surface, 20 measuring points are evenly distributed on the substrate, and the pressure-time curve of the measuring point is shown in Figure 5(b). At an impact velocity of 600 m/s, the welding effect is best, the maximum pressure at the joint reaches 30GPa, the average maximum pressure is 15 GPa, and the maximum pressure duration is short.

(a)The pressure cloud diagram (b)The pressure-time curve

Figure 5. The collision angle is 15 degrees, impact speed is 600m/s,the pressure cloud diagram of combined surface at different impact velocities.

Extracting the average maximum pressure under all conditions, as shown in Table 4. It can be found that the average maximum pressure decreases as the collision angle increases and the impact speed decreases.

Table 4. The maximum pressure summary of Joint surface average (Pa)

	Impact speed (m/s)		
Collision angle (°)	400	600	800

10	7.00E+06	1.75E+07	2.50E+06
15	3.50E+06	0.75E+07	1.50E+07
20	3.00E+06	4.00E+06	1.25E+06

4.2.2 Substrate speed analysis. In the existing literature analysis, most of them focus on the response speed of the composite plate under the impact of explosives. This paper analyses the feedback speed of the substrate as shown in the figure. It can be seen that in the same direction, the velocity curve of the substrate exhibits the same trend as the pressure curve, and its variation with the impact velocity and the collision angle is also the same as the pressure variation. This is because the speed of the substrate is due to the strong action of the composite plate, and the substrate material collides with the composite plate to cause severe plastic deformation, which is reflected in the simulation is the change in the velocity of the substrate particles.

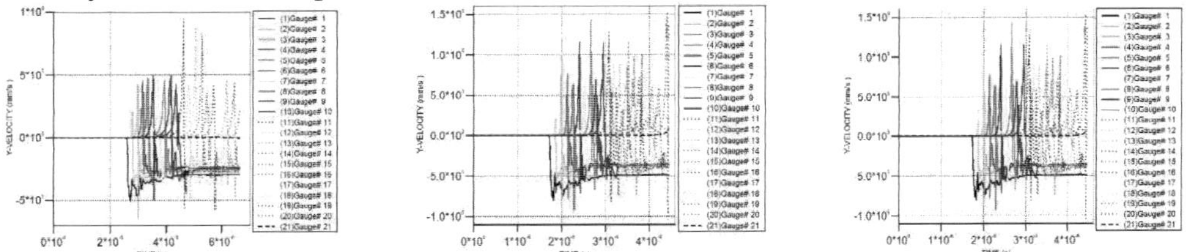

(a)Collision angle:10; Impact speed:400m/s. (b) Collision angle:10; Impact speed:600m/s (c)Collision angle:15; Impact speed:600m/s

5. Conclusion

The experimental results show that: (1) By simulating the welding process under different impact speeds, it is found that there is no obvious jet phenomenon when the impact velocity is small, and the welding effect of the joint surface is not ideal. When the impact velocity is too large, the cracking phenomenon may occur near the bonding surface of the substrate due to the excessive impact force. Under this experimental condition, 600 m/s is a suitable impact velocity.(2) By simulating the welding under different impact angles, it is found that the average maximum pressure between the substrate and the composite plate decreases with the increase of the collision angle, but the tightness of the welding increases and the spallation phenomenon decreases.(3) In actual manufacturing, it is necessary to select the appropriate impact speed and collision angle to achieve the best welding results.

6. Acknowledgments

This paper is funed by the International Exchange Program of Harbin Engineering University for innovation-oriented Talents Cultivation; Supported by National Natural Science Foundation of China(11602069,51779056); Supported by the Heilongjiang Natural Science Fund(E2017026).

7. References

[1] Wu Tairan, Fan fei. Application and Research Status of Explosive Welding Technology[J]. Coal mine Blasting,2018(05):20-24,32.
[2] Liu Jiang, Zheng Yuanyuan, Shen Zongbao, et al. Simulation of Explosive Welding Process Based on SPH Method[J].Welding Technol0gy,2013,42(12):17-20.
[3] Yan Jian-wen, Lei Fang-chao, Yu Qi-yang, Yang Ke-yin. Research of explosive charge impact on explosive welding interface wave [J].Engineering Blasting,2018,24(06):10-17.
[4] Huang Xingli, Wang Yang, Xu Peng, Hao Hongwei. Research on Property of Aluminum/Titanium/Steel Explosive Clad Plate [J].Material&Heat Treatment, ,2011,40(16):103-105.
[5] Li Xiaojie, Yang Wenbin. Xi Jinyi, et al. The Lower Limit of Explosive Welding Parameter Window for Bimetal [J].Explosive Materials, 1999 (3) :22-26.

The Study on Effect of Lightweight Concrete Block by Water Hyacinth Adding

PRANG SUBPA-ASA[1], LAEMTHONG LAOKONGTHAVORN[2], SHIGEYUKI DATE[3]

[1]Graduate student, Department of Civil Engineering, Graduate School of Science and Technology, Tokai University, Hiratsuka-pref., Japan, 259-1292
[2]Associate professor, Civil Engineering Department, Faculty of Engineering, King Mongkut's Institute of Technology Ladkrabang, Bangkok, Thailand, 10520
[3]Professor, Civil Engineering Department, Faculty of Engineering, Tokai University, Kanagawa-pref., Japan, 259-1292

[a]prangsub@gmail.com

Abstract: The objective of this study is to examine the pattern of water hyacinths which is a plant that is suitable for making lightweight concrete and can be used to replace sand in lightweight concrete production. The proportion of cement: sand: water: foam is 1: 1: 0.65: 44.08 and the proportion of water hyacinths that will replace the sand by the cement's weight is 0%, 2.5%, 5%, 7.5% and 10%. The density, water absorption, compressive strength (by the standard of TIS 1505-2541) and heat conduction of the lightweight concrete at 14 days will be studied. The results of the compressive strength of the lightweight concrete (LC) and the cellular lightweight concrete (CLC) are 167.56 ksc and 284.94 ksc, respectively. Both the compressive strength of the cellular lightweight concrete mixed water hyacinth (CLCH) and the lightweight concrete mixed water hyacinth (LCH) were investigated. The compressive strength of the CLCH and the LCH are 156.37 ksc and 172.45 ksc, respectively. The results are that the percentage loss of the CLC and the LC compressive strength are 6.68% and 39.83%. As for the physical properties, which are density and thermal conductivity, decreased. In contrast, water absorption increased by the percentage of water hyacinth by sand replacement. The optimal water hyacinth ratio in this study is 5%. Moreover, the CLC block is specified by the Thai Industrial Standards Institute 1505-2541 and is classified as C16 of the CLC block.

1. Introduction

The growing demand for construction in Thailand has increased the usage of construction materials like lightweight concrete blocks which are used in many types of buildings like residences and warehouses. The benefits of using lightweight aggregate concrete include reduced dead loads which makes savings in foundations and reinforcement, improved thermal properties, improved fire resistance, lessened transporting and handling precast units on-site and a decrease in formwork and propping. On the other hand, convention concrete making is not completely environmentally friendly and this has enthused research on seeking greener alternative material for concrete. [1]

Water hyacinth (WH) is known as one of the fastest-growing plants. Its primary means of reproduction is by way of runners or stolons, which eventually form daughter plants. It also produces large quantities of seeds that are viable for up to 30 years. With the fast rate of reproduction of water hyacinths, it has been found that populations often double in size in just two weeks. [2] They flourish in warmer climates and have broad, glossy and thick leaves that are 10-20 cm across. The hanging roots of the plant are feathery and purplish-black.

WH is classified as an aggressive invasive species. It has been named one of the ten worst weeds and the most widespread and damaging aquatic plant species in the world. Its growth rate is among the highest in the among plants, as it reproduces vegetation by having short runner stems that branch out from the base of the plant. This means that it has a huge effect on the life cycle of the environment. Thus, using this weed as cement replacement or concrete aggregate which supports concrete properties is suitable. [3]

Fiber cement material is made of cement paste containing coconut coir fibers and oil palm residues which are used as roof sheets and siding materials to reduce heat transfer through buildings and for energy conservation. The investigation focused mainly on the effects of both cellulose fibers on the physical-mechanical and thermal properties of the products. The results are that, the mixtures of fiber cement products containing 5% of both cellulose fibers by weight of portland cement yielded optimal physical and mechanical properties. The thermal conductivity of the fiber cement paste was 66% less than that of the control specimens. [4, 5]

Similarly, the research investigated the behavior of concrete with rice husk ash after exposure to different high temperatures (200, 300 and 400◦C). Then, samples were used to measure the compressive strength and water permeability of the concrete. The results showed that the strength of the concrete reduced after exposure to high temperatures compared with the control concrete and tended to decrease with an increasing percentage of replacement of Portland. [6]

Specifically, the purpose of this investigation is to examine the properties of lightweight concrete blocks mixed water hyacinth by focusing on their compressive strength and various mixing.

2. Methodology

2.1. Materials and mix properties

The water hyacinth was located in Ladkrabang district, a suburb area in Bangkok, Thailand. The water hyacinth was cut into about 5 mm in length using a long knife and then dried in a hot air oven at 60°C for 48 hours. [8] Then it was divided into two parts where some was blended through a blender and some was not. They were classified into three classes 1) 0.5 cm of length, 2) were not passed through sieves No.50 and 3) passed through sieves No.50.

2.1.1. Compressive strength testing by Trial size of water hyacinth.

In the first part of the research, to find the optimal size of water hyacinth, dried specimens were used to determine the trend of maximum compressive strength. The water hyacinth specimens were tested in three sizes as shown in Figure 1 and mixed proportion show in Table 1.

The result was found that A, B1-B5, C1-C5, D1-D5 are 369.02, 233.27, 216.15 and 165.39 ksc of compressive strength, respectively and were presented in Table 2. The dried powder water hyacinth that was passed through Sieve No.50 was the optimal size for this part of the research.

2.1.2. The experiment for finding optimal ratio of water hyacinth dried powder pass through sieves NO.50.

In the second part, after the optimal size of the dried powder water hyacinth was found, it was used to determine the trend of maximum compressive strength for the research. The ratio of water hyacinth by sand replacement was classified in five states, as 0%, 2.5%, 5.0% 7.5% and 10%.

Part 2.1 The sample of the mortar with dried powder water hyacinth passed through Sieve NO.50 adding: Mixed proportion was shown in table 3.

Part 2.2 Sample of the mortar with dried powder hyacinth passed through Sieve No.50 and CLC foaming agent adding: Mixed proportion is shown in Table 4.

Table 1. Mixed proportions.

	Description	Ratio			
		C	S	WH	W
A	Cement plate	1.00	2.00	-	0.60
Adding Water Hyacinth 3% by mass of cement					
B1-B5	Cement plate with water hyacinth dried powder pass through sieves NO.50	1.00	2.00	0.03	0.60
C1-C5	Cement plate with water hyacinth dried powder do not pass through sieves NO.50	1.00	2.00	0.03	0.60
D1-D5	Cement plate with water hyacinth dried 0.5 cm of length	1.00	2.00	0.03	0.60

Table 2. Result of compressive strength Test by Trial size of water hyacinth.

Description	Weight(g)	Cross-sectional Area (cm^2)	Age (days)	Compressive Strength (ksc)
Mortar by non-water hyacinth adding	258.01	25.10	14	369.02
Mortar with water hyacinth dried powder pass through sieves NO.50 adding	251.52	25.47	14	233.27
Mortar with water hyacinth dried powder do not pass through sieves NO.50 adding	252.08	25.35	14	216.58
Mortar with water hyacinth dried 0.5 cm of length adding	247.29	25.32	14	165.39

Table 3. Mixed proportions of mortar with water Hyacinth dried powder pass sieves NO.50 adding.

	Description	Ratio			
		C	S	WHSR*	W
A	Mortar	1.00	1.00	0%	0.65
B1-B5	Mortar with water hyacinth dried powder pass through sieves NO.50 adding	1.00	1.00	2.5%	0.65
C1-C5		1.00	1.00	5.0%	0.65
D1-D5		1.00	1.00	7.5%	0.65
E1-E5		1.00	1.00	10%	0.65

Table 4 Mixed proportions of mortar with water Hyacinth dried powder pass through sieves NO.50 adding and CLC foaming agent.

	Description	Ratio				
		C	S	WHSR*	W	CLCFA*
A	Mortar	1.00	1.00	0%	0.65	44.08%
B1-B5	Mortar with water hyacinth dried powder pass through sieves NO.50 and CLC foaming agent adding	1.00	1.00	2.5%	0.66	44.08%
C1-C5		1.00	1.00	5.0%	0.67	44.08%
D1-D5		1.00	1.00	7.5%	0.68	44.08%
E1-E5		1.00	1.00	10%	0.69	44.08%

3. Physical Properties Testing

3.1. Compressive strength testing

The sample (after 14 days), was tested for compressive strength according to TIS. 1505-2541 by ensuring the compressive load was controlled between 0.05 - 0.20 N/mm2/sec until failure.

3.2. Dry density testing

The dry density of the mortar testing was according to TIS. 1505-2541 by measuring the sample volume then dried in a hot air oven at 105±5 °C for 24 hours.

3.3. Water absorption testing

Water absorption testing also followed by institute TIS. 1505-2541 where the sample was dried in a hot air oven until constant weight is reached in more than 24 hours at 105±5 °C. Then after more than four hours, it cooled down to room temperature, so its mass was measured. Once the sample was cured for 24 hours in water, its surface was wiped and the hardened mortar mass was measured in three minutes.

4. Result

4.1. Compressive strength testing of WHSR mortar and WHSR and CLCFA mortar

The representative of the compressive strength of the WHSR mortar was measured by the compressive load test shown in Table 3, where the mass of cement and water cement ratio were controlled by mortar cement ratio 1 :1 The non-WHSR is 284.94 ksc in compressive strength. Figure 1 presents the results of this study as the WHSR ratio of 2.5%, 5%, 7.5% and 10% are 225.99 ksc, 171.45 ksc, 6.09 ksc and 4.23 ksc in compressive strength, respectively.

On the other hand, the CLCFA non-WH is 167.56 ksc in compressive strength. WHSR and CLCFA mortar shown in Figure 2 as WH with sand replacement of the ratio 2.5%, 5%, 7.5% and 10% and CLCFA are 136.85, 156.37, 5.39 and 6.26 ksc in compressive strength, respectively.

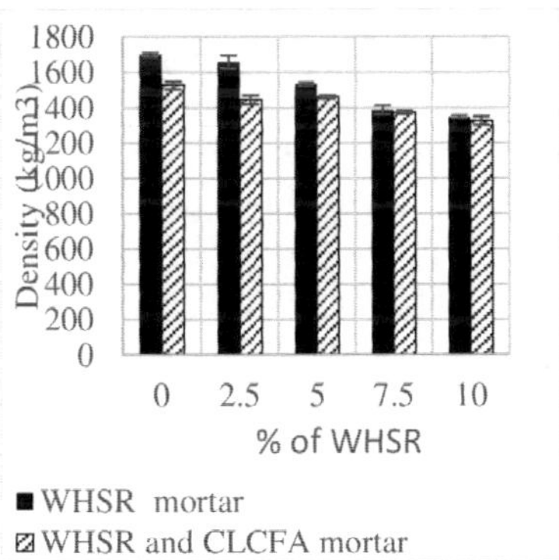

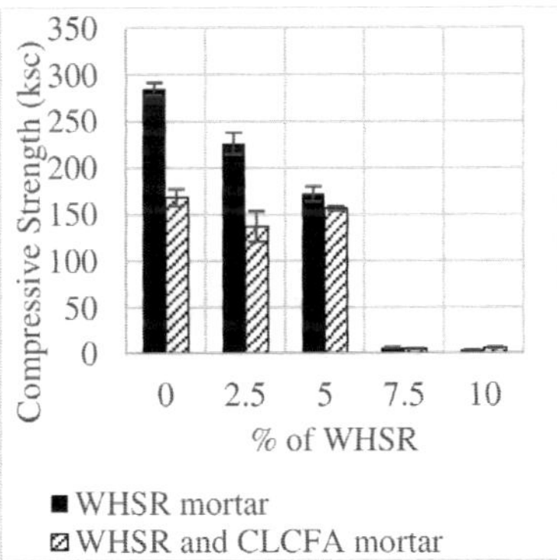

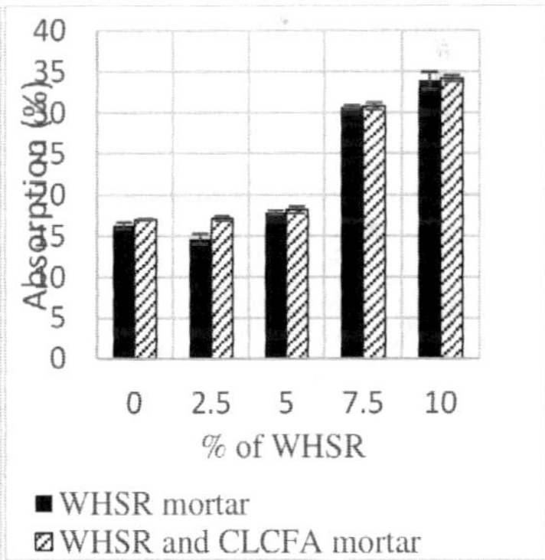

Figure 1. Compressive strength Test. **Figure 2.** Density Test. **Figure 3.** Water absorption Test.

4.2. Dry density testing of WHSR mortar and WHSR and CLCFA mortar

The representatives of the dry density of the WHSR mortar were measured and followed by equation 1.1. The non-WHSR is 1696.57 kg/m3 in dry density. Figure 2 shows the results of this test as the WH sand replacement ratio of 2.5%, 5%, 7.5% and 10% are 1656.59, 1533.52, 1388.49 and 1347.14 (kg/m3) of dry density, respectively.

In the same way, the CLCFA mortar (non-WHSR) is 1526.98 (kg/m3) in dry density. Figure 3 presents the results of this test as the WH sand replacement ratio of 2.5%, 5%, 7.5% and 10% are 1445.57 , 1460.66, 1373.41 and 1324.31 (kg/m3) in dry density, respectively.

4.3. Water absorption testing of WHSR mortar and WHSR and CLCFA mortar

The representatives of the water absorption of the WHSR mortar were measured and followed by equation 1.2. The general mortar (non-WHSR) has a 16.29% rate of water absorption. Figure 3 shows the results of this test as the WH sand replacement ratio of 2.5%, 5%, 7.5% and 10% are 14.64%, 17.84%, 30.73% and 33.89% of water absorption, respectively.

Similarly, the CLCFA mortar (non-WHSR) has a 16.95% rate of water absorption. Figure 3 presents the results of this test as the WH sand replacement ratio of 2.5%, 5%, 7.5% and 10% has a 17.09%, 18.15%, 30.8% and 34.2% rate of water absorption, respectively.

5. Conclusion

The study has investigated the possibility of making a LC with WH, which was qualified from TIS, and its physical and mechanical properties. The main results can be concluded as follows:

1) The compressive strength decreased when the percentage of the WHSR was increased. It was significantly reduced by the WHSR on 7.5. the optimum percentage of the WHSR ratio, both of the WHSR mortar and the WHSR and CLCFA mortar at 5% of water WHSR are 171.45 and 156.37 ksc in compressive strength, according to the TIS 2601-2556.

2) The percentage of WHSR increased which then caused the density to decrease. The lowest density was obtained from the WHSR mortar, the WHSR and CLCFA mortar on 10% as 1347.14 and 1342.21 (kg/m3) respectively. Nonetheless, this did not pass the standard of the TIS 2601-2556. Hence, the optimum percentage of the WHSR ratio was 5% as the WHSR mortar, the WHSR and CLCFA mortar, with a density of 1533.52 and 1460.66 kg/m3, respectively which can be classified in the lightweight concrete block type C16 according to the TIS 2601-2556.

3) Water absorption increased as the percentage of dried water hyacinth sand replacement increased. The lowest water absorption rate was obtained from the WHSR mortar, the WHSR and CLCFA mortar on 2.5% which are 14.64% and 17.09% respectively, which is also lower than the mortar. For this result, the optimum percentage of dried WH sand replacement ratio is 5% as the WHSR mortar and the WHSR and CLCFA mortar, has a density of 17.84% and 18.15%, respectively which is suitable according to the TIS 2601-2556.

This study investigated the mixing of WHSR ratio, the size of WHS in which compressive strength testing, dry density and water absorption testing were carried out. This was done in a ratio where the optimal size and ratio of dried powder water hyacinth which is passed through Sieve No.50 can be summarized. As a result, the compressive strength is much higher than the standard of the TIS 2601-2556 which is 20.4 ksc, the WHSR mortar which have 7.5 and 10% of WHSR did not pass the standard. Hence, 2.5 and 5.0% of WHSR are the optimal conditions for producing lightweight concrete blocks. The mortar which contained 2.5% of the WHSR is 225.99 ksc in compressive strength, which is the highest in this study compared to the 5.0% of WHSR and CLCFA which is 156.37 ksc in compressive strength. In contrast, the density and the water absorption of the 5.0% of WHSR and CLCFA are 1460.66 kg/m3 and 18.15% which is the best condition. The optimal condition does not necessarily have to have a high compressive strength load but it has to be over the qualification of the TIS and focus on eliminating more WH volume. As a result, 5.0% of WHSR and CLCFA mortar is the optimal condition. Furthermore, it would reduce water hyacinth in 1 m2 in the canal by the standard size of lightweight concrete block production which is an environmentally-friendly material.

References

[1] Thailand Institute Standard,2556, Aerated lightweight concrete elements, vol 4562 (2556) Thai industrial standard institute, Ministry of industry (TIS.2601-2556)

[2] Wannee Ekasilp and Chairat Boonthanomwong, Thermal Insulator Performance of Water Hyacinth and Sawdust Hollow Concrete Block, National Research Conference 2014.

[3] Tharaporn Budnumpecthm, Nipon Tanpaiboonkul 2016 Properties of lightweight concrete containing crumb rubber and cement replacing with fly ash from coal power plant, *Veridian E-J. Sci. Technol. Silpakorn Univ.* **3(4)** SN 2408 – 1248.

[4] Pusit Lertwattanaruk, Anchisa Suntijitto 2015 Properties of natural fiber cement materials containing coconut coir and oil palm fibers for residential building applications, *Construct. Build. Mater. Elsevier,* **94** 664-9.

[5] Pisit Lertwattanaruk, Kanokwan Masuwan Investigation of sound Insulation Properties of Fiber Cement Board Containing Natural Fiber, *KMUTT res. develop. J.* **38(1)**.

[6] Natthapong Ladbat, Maneerat Ongwandee, Sahalaph Homwuttiwong, Compressive strength and permeability of concrete with rice-husk ash exposed to various temperatures, Esearch Facilitation And Dissemination Mahasarakham University.

[7] Roy D G, Mehrotra S P, Kapur P C 1984 Lightweight masonry blocks from fly ash pellets, *Resources Conserv.* **11(1)** 63-74.

[8] Ouensanga A 1989 Variation of fiber composition in sugarcan stalks, *Wood fiber sci.* **21(2)** 105-11.

IOP Conf. Series: Materials Science and Engineering **811** (2020) 012015 doi:10.1088/1757-899X/811/1/012015

Progress on Pine Derivative Products as Fuel Source in Indonesia

H Y Setyawan*[1], N M S Sunyoto[1], S Wijana[1], D Pranowo[1]

[1]Department of Agro-industrial Technology, Universitas Brawijaya, Malang-Indonesia

Abstract. Pine resin is a non-wood forestry product in Indonesia that can be converted into a much higher value product. Pine resin traditionally is converted into rosin and turpentine. Due to the scarcity supply of fossil fuel, turpentine has been studied as fuel for diesel substitution. Pinene, the highest component found in pine resin is the primary fuel source candidate. Pinene dimer, modified pinene using catalysts or ozone have a heating value similar to that of JP-10, a tactical fuel for rockets. This paper discussed the technological development of pinene and pinene dimers as a fuel. The opportunity and challenge for producing pine based fuel in Indonesia were also addressed.

1. Introduction

Oleoresin is one of the potential non-woods commodities in Indonesia, and raw pine resin is the main product derived. In 2018, the total production of raw pine resin in Indonesia was over 50,000 tons [1]. Figure 1 shows the spread of location of pine resin production. Three main pine producers in Indonesia are Java island, Sumatra and Celebes. It is clear that in 2015, 32.78 thousand tons (73%) of pine resin was produced in Java island, followed by 6.89 thousand tons (15%) in Sumatra Island. The remaining 5.23 thousand tons (11%) was produced in Celebes island [2].

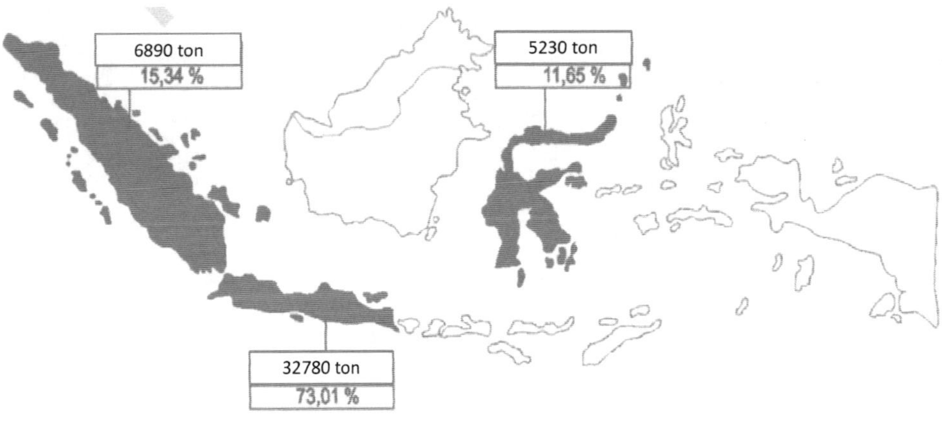

Figure 1. Raw pine resin production by location in Indonesia [2].

Indonesia is among the highest resin exporters [3]. France, Sudan, India and Indonesia are the top four exporter. However, only Sudan and Indonesia are actually not among the highest importers. The high production of raw pine resin in Indonesia is an opportunity to be converted into a much higher value product. One of the high-value products is pinene, which can be used as a fuel and has a high energy density for fuelling jet engine. This paper will address the progress of pine resin research and development, especially the technological development of converting pinene into fuel. This paper will also discuss the opportunity and challenge the development of fuel from pine derivative product in Indonesia.

2. Pine Resin, Rosin and Turpentine Oil

Pine or *Pinus merkusii* is a tropical pine species with high economic importance. Pine is harvested for its wood and non-wood products. The non-wood product of pine is raw pine resin, which could be used in the production of soap, paints, adhesives, printing inks, coatings, paper-sizing, varnishes, perfume, disinfectants and cleaning agents [4]. The raw pine resin is a yellowish, sticky, opaque mass. The raw pine resin can be processed into rosin and turpentine via distillation. The raw pine resin, distilled at 3-30 mbar and 170-290 °C, produced 10-15% turpentine, 20-40% fatty acid, 25-35% rosin, and 20-30% pitch residue [4, 5].

Rosin is the heavy/solid fraction of pine resin distillation. Rosin is transparent, insoluble in water but soluble in many solvents. The turpentine oil is a colourless, volatile liquid with a specific odour. Table 1 shows the chromatogram profile of neutral fraction and turpentine oil. The oil contains α-pinene, *d*-camphene, ß -pinene, Myrcene, α- phellandrene, △-carene, *p*-cymene, and *d*-limonene. The highest component of turpentine oil is pinene, followed by carene [6].

Table 1. The composition of neutral fraction and turpentine oil [6].

No	Constituent	Neutral Fraction	Turpentine oil
1	α-pinene	73.1	82.9
2	*d*-camphene	0.8	0.9
3	ß -pinene	1.8	2.2
4	Myrcene	0.7	0.4
5	α- phellandrene	0.2	0.4
6	△-carene	16.0	11.0
7	*p*-cymene	0.8	1.1
8	*d*-limonene	1.9	1.3

Pinene, $C_{10}H_{16}$, is a bicyclic terpene hydrocarbon consisting of two isomeric forms, alpha- and beta-pinene. However, pinene described usually means alpha-pinene. Recent research on pinene is ranging from the production of pinene through a microbial process [7] to its utilisation for antibacterial [8], food flavour [9], and energy [10].

3. Turpentine Oil as a Fuel

Table 2 shows the thermodynamic properties of turpentine oil [5]. Turpentine oil can be employed as an alternative fuel. Table 2 shows that the distilled turpentine oil has a heating value of 46-47MJ/kg, slightly higher compared to the other fuels such as diesel (44 MJ/kg). Compared to diesel fuel, turpentine oil has lower autoignition temperature and comparable specific heat capacity and thermal conductivity. However, turpentine oil poses some challenges to be used as fuel. The low cetane number of turpentine oil makes it difficult/take longer time to ignite, and it has much higher viscosity which promotes difficulties in fuel injection.

Mixing fuel with a fuel that has better combustion properties can effectively increase the performance of fuel [11, 12]. The mixture of turpentine oil and diesel oil has also been a subject of interest [13]. Improved performance and lower emission of combustion of diesel blended with turpentine oil have been demonstrated [14]. The best mixture was 20% turpentine oil and 80% diesel, producing better performance and lower emission values compared to diesel fuel [14]. The mixing turpentine oil and linseed oil could substitute the energy produced from diesel fuel at high turpentine oil proportion. However, on the high linseed oil proportion, the power is decreased, and it consumed more fuel [13].

Table 2. Thermodynamics properties of turpentine oil [3, 15].

Parameters	Turpentine oil	Biodiesel	Diesel
Molecular weight	136–140	~308	~200
Specific gravity @ 20°C	0.84–0.87	-	0.83
Boiling point (K)	426.8–443	588-630	433-633
Composition (% wt)	C: 88.2 H: 11.8	-	C: 87 H: 16
Vapor pressure (kPa)	4	-	2
Flash point (K) (cc)	303–319	422	328
Autoignition temperature (K)	493–528	646	588
Heating value ($MJkg^{-1}$)	46.2 - 47.2	42.2	44.8
Viscosity (mPa.s)	3.89	3.36 – 3.68	2.20 – 2.59
Specific heat capacity ($kJ\ kg^{-1}$)	1.72	2.2	1.8
Thermal conductivity ($Wm^{-1}K^{-1}$)	0.128	0.1490	0.1169
Adiabatic flame temperature (K@973K)	-	2291	2413
Flammability limit (% volume)	0.8	-	1.0
Cetane number	20-25	51.7	47.4-63.9
Explosive limit	LEL=0.8%;UEL: 6%		
Solubility in water	Insoluble	Insoluble	Insoluble

4. High Density Energy Fuel from Pinene Dimers

The heating value of conventional fuel/biofuel is too low to power energy-intensive equipment, such as rockets, missiles, and aircraft. Pinene dimers, however, shown to have a heating value similar to that of JP-10, a tactical fuel for rockets [16]. The pinene dimers, a modification of pure pinene structure, have similar thermodynamics properties close to that of tactical fuel. Figure 2 shows the comparison of heating value as a function of the density of conventional fuels to pinene dimers and JP-10 and RJ-5.

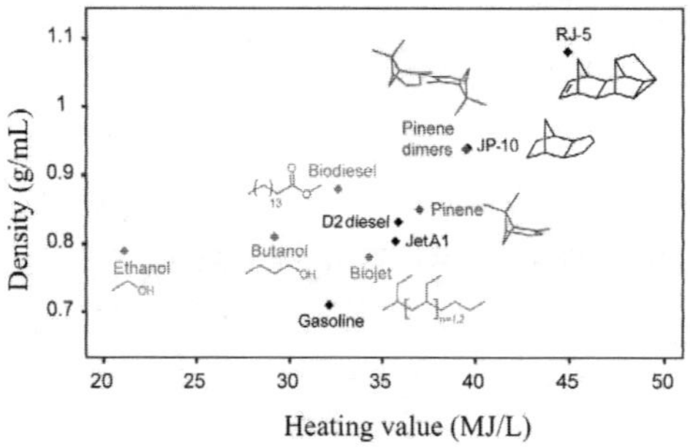

Figure 2. Comparison of energy density various fuels [16].

Therefore, research on α-pinene as fuel is mainly pursuing pinene dimer production. Pinene dimer is created by dimerization of pinenes. The dimerization process can significantly increase the fuel density of pinene, which already has a high heating value due to its ring strain. There are many ways of pinene dimerization, such as using catalyst and ozone. One of the successful dimer production attempt is able to convert pinene into a mixture containing about 60% pinene dimers and 40% isomers via dimerization/isomerization over a chemical catalyst and porous catalyst [17, 18].

The catalytic process of pinene into pinene dimers showed that the catalyst pore size of 20-50μm and catalyst surface treatment using organic ligand was a better mean to produce pinene dimers [19]. In the ozone process, the effect of hydroxyl radicals, temperature, and relative humidity has also been investigated [20].

Pinene dimers provide an opportunity to produce high intensive energy from Indonesian pine. However, research on pinene dimers in Indonesia is scarce. Research on pinene in Indonesia is still in the first generation where it focussed on the production of raw pine resin and its utilization. However, the downstream process and product has yet to be considered [6].

5. Opportunity and Challenge

Biofuel based pinene dimer derived from pine has a bright future in Indonesia due to the high production of pin resin. The government policies to integrate biofuel in energy resource is also encouraging as stated by Government Regulation No. 79 of 2014 on National Energy Policy, new and renewable energy mix target is at least 23% by 2025 and 31% by 2050 [21]. However, due to government subsidy, the fossil fuel price is lower than that of biofuel, causing the bioenergy industry is not competitive yet. Therefore, research and production of biofuel such as pinene dimer require serious support from the government.

6. Conclusion

Pine resin is one of the precious non-wood forestry products in Indonesia, who among the leaders of the production of pine resin. Pine resin traditionally is converted into rosin and turpentine. Due to the scarcity supply of fossil fuel, turpentine has been studied as fuel as a diesel substitute. Pinene, the highest component found in pine resin has been studied as a renewable energy source. Pinene dimer, modified pinene using catalysts or ozone have a heating value similar to that of JP-10, a tactical fuel for rockets. However, despite the abundance production of pinene and government policies, more research support from Indonesia government is required to create a full-scale production of pinene based fuel.

Reference

[1] United Nations Commodity Trade Statistics Database 2019 *Lac; gums, resins and other vegetable saps and extracts* UN Comtrade.
[2] Statistics of Forestry Production 2015 *BPS-Statictics Indonesia*: Jakarta. p. 146.
[3] Alma M H, Salan T 2017 A review on novel bio-fuel from turpentine oil *Process of Petrochemistry and Oil Refinery* **18(1)** 12.
[4] Coppen J J W, Hone G A 1995 Gum naval stores: turpentine and rosin from pine resins, In: *Non-Wood Forest Product* **2** FAO *Rome*
[5] Pohanish R P 2017 *Sittig's Handbook of Toxic and Hazardous Chemicals and Carcinogens (Seventh Edition)* William Andrew Publishing 3012-223.
[6] Wiyono B, Tachibana S, Tinambunan D 2006 Chemical compositions of pine resin, rosin and turpentine oil from West Java *J. Forestry Res.* **30(1)** 10.
[7] Niu F-X, He X, Wu Y-Q, Liu J-Z 2018 Enhancing production of pinene in escherichia coli by using a combination of tolerance, evolution, and modular co-culture engineering *Eng. Front. Microbiol.* **9** 1623.
[8] Mercier B, Prost J, Prost M 2009 The essential oil of turpentine and its major volatile fraction (α- and β-pinenes): A review *Int. J. Occupational Medicine Environ. Health* **22(4)** 331-42.
[9] Vespermann K A C, Paulino B, Barcelos M, Pessôa M, Pastore G, Molina G 2017 Biotransformation of α- and β-pinene into flavor compounds *Appl. Microbiology Biotechnol.* **101(5)** 1805-17.
[10] Romonosky D E, Li Y, Shiraiwa M, Laskin A, Laskin J, Nizkorodov S A 2017 Aqueous photochemistry of secondary organic aerosol of α-pinene and α-humulene oxidized with ozone, hydroxyl radical, and nitrate radical *J. Physical Chem.* **121(6)** 1298-309.
[11] Setyawan H Y, Zhu M, Zhou W, Zhang D 2013 *Effect of Methanol Addition on Combustion Characteristics of Single Droplets of Glycerol* Australian Combustion Symposium Perth Australia.
[12] Zhu M, Zhang Z, Zhang Y, Setyawan H, Li J, Zhang D 2017 Ignition and combustion characteristics of single particles of Zhundong lignite: Effect of water and acid washing *Proceedings of the Combustion Institute* **36(2)** 2475-82.
[13] Yadav N K, Gupta R 2016 Emission reduction by using linseed turpentine oil blend as alternative fuel in diesel engine *Int. J. Eng. Res. Technol.* **5(5)** 5.
[14] Anand B P, Saravanan C G, Srinivasan C A 2010 Performance and exhaust emission of turpentine oil powered *Renew. Energy* **35** 5.

[15] Setyawan H Y, Zhu M, Zhang Z, Zhang D 2016 Ignition and combustion characteristics of single droplets of a crude glycerol in comparison with pure glycerol, petroleum diesel, biodiesel and ethanol *Energy* **113** 153-9.

[16] Sarria S, Wong B, Garcia Martin H, Keasling J D, Peralta-Yahya P 2014 Microbial synthesis of pinene. *ACS Synth. Biol.***3(7)** 466-75.

[17] Becerra J A, Villa A L 2018 Thermodynamic Analysis of α-Pinene and Limonene Allylic Oxidation over a FePcCl16-NH_2-SiO_2 Catalyst *Chem. Eng. Technol.* **41(1)** 124-33.

[18] Shcherban N D, Barakov R Y, Mäki-Arvela P, Sergiienko S A, Bezverkhyy I, Eränen K 2018 Isomerization of α-pinene oxide over ZSM-5 based micro-mesoporous materials *Appl. Catalysis A, General* **560** 236-47.

[19] Jung J K, Lee Y, Choi J-W, Jae J, Ha J-M, Suh D J 2016 Production of high-energy-density fuels by catalytic β-pinene dimerization: Effects of the catalyst surface acidity and pore width on selective dimer production *Energy Convers. Manag.* **116** 72-9.

[20] Kristensen K, Cui T, Zhang H, Gold A, Glasius M, Surratt J D 2014 Dimers in α-pinene secondary organic aerosol: effect of hydroxyl radical, ozone, relative humidity and aerosol acidity *Atmospheric Chem. Physics* **14** 17.

[21] Siswanto D, Mujiyanto S, Suharyati, Pambudi S H, Wibowo J L, Pratiwi N I 2019 *Indonesia energy outlook* Editors. Secretariat General National Energy Council: Jakarta. 73.

IOP Conf. Series: Materials Science and Engineering **811** (2020) 012016 doi:10.1088/1757-899X/811/1/012016

Research on Modular Coordination of Indemnificatory Finished Product Housing in Severe Cold Region of China ——A Case Study of an Experimental Residential Project in Harbin

Yu Shao[1, 2], Yizhou Mo[1, 2]

[1]School of Architecture, Harbin Institute of Technology, Harbin 150001, China
[2]Key Laboratory of Cold Region Urban and Rural Human Settlement Environment Science, Ministry of Industry and Information Technology, Harbin 150001, China

shaoyuu@163.com, Yizhoumo0@outlook.com

Abstract. This paper study the case of an experimental house built in Harbin and analyses the modular coordination of indemnificatory housing in severe cold area. Using case analysis and literature research methods, the modular coordination system of the experimental house is analysed, and this conclusion is elaborated from three aspects: component position, space optimization size and modular grid. The research shows three key problems in modular coordination of indemnificatory finished product housing in severe cold region: how to select the positioning mode of building components; how to select the optimal size according to the infill parts; and how to establish modular grid according to the spatial organization.

1 Introduction

The indemnificatory finished product housing in severe cold area refers to the indemnificatory housing delivered in the form of finished house. The concept of finished housing delivery refers to the provision of new industrial housing with direct use function. The delivery standard of the finished residence is that the internal covering and finishing are completed, the equipment and pipeline ports are complete, and the door parts, window parts, kitchen and bathroom parts are also installed in place.

The concept of indemnificatory finished product housing in severe cold area is based on the industrialization of housing, and the construction system of "SI house"[1] which separates support and infill parts in Japan is used for reference, including the Infill, equipment and pipeline system, building main structure system and exterior enclosure system [2]. The professional cooperation among architectural design, structural design and electromechanical design is the premise to ensure the coordination of each subsystem. This kind of coordination needs to establish a rule to solve the problem of size coordination, that is, to establish a modular coordination system belonging to the finished product housing.

The modular coordination system for the laboratory includes three aspects: component positioning mode, optimal size of space and module grid setting.

1.1. Modular Grid

Building modular grid is the systematic expression of modular coordination results. The setting needs to meet the needs of component installation and spatial organization. Modular grid consists of a series of datum lines, and the relationship between building components and datum lines is also known as its positioning mode. And the distance between successive datum lines should be selected from the preferred size series conforming to modulus [3].

1.2. Two Ways of Component Positioning

The two positioning modes of components are shown in figure 1 and figure 2. Among them:

The way of axis positioning is more commonly used, which is conducive to structural safety calculation and determine the size of prefabricated structural members.

The way of interface positioning is conducive to ensure that the clear space of the building conforms to the module, and is conducive to the installation of the Infill parts, especially the integrated infill parts.

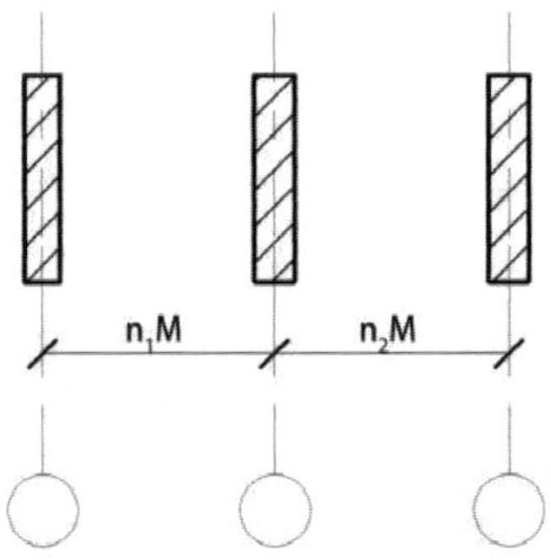

Figure 1. Axis positioning.

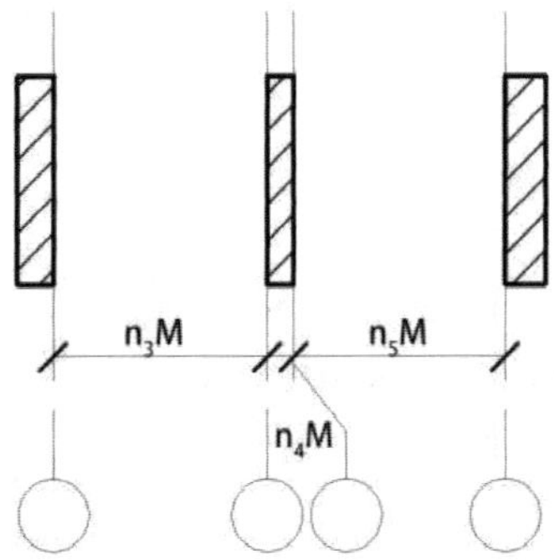

Figure 2. Interface positioning.

1.3. Space optimal Size and Preferred Size

Under certain type of positioning ways， we can tell preferred size from space optimal size. The selection of space optimal size is related to architecture design. The goal of optimal size design of indemnificatory house is to obtain intensive space [4]. It is necessary to conduct quantitative research on the size of each space to find a balance between space economy and human use.

This paper aims to investigate and summarize the current situation of indemnificatory housing in Harbin. Using qualitative and quantitative research methods, the paper puts forward the modular coordination strategy and key points of the experimental house.

2 Basic Information of the Studied Building

2.1. Climate Condition of the Building Environment

According to the Code for Design of Civil Buildings (GB 50352-2005), Severe Cold Region is one of five different climate zones in China, and Harbin is a typical city in severe cold area (table 1). Harbin lies between 125 ° 42 ′ - 130 ° 10 ′ E and 44 ° 04 ′ - 46 ° 40 ′ n. The climate condition of Harbin is presented in table 1.

Table 1. Climate condition of Harbin.

Climate Region	Temperature	
	Hottest	Coldest
Severe Cold	≤ 25°C	≤– 10°C

Data Source: Code for Design of Civil Buildings (GB 50352-2005).

IOP Conf. Series: Materials Science and Engineering **811** (2020) 012016 doi:10.1088/1757-899X/811/1/012016

2.2. Details of the Studied Building

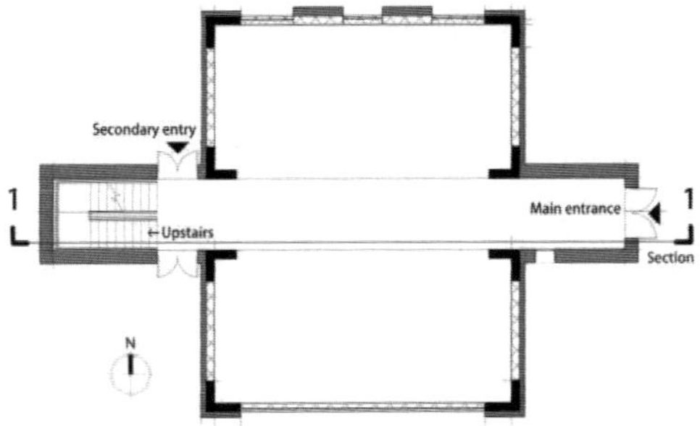

(**a**) Ground floor plan of the building

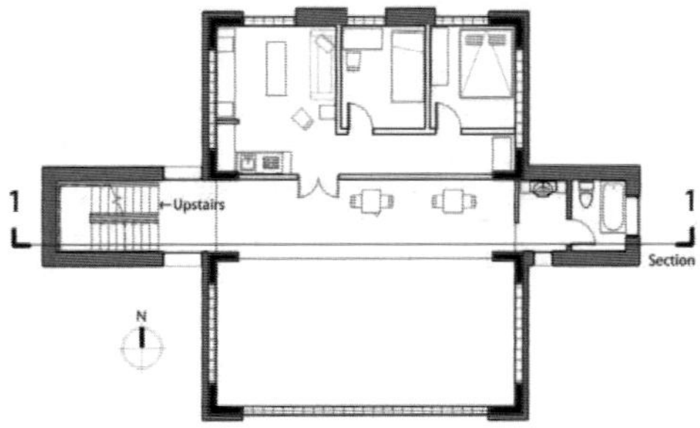

(**b**) First floor plan of the building

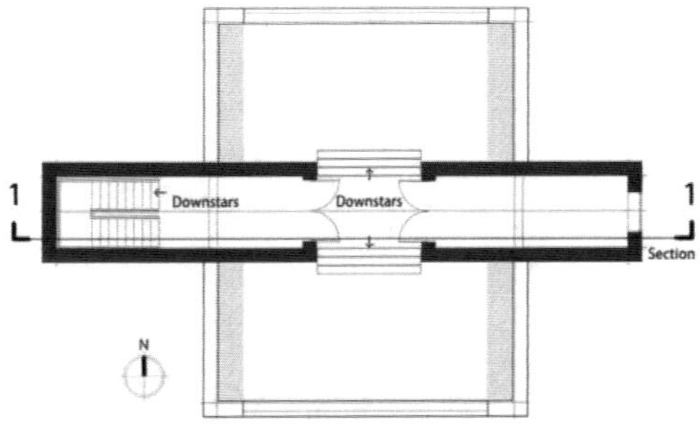

(**c**) Top floor plan of the building

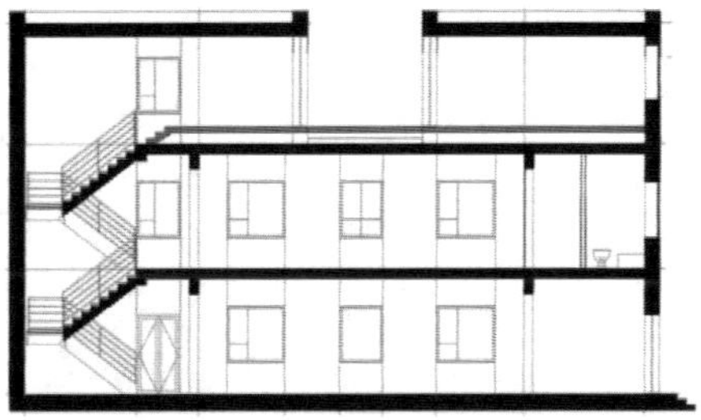

(**d**) Section of the building

Figure 3. Floor plan and section of the experimental house.

The experimental house built in Harbin has three floors, with a total floor area of 224 m^2. The total height of the building is 10.6m, and the standard height of each floor is 3.4m. Figure 3 shows the floor plan of the house. The first floor of the house is the exhibition hall. The second floor is mainly composed of the indoor space and corridor of the residence. It can be arranged as an apartment with two rooms, and the third floor is the roof auxiliary facilities.

The building main structure subsystem of the experimental house is composed of shear wall and coupling beam. The load-bearing member is a group of Cast-in-place Short-leg Shear Walls with a thickness of 200 mm. The shear walls are connected by cast-in-place or prefabricated coupling beams. The main structure does not need to use prefabricated components.

The interior components of the laboratory are mainly included in the Infill parts subsystem, equipment subsystem and pipeline subsystem. Infill is the key to realize the "delivery in the form of finished products" of finished housing. The interior decoration of the experimental house includes integrated bathroom, integrated kitchen, overhead ground system, lightweight partition system and other equipment, etc. In order to meet the regional demand, the raised flooring systems and the under-floor heating system are arranged in combination.

The exterior enclosure subsystem of the experimental house needs to meet the regional needs. The subsystem consists of light embedded enclosure subsystem and mounted external enclosure subsystem, and adopts filled wall, insulation board, facade curtain walls and other modules.

3 Method

In this paper, *SPSS Statistics 25.0* is used to analyse the data that obtained from the case investigation. The original data source is presented in table 2.

Table 2. Basic information of public rental housing project.

Project name	Type	Floor area（㎡）	Sample size
HuaRunKaiXuanMen	High-rise Apartment	49~60	7
LvHaiHuaYuan	High-rise Apartment	36~41	5
ZhongShanLongJun	High-rise Residential Building	53~58	3
DongDuGongYuan	High-rise Residential Building	40-49	4
GeLanYunTian	High-rise Residential Building	38~59	12
MingGuangShuiAn	High-rise Apartment	48~50	7
HengDaMingDu	High-rise Residential Building	48~49	3
HuaYuanShuiMuQingHua	High-rise Apartment	43~53	5
JinRuiLinCheng	High-rise Residential Building	48~50	8
LvHaiHuaTing	High-rise Apartment	47~49	20
ZhongTianFuCheng	High-rise Residential Building	47~49	4
YuanChuangYueFu	High-rise Residential Building	52~63	12

Through case analysis, this paper summarizes the modular data of the space size of indemnificatory housing in severe cold area, and puts forward the positioning ways of the components through comprehensive evaluation, obtains the space optimal size series of the housing space and draws the modular grid plan.

Due to the timeliness of the construction demand, this paper investigates 30 newly-built indemnificatory housing projects in Harbin, and obtains 90 sets of samples by random sampling survey. The information of component and space size of indemnificatory housing are sorted into database. Through quantitative description and qualitative analysis, this paper summarizes the spatial characteristics and functional requirements of indemnificatory housing, and provides the basis for the spatial design and component selection of experimental room.

According to relevant standards, the floor area of single indemnificatory housing is controlled within 50 square meters. In the severe cold area, the floor area is large under the same use area, so the floor area of the investigated case is between 40 to 60 square meters. The housing types investigated include ordinary housing, apartments and affordable housing. The project nature, quantity and other information of the case are also shown in table 2.

4 Results and Analysis

4.1. Optimal Size Series

The data of spatial layout and spatial size of indemnificatory housing in severe cold area are obtained and BIM database is formed. The design of space optimal size series needs to meet three conditions: first, it can realize the intensive layout of space; second, it can cover the commonly used size range; third, it can adopt appropriate multi-module [5].

The analysis of spatial dimension database can summarize the common spatial layout and the range of spatial dimension. IN the process of selecting and expanding modulus, the factors considered are coordination, applicability and economy. The multi-module is coordinated with the whole module system. The size formed by multi-module can satisfy various spatial layout modes. The modulus value is beneficial to the economy of component construction [6].

Through the analysis of the above three conditions, the optimal size series of indemnificatory housing space in severe cold area is obtained as shown in figure 4 [7].

Figure 4. The optimal size series of bedrooms, kitchens, living rooms and toilets.

4.2. Component Positioning

The experimental house can be divided into four subsystems. The main structure system mainly consists of load-bearing structural components, the Infill, equipment and pipeline system mainly consists of Infill parts, and the exterior enclosure system mainly consists of exterior parts.

The positioning mode of main structure components is shown in figure 5. It has the following advantages:

It can choose more kinds of thermal insulation materials and adopt passive technology to deal with regional problems, excluding the interference of different materials, forms and thicknesses of the external enclosure system on the other systems; The positioning mode contributes a complete internal clear space that conforms to the module. It is conducive to the organization of parts in the interior space.

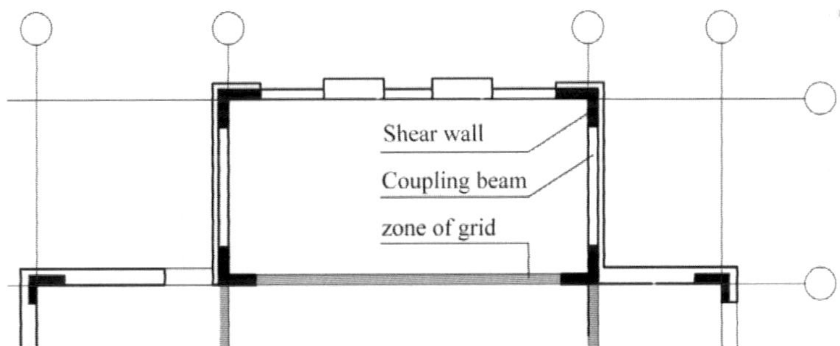

Figure 5. the positioning of components of main structure system.

After comprehensive consideration, the way of interface positioning is adopted for the Infill parts. There are two subtypes:

Interface positioning method are adopted to integrated bathroom and integrated kitchen unit (figure 6); Residential living room, bedroom, corridor space adopted interface positioning method and set the zone of grid (figure 7).

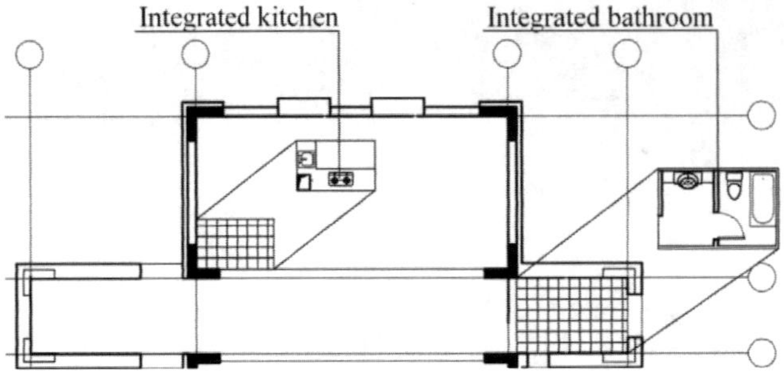

Figure 6. Integrated bathroom layout with interface positioning.

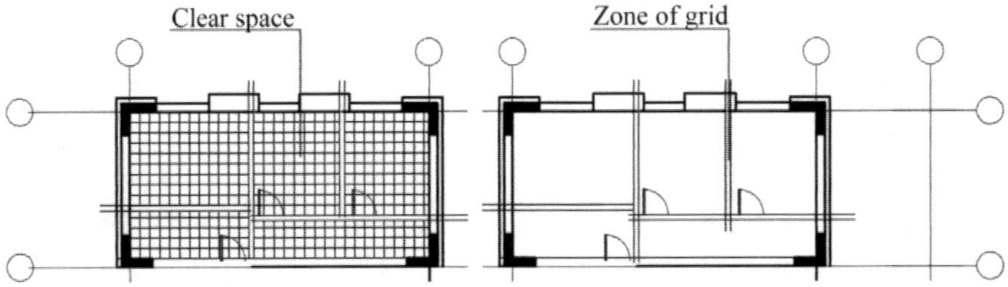

Figure 7. Interface positioning of bedroom and corridor unit.

Similar to the poisoning of main structure subsystem, the exterior enclosure subsystem uses the way of interface positioning to avoid the interference to other building components. The following two subtypes are shown in figure 8:

Light embedded enclosure system, ALC（Autoclaved Lightweight Aerated Concrete Panel）is used to fill the enclosure structure; The light external enclosure system is located in the outer layer of the main structure, and the auxiliary datum line is used to locate the components. The mounted external enclosure system adopts vacuum insulation board and other related materials to meet the insulation performance requirements in the severe cold area, and the mounted external curtain wall is used as the façade (figure 9).

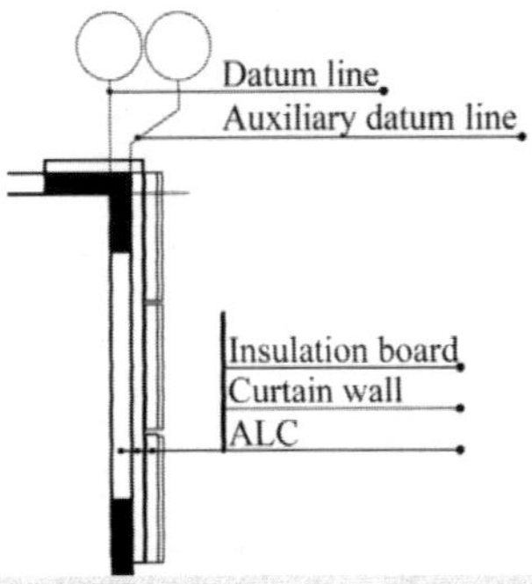

Figure 8. The positioning mode of the light embedded enclosure system and the light external enclosure system.

Figure 9. ALC block products, polystyrene thermal insulation layer and mounted external concrete curtain wall.

4.3. Modular Grid

The component positioning ways provides the qualitative conclusion, and the optimal size series provides the quantitative conclusion of the modular coordination. On this basis, a modular grid suitable for the experimental house can be proposed. Among them, the module coordination problem of the second floor is the most typical, and the module grid plan is shown in figure 10.

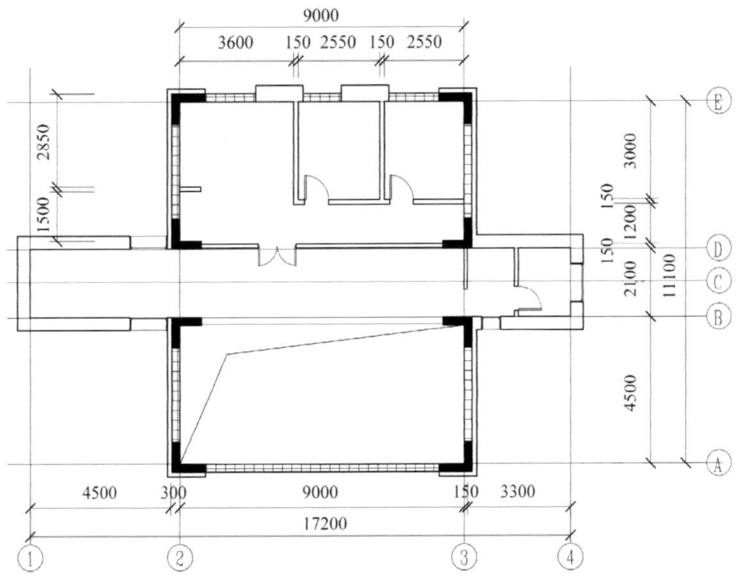

Figure 10. modular grid plan of the second floor of experimental house.

5 Conclusion

The finished product housing provides a new construction mode for indemnificatory housing in severe cold area. The main problems in the application of modular coordination system are as follows:

First, in the process of building modular grid, reasonable and optimal size should be selected. The selection of optimal size should consider the economic and human needs, and the space corresponding to the modular grid should be intensive.

Second, the component positioning mode shall meet the regional requirements. In this case, the passive means to increase the thermal performance of the enclosure structure is selected, and the active heating mode is provided by the overhead ground system and the under-floor heating system. It has a significant impact on the positioning mode of components.

Third, in the process of building module grid, the zone of grid shall be set reasonably to ensure the coordinated installation of the Infill parts, especially the Integrated infill parts, so as to achieve the purpose of the delivering of finished product housing.

Acknowledgment

This work was supported by the National Natural Science Foundation of China (51878199).

References

[1] Seiichi F, Xinxin G The history of developments toward Open Building in Japan 2011 *J. New Architecture.***6** 14-7(in Chinese).

[2] Cuperus Y 2001 An introduction to Open Building *9th Annual Conf. of the Int. Group for Lean Construction* (Singapore: Ballard G & Chua D).

[3] Xin Z, Wei X 2013 The application research of Europe and the United States construction module in residential industrialization system *J. Architecture & Culture* **2** 82-5 (in Chinese).

[4] Hasim A, Nicola G, Sam M and Antonio F 2015 Redesigning terraced Social Housing in the UK for flexibility using building energy simulation with consideration of passive design *J. Sustainability* **7(5)** 5488-5507.

[5] Xiaohong Z, Lin L, Jishou Z 2012 Development and application of modern building modulus theory *J. Architectural Jl.***4** 30(in Chinese).

[6] Yoshiya U General System of Construction Industrialization 1983 *Build. Construct. Syst.* (Shanghai Scientific & Technical Publishers) (in Chinese).

[7] Hua L 2015 Secondary modular system and its use in housing construction *J. Architectural J.* **1** 237-41 (in Chinese).

Global Existence of a 2-D Heat-Conductive Incompressible Viscous Fluid

P. Yu[1]

[1]School of Mathematics Science, University of Electronic Science and Technology, Xiyuan Ave 2006, Chengdu, 611731, China

yp9106@gmail.com

Abstract. In this paper, we consider the Cauchy problem of non-stationary heat conducting problem coupled with viscous incompressible fluid. We construct the global well-posed ness of this equation and improve coefficient range.

1. Introduction

In this paper, we consider the following Cauchy problem of a heat-conductive incompressible viscous fluid system posed in $\mathbb{R}^2 \times \mathbb{R}^+$

$$\begin{cases} \mathbf{u}_t + (\mathbf{u}\cdot\nabla)\mathbf{u} - \operatorname{div}(\mu(\theta)\nabla\mathbf{u}) + \nabla P = 0, \\ \theta_t + (\mathbf{u}\cdot\nabla)\theta - \operatorname{div}(\kappa(\theta)\nabla\theta) - \mu(\theta)\,|\nabla\mathbf{u}|^r = 0, \\ \operatorname{div}\mathbf{u} = 0, \\ \mathbf{u}\big|_{t=0} = \mathbf{u}_0, \quad \theta\big|_{t=0} = \theta_0, \end{cases} \tag{1.1}$$

for any $r \in [3/2, 2)$. Here, u, P, θ are the unknowns denote fluid velocity, temperature and the pressure of fluid, respectively. The temperature-dependent functions $\mu(\theta)$, $\kappa(\theta)$ are the viscosity of fluid and heat conductivity, which are assumed to be

$$\begin{aligned} &\mu(\cdot), \kappa(\cdot) \in C^3([0,\infty)), \quad 0 < \underline{\mu} \le \mu(\theta) \le \overline{\mu}, \quad |\mu'(\theta)| \le \mu_1, \\ &0 < \underline{\kappa} \le \kappa(\theta) \le \overline{\kappa}, \quad |\kappa'(\theta)| \le \kappa_1, \quad 1 < \overline{\kappa}\underline{\kappa}^{-1} < \kappa_2. \end{aligned} \tag{1.2}$$

The model (1.1) coupled with conditions (1.2) is widely studied for decades. The results of well-posedness mostly concern with the case of $r = 2$. Consider the non-convection case, Shillkin [2] showed that the convection-free model, where (1.1) degenerates to a nonlinear parabolic system

$$\begin{cases} \mathbf{u}_t - \operatorname{div}(\mu(\theta)\nabla\mathbf{u}) = f, \\ \theta_t - \kappa(\theta)\Delta\theta = \mu(\theta)\kappa(\theta)\,|\nabla\mathbf{u}|^2, \end{cases} \tag{1.3}$$

processes a unique global weak solution in spatial dimensional three and a unique strong solution in spatial dimensional two. Without ignorant of the convective term, Ye [3] and Ye-Zhu [4], established the global existence of global well-posedness results of 2D and 3D Cauchy problem. Recently, Ye [4] deduced the boundary layer profile of half-space problem.

This paper concerns about the global existence of strong solution to (1.1)-(1.2) with small assumptions on initial data and $r=\frac{3}{2}$. To this end, we give the local-in-time existence result which can be deduced via the well-known Banach fixed point theorem and the continuity method [1].

Theorem 1.1 (Local-in-time existence) Assume that $(\mathbf{u}_0,\theta_0)\in\left(H^3(\mathbb{R}^2)\right)^2\times H^3(\mathbb{R}^2)$ and $\theta_0\geq 0$. Then there exist a maximum existing time T and a unique classical solution $(\mathbf{u},\theta)$ to (1.1) such that

$$(\mathbf{u},\theta)\in L^\infty\left(0,T;H^3\left(\mathbb{R}^2\right)\right),\quad (\nabla\mathbf{u},\nabla\theta)\in L^2\left(0,T;H^3\left(\mathbb{R}^2\right)\right). \tag{1.4}$$

Moreover, the solution (u,θ) blows up if and only if

$$\int_0^t\left\|\nabla\mathbf{u}(s)\right\|_{L^4}^3\,\mathrm{d}s\to+\infty, \tag{1.5}$$

as $t\to T$.

We claim that the obtained solution which exists local-in-time will be extended globally by some *a priori* estimates. We illustrate the main result of global well-posedness as follows:

Theorem 1.2 (Global-in-time existence) Assume that $(\mathbf{u}_0,\theta_0)\in\left(H^1(\mathbb{R}^2)\right)^2\times H^1(\mathbb{R}^2),\ \theta_0\geq 0$. If (1.2) holds and $\|\nabla\mathbf{u}_0\|_{L^2}^2$ is small enough, then there exists a global-in-time solution $(\mathbf{u},\theta)$ to (1.1) such that

$$(\mathbf{u},\theta)\in L^\infty\left(0,\infty;H^1\left(\mathbb{R}^2\right)\right),\quad (\nabla\mathbf{u},\nabla\theta)\in L^2\left(0,\infty;H^1\left(\mathbb{R}^2\right)\right) \tag{1.6}$$

as well as

$$\|(\nabla\mathbf{u},\nabla\theta)(t)\|_{L^2}^2\leq C_1(1+t)^{-1}. \tag{1.7}$$

2. A Priori Estimates

First of all, the following energy estimate of u holds.

Lemma 2.1 Assume that $(\mathbf{u},\theta)$ is a smooth solution to (1.1)-(1.2). Then u satisfies

$$\sup_{0\leq t\leq T}\|\mathbf{u}\|_{L^2}^2+\frac{\mu}{2}\int_0^T\|\nabla\mathbf{u}\|_{L^2}^2\mathrm{d}t\leq\|\mathbf{u}_0\|_{L^2}^2. \tag{1.8}$$

Proof: We can easily obtain the estimate by multiplying u to the first sub equation to (1.1) and integrating by parts.

Next, we introduce an important variable

$$\hat{\theta}=\int_0^\theta k(z)\mathrm{d}z,$$

comes from ordinary differential equation theory to gather the estimate on θ.

Lemma 2.2 Assume that $(\mathbf{u},\theta)$ is a smooth solution to system (1.1)-(1.2) satisfying

$$\int_0^T\|\nabla\mathbf{u}\|_{L^4}^3\,dx\leq M. \tag{1.9}$$

Then

$$\sup_{0\leq t\leq T}\|\nabla\theta\|_{L^2}^2\leq C_1 \tag{1.10}$$

as well as

$$\int_0^T \|\Delta\theta\|_{L^2}^2 \mathrm{d}t \le C_2, \tag{1.11}$$

where C_1, C_2 depend only on $\underline{\kappa}$, $\bar{\mu}$, κ_1, κ_2 and M.

Proof: By a direct calculation, the second sub problem of (1.1) could be rewritten as follows

$$\begin{cases} \hat{\theta}_t + \mathbf{u}\cdot\nabla\hat{\theta} = \kappa(\theta)\Delta\hat{\theta} + \mu(\theta)\kappa(\theta)|\nabla u|^{\frac{3}{2}}, \\ \hat{\theta}(x,0) = \int_0^{\theta_0(x)} k(z)\mathrm{d}z \,\square\, \hat{\theta}_0(x). \end{cases} \tag{1.12}$$

Therefore, by multiplying $-\Delta\hat{\theta}$ to each side of (1.12) and integrating the resulted equation by parts on $\mathbb{R}^2$, we can easily get

$$\begin{aligned} &\frac{1}{2}\frac{\mathrm{d}}{\mathrm{d}t}\left\|\nabla\hat{\theta}\right\|_{L^2}^2 + \left\|(\kappa(\theta))^{1/2}\Delta\hat{\theta}\right\|_{L^2}^2 \\ &\le \left\|\mathbf{u}\nabla\hat{\theta}\Delta\hat{\theta}\right\|_{L^1} + \left\|\mu(\theta)\kappa(\theta)|\nabla\mathbf{u}|^{\frac{3}{2}}\Delta\hat{\theta}\right\|_{L^1} \\ &\le \frac{2}{\underline{\kappa}}\|\mathbf{u}\|_{L^4}^2\left\|\nabla\hat{\theta}\right\|_{L^4}^2 + 2\bar{\mu}^2\bar{\kappa}\kappa_2\|\nabla\mathbf{u}\|_{L^4}^3 + \frac{\bar{\kappa}}{2}\left\|\Delta\hat{\theta}\right\|_{L^2}^2, \end{aligned} \tag{1.13}$$

where we used the Young's inequality and the conditions of coefficients (1.2). Moreover, we could use the Gagliardo-Nirenberg-Sobolev inequality to show that

$$\begin{aligned} &\frac{1}{2}\frac{\mathrm{d}}{\mathrm{d}t}\left\|\nabla\hat{\theta}\right\|_{L^2}^2 + \underline{\kappa}\left\|\Delta\hat{\theta}\right\|_{L^2}^2 \\ &\le \frac{C_2}{\underline{\kappa}}\|\mathbf{u}\|_{L^2}\|\nabla\mathbf{u}\|_{L^2}\left\|\nabla\hat{\theta}\right\|_{L^2}\left\|\Delta\hat{\theta}\right\|_{L^2} + 2\bar{\mu}^2\bar{\kappa}\kappa_2\|\nabla\mathbf{u}\|_{L^4}^3 + \frac{\bar{\kappa}}{2}\left\|\Delta\hat{\theta}\right\|_{L^2}^2 \\ &\le \frac{C_2}{\underline{\kappa}^2}\|\mathbf{u}\|_{L^2}^2\|\nabla\mathbf{u}\|_{L^2}^2\left\|\nabla\hat{\theta}\right\|_{L^2}^2 + 2\bar{\mu}^2\bar{\kappa}\kappa_2\|\nabla\mathbf{u}\|_{L^4}^3 + \frac{3\bar{\kappa}}{4}\left\|\Delta\hat{\theta}\right\|_{L^2}^2. \end{aligned} \tag{1.14}$$

We can apply the Gronwall's inequality to obtain that

$$\begin{aligned} &\sup_{0\le t\le T}\left\|\nabla\hat{\theta}\right\|_{L^2}^2 + \frac{\kappa}{2}\int_0^T\left\|\Delta\hat{\theta}\right\|_{L^2}^2\mathrm{d}t \\ &\le \exp\left\{\frac{2C_3}{\underline{\kappa}^2}\|\mathbf{u}_0\|_{L^2}^2\int_0^T\|\nabla u\|_{L^2}^2\mathrm{d}t\right\}\left(\left\|\nabla\hat{\theta}_0\right\|_{L^2}^2 + 4\bar{\mu}^2\bar{\kappa}\kappa_2\int_0^T\|\nabla\mathbf{u}\|_{L^4}^3\mathrm{d}t\right) \\ &\le C_4\exp\left(\|\mathbf{u}_0\|_{L^2}^4\right)\left(\left\|\nabla\hat{\theta}_0\right\|_{L^2}^2 + M\right), \end{aligned} \tag{1.15}$$

where C_4 depends only on coefficients in (1.2), M and the initial data. Thus, by a simple calculation of relationship between $\hat{\theta}$ and θ we get (1.10)-(1.11) and complete the proof.

Lemma 2.3 Assume that $(\mathbf{u},\theta)$ is a smooth solution to (1.1)-(1.2) and the initial data $\|\mathbf{u}_0\|_{L^2}^2$, $\|\nabla\mathbf{u}_0\|_{L^2}^2$ are small enough. There exists positive constants K_1 and K_2 depending only coefficients in (1.2) and initial data such that

$$\sup_{0\le t\le T}\|\nabla \mathbf{u}(t)\|_{L^2}^2+\int_0^T\|\nabla^2\mathbf{u}\|_{L^2}^2 dt\le K_1 \tag{1.16}$$

and

$$\int_0^T\|\nabla \mathbf{u}\|_{L^4}^3 dt\le K_2. \tag{1.17}$$

Proof: By multiplying $-\Delta\mathbf{u}$ to the first sub problem of (1.1), integrating the resulted equation by parts, we get

$$\frac{1}{2}\frac{d}{dt}\|\nabla \mathbf{u}\|_{L^2}^2 dx+\frac{\mu}{2}\|\Delta \mathbf{u}\|_{L^2}^2\le\|\nabla\mu(\theta)\nabla\mathbf{u}\|_{L^2}\|\Delta\mathbf{u}\|_{L^2}\le|\mu'(\theta)|\|\nabla\theta\|_{L^4}\|\nabla\mathbf{u}\|_{L^4}\|\Delta\mathbf{u}\|_{L^2} \tag{1.18}$$

By using Young's inequality and Gagliardo-Nirenberg-Sobolev inequality, we obtain that

$$\frac{1}{2}\frac{d}{dt}\|\nabla \mathbf{u}\|_{L^2}^2+\frac{\mu}{2}\|\Delta \mathbf{u}\|_{L^2}^2\le\frac{\kappa_2}{2}\|\nabla\theta\|_{L^2}\|\Delta\theta\|_{L^2}\|\nabla\mathbf{u}\|_{L^2}. \tag{1.19}$$

Therefore, an application of Gronwall's equation again yields that

$$\sup_{0\le t\le T}\|\nabla \mathbf{u}\|_{L^2}^2+\underline{\mu}\int_0^T\|\Delta \mathbf{u}\|_{L^2}^2 dt\le\|\nabla\mathbf{u}_0\|_{L^2}^2\exp\left(\exp\left(\|\nabla\mathbf{u}_0\|_{L^2}^4\right)P_1\left(\|\nabla\theta_0\|_{L^2}^2\right)+M\right), \tag{1.20}$$

where $P_1(\cdot)$ is a polynormal satisfying $P_1(0)=0$. Moreover, we take the Gagliardo-Nirenberg-Sobolev inequality again to show

$$\int_0^T\|\nabla \mathbf{u}\|_{L^4}^3 dt\le 2\sup_{0\le t\le T}\|\mathbf{u}\|_{L^2}^2\int_0^T\|\nabla^2\mathbf{u}\|_{L^2}^2 dt. \tag{1.21}$$

Thus, by a direct observation on continuous function on the right-hand-side of (1.20), we find that if the initial data is small enough, then the left-hand-side of (1.20) could be estimated as

$$\sup_{0\le t\le T}\|\nabla \mathbf{u}\|_{L^2}^2+\underline{\mu}\int_0^T\|\Delta \mathbf{u}\|_{L^2}^2 dt\le\|\nabla\mathbf{u}_0\|_{L^2}^2\exp\left(f\left(\|\mathbf{u}_0\|_{L^2}^2,\|\nabla\theta_0\|_{L^2}^2\right)+M\right), \tag{1.22}$$

where $f(\cdot,\cdot)$ is a nonnegative continuous function satisfying $f(0,0)=0$. Therefore, we obtain (1.16) and (1.17) which complete the proof.

Lemma 2.4 Assume that $(\mathbf{u},\theta)$ is a smooth solution to (1.1)-(1.2). If the initial data $\|\mathbf{u}_0\|_{L^2}$ is small enough, then

$$\sup_{0\le t\le T}\|\theta\|_{L^2}^2+\int_0^T\|\nabla\theta\|_{L^2}^2 dt\le C_5, \tag{1.23}$$

for some positive constant C_5 depending only on coefficients in (1.2) and initial data.

Proof: We take L^2 inner product of θ and the second sub problem of (1.1) to obtain that

$$\frac{1}{2}\frac{d}{dt}\|\theta\|_{L^2}^2+\underline{\kappa}\|\Delta\theta\|_{L^2}^2\le C_6\|\nabla\mathbf{u}\|_{L^2}^2\|\theta\|_{L^2}^2+\frac{\underline{\kappa}}{2}\|\Delta\mathbf{u}\|_{L^2}^2. \tag{1.24}$$

Now we apply the Gronwall's inequality to the above differential equation to gather

$$\sup_{0\le t\le T}\|\theta\|_{L^2}^2+\int_0^T\|\nabla\theta\|_{L^2}^2 dt\le C_2. \tag{1.25}$$

This completes the proof.

In this moment, we have the uniform estimates

$$\sup_{0\le t\le T}\|\mathbf{u}\|_{L^2}^2,\ \sup_{0\le t\le T}\|\nabla\mathbf{u}\|_{L^2}^2,\ \sup_{0\le t\le T}\|\theta\|_{L^2}^2,\ \sup_{0\le t\le T}\|\nabla\theta\|_{L^2}^2<+\infty \tag{1.26}$$

and

$$\int_0^T\|\nabla^2\mathbf{u}\|_{L^2}^2\,\mathrm{d}t,\int_0^T\|\nabla^2\theta\|_{L^2}^2\,\mathrm{d}t<+\infty. \tag{1.27}$$

3. Proof of Theorem 1.1

These uniform boundedness (1.26) and (1.27) ensure Theorem 1.1 can be deduced as follows. By multiplying (1.19) by $1+t$, we find that

$$\frac{\mathrm{d}}{\mathrm{d}t}\left[(1+t)\|\nabla\mathbf{u}\|_{L^2}^2\right]+(1+t)\|\Delta\mathbf{u}\|_{L^2}^2\le C_5\|\nabla\theta\|_{L^4}^4(1+t)\|\nabla\mathbf{u}\|_{L^2}^2+\|\nabla\mathbf{u}\|_{L^2}^2, \tag{1.28}$$

which implies

$$\sup_{0\le t\le T}\|\nabla\mathbf{u}\|_{L^2}^2\le C_7(1+t)^{-1}. \tag{1.29}$$

We couple the above estimate with the resulted inequality of (1.24) multiplied by $1+t$

$$\frac{\mathrm{d}}{\mathrm{d}t}\left[(1+t)\|\nabla\theta\|_{L^2}^2\right]+(1+t)\|\Delta\theta\|_{L^2}^2\le C_5\|\mathbf{u}\|_{L^4}^4(1+t)\|\nabla\theta\|_{L^2}^2+(1+t)\|\nabla\mathbf{u}\|_{L^2}^2\|\Delta\mathbf{u}\|_{L^2}^2 \tag{1.30}$$

to obtain

$$\sup_{0\le t\le T}\|\nabla\theta\|_{L^2}^2\le C_8(1+t)^{-1}. \tag{1.31}$$

By using the blow-up criteria (1.5), we complete the proof

□

4. Global Existence of Solutions

Using these uniform boundedness results, we can give the proof of Theorem 1.1:

Proof of Theorem 1.1: By multiplying (1.19) by $1+t$, we find that

$$\frac{\mathrm{d}}{\mathrm{d}t}\left[(1+t)\|\nabla\mathbf{u}\|_{L^2}^2\right]+(1+t)\|\Delta\mathbf{u}\|_{L^2}^2\le C_5\|\nabla\theta\|_{L^4}^4(1+t)\|\nabla\mathbf{u}\|_{L^2}^2+\|\nabla\mathbf{u}\|_{L^2}^2, \tag{2.1}$$

which implies

$$\sup_{0\le t\le T}\|\nabla\mathbf{u}\|_{L^2}^2\le C_6(1+t)^{-1}. \tag{2.2}$$

If we couple the above estimate with the resulted inequality of (1.24) multiplied by $1+t$,

$$\frac{\mathrm{d}}{\mathrm{d}t}\left[(1+t)\|\nabla\theta\|_{L^2}^2\right]+(1+t)\|\Delta\theta\|_{L^2}^2\le C_5\|\mathbf{u}\|_{L^4}^4(1+t)\|\nabla\theta\|_{L^2}^2+(1+t)\|\nabla\mathbf{u}\|_{L^2}^2\|\Delta\mathbf{u}\|_{L^2}^2 \tag{2.3}$$

to gather

$$\sup_{0\le t\le T}\|\nabla\theta\|_{L^2}^2\le C_6(1+t)^{-1}. \tag{2.4}$$

(2.2) and (2.4) yield that $\|\nabla\mathbf{u}\|_{L^2}^2$ and $\|\nabla\theta\|_{L^2}^2$. By the above analysis, we find that.

IOP Conf. Series: Materials Science and Engineering **811** (2020) 012017 doi:10.1088/1757-899X/811/1/012017

References

[1] Evans L C 1998 *Partial differential equations* (New York: American Mathematical Society).
[2] Shilkin T 2005 Classical solvability of the coupled system modelling a heat-convergent Poiseuille-type flow *J. Mathematical Fluid Mech.* **7(1)** 72-84.
[3] Ye X 2019 Boundary-layer phenomena for the boundary layer for the heat-conductive incompressible viscous fluids in the half space *Comput. Mathematics Appl.* **78(6)** 1889-96.
[4] Ye X, Zhu M 2016 Global Well-Posedness for the Heat-conductive Incompressible Viscous Fluids *Mathematical Physics Anal. Geometry* **19(3)** 19
[5] Ye X 2017 Global existence of heat-conductive incompressible viscous fluids *Acta Applicandae Mathematicae* **148(1)** 61-9.

IOP Conf. Series: Materials Science and Engineering **811** (2020) 012018 doi:10.1088/1757-899X/811/1/012018

The Potential Use of Waste Cement as a Filler in Polymer Composites

A Rattanapan[1], T Sritapunya, S Tuampoemsab, P Sapsrithong

[1]Division of Polymer Engineering Technology, Department of Mechanical Engineering Technology, College of Industrial Technology, King Mongkut's University of Technology North Bangkok, Bangkok, 10800, Thailand

apaipan.r@cit.kmutnb.ac.th

Abstract. This study explores the potential to use waste cement particles from ready-mixed concrete industry as a low cost filler in polypropylene. Its effects on mechanical and morphological properties were investigated in comparison with commercial cement and calcium carbonate. Mixing and shaping procedures of the polymer composites specimen were conducted by two-roll mill and injection molding, respectively. Next, the effect of filler loading (0, 10, 20, 30, 40, and 50 phr) and filler type on the mechanical and morphological properties of the polymer composites were studied. The results show that when filler loadings increased, the young's modulus, flexural modulus, impact strength, and hardness of the polymer composites also increased accordingly, while the tensile strength and flexural strength decreased. Moreover, the results show that the waste cement had a greater effect on the enhanced mechanical properties compared to commercial cement and calcium carbonate. Furthermore, the polymer composites' morphology was studied using scanning electron microscopy. A microstructure analysis was performed to verify and provide a better assessment of the performance properties of the polymer composites.

1. Introduction

Recently, ready-mixed concrete has become widespread and several construction industries in Thailand utilize ready-mixed concrete in building construction. Customers normally receive ready-mixed concrete from the producer using rotating-drum truck mixers on demand and in the specific quantity required. Occasionally these trucks return residues of the ready-mixed concrete back to the producer. That sand, gravel, and cement are then separated and the sand and gravel is reused for ready-mixed concrete processing. However, waste cement is eliminated by landfill, which pollutes the environment and increases costs for the producer. Due to the rise in the amount of waste cement, this study generates alternative uses for waste utilization, specifically proposing waste cement as a filler for polymer composites.

Polymer composites have gained greater attention over recent decades in different technological applications due to their light weight, cost reduction, and improved mechanical properties. Normally, fillers such as calcium carbonate, clay, and talc have been widely selected for use in improving properties and reducing costs in the plastics industry [1-4]. A number of studies have investigated polymer composites prepared using waste material as a filler. Žlebeka T et al. studied the suitable use of waste glass in polymer anchor materials based on epoxy resin. The results led to the selection of a formulation that yielded the best quality of polymer anchor [5]. Meanwhile, Çınar M E and Kar F studied the potential for using marble dust as a reinforced element in composite materials. They reported that increasing the marble dust ratio improved the bending strength, tensile strength, hardness, and thermal conductivity [6]. Further, Aman A et al. examined the possibility of waste material as a filler for polymeric composite insulation to be environmentally and economically beneficial. Seashells ($CaCO_3$) and waste glass (SiO_2) were selected to produce an artificial wollastonite (calcium silicate-$CaSiO_3$) to use as a filler in a polypropylene (PP) matrix. It was found that the chemical compositions of the artificial wollastonite was similar to natural wollastonite [7].

The main aim of the present study was to explore the potential for using waste cement as a filler in polymer composites. Polypropylene (PP)—a commodity plastic—is used as a matrix. The effect of filler content on the mechanical properties were examined. Commercial cement and calcium carbonate were selected for use as a filler to compare their properties with waste cement. The study results show the potential for using waste cement in the plastics industry.

2. Experimental

2.1. Materials

Polypropylene, 1100NK was purchased from IRPC Public Company Limited (Thailand). This had a melt flow index of 11 g/10 min (190°C, 2.16 kg). Waste cement powder derived from ready-mixed concrete was selected from Concrete Products and Aggregate Company Limited (CPAC) (Thailand), which was filtered with a screen panel of 400 meshes. Commercial cement (Ordinary Portland cement-grade) was obtained from Siam Cement Group, Thailand. Calcium carbonate which selected as the commercial filler and was supplied by Kij Paiboon Chemical Limited Partnership (Bangkok, Thailand).

2.2. Preparation of polymer composites

The polypropylene filled with different filler, i.e. waste cement, commercial cement, and calcium carbonate were added at 0, 10, 20, 30, 40, and 50 phr. All the polymer composites were melt compounded in two-roll mill at 190°C for 12 minutes and then pelletized and oven dried.

2.3. Mechanical analysis

An injection molding process was used to prepare the specimens for tensile, flexural, impact, and hardness testing. Tensile testing was conducted using a universal testing machine (model M500-25AT, Testometric). The specimens were made in accordance with ASTM D 638. The instrument was operated at a crosshead speed of 5 mm/min using a 2,500 kgf load cell. The flexural modulus and flexural strength testing of the composite specimens were carried out on a universal testing machine (model M500-25AT, Testometric) in accordance with ASTM D 790 at a crosshead speed of 1.3 mm/min. The notched Izod impact test was determined according to ASTM D 256 using a Ceast impact testing model 6957. The hammer was used to strike the sample with 2.75 J. Hardness was tested by a Shore hardness tester (model Shore D, Desik) in accordance with ASTM D 2240. In each case, the tests were performed at room temperature, with five specimens tested to obtain an average value.

2.4. Scanning electron microscopy (SEM)

The polymer composites' dispersion morphology was observed by scanning electron microscopy (model S-4800, Hitachi). The samples were soaked in liquid nitrogen and before being fractured. The test samples were mounted on stubs and then gold coated to avoid electric charge accumulation during examination.

3. Results and Discussion

Figure 1 depicts the tensile properties of polypropylene filled with various filler types at different filler loadings. It is found that the tensile strength decreased with increasing filler content. This means that the addition of organic filler in polypropylene affected the tensile strength, but polypropylene filled with waste cement had an effect that was less than polypropylene filled with commercial cement and calcium carbonate. This may be because in this study, the polypropylene was filled with organic fillers without any compatibilizer, incompatibility between polypropylene, which non-polar polymer and fillers, which polar material. Therefore, this had a weak interaction effect on the tensile strength reduction. Similar to the tensile strength results, the strain at break decreased with increasing filler content, indicating that the polymer composites became somewhat brittle compared with neat polypropylene. Similar findings are shown in the literature and have been explained by a failure mechanism [8-10]. Under tensile stress, micro-cracks start at the end of the surface of filler particles, due to higher stress concentration at these points. These micro-cracks then move along the particle and spread across the matrix. When the cracks grow to a critical size, fracturing of the composite occurs. This result shows that the failure is closely related to the content of the particle and explains why increasing filler content decreases tensile strength and strain at break values. Although the use of fillers shows no positive effect on the tensile strength of the polypropylene composites, the addition of fillers enhances the tensile modulus. It was found that the tensile modulus of polymer composites increased with increasing filler content from 0 to 50 phr. Interestingly, the comparison between polypropylene filled with different filler i.e. waste cement, commercial cement, and calcium carbonate indicates that the trend of tensile strength and tensile modulus value is the same for polypropylene filled with waste cement and calcium carbonate.

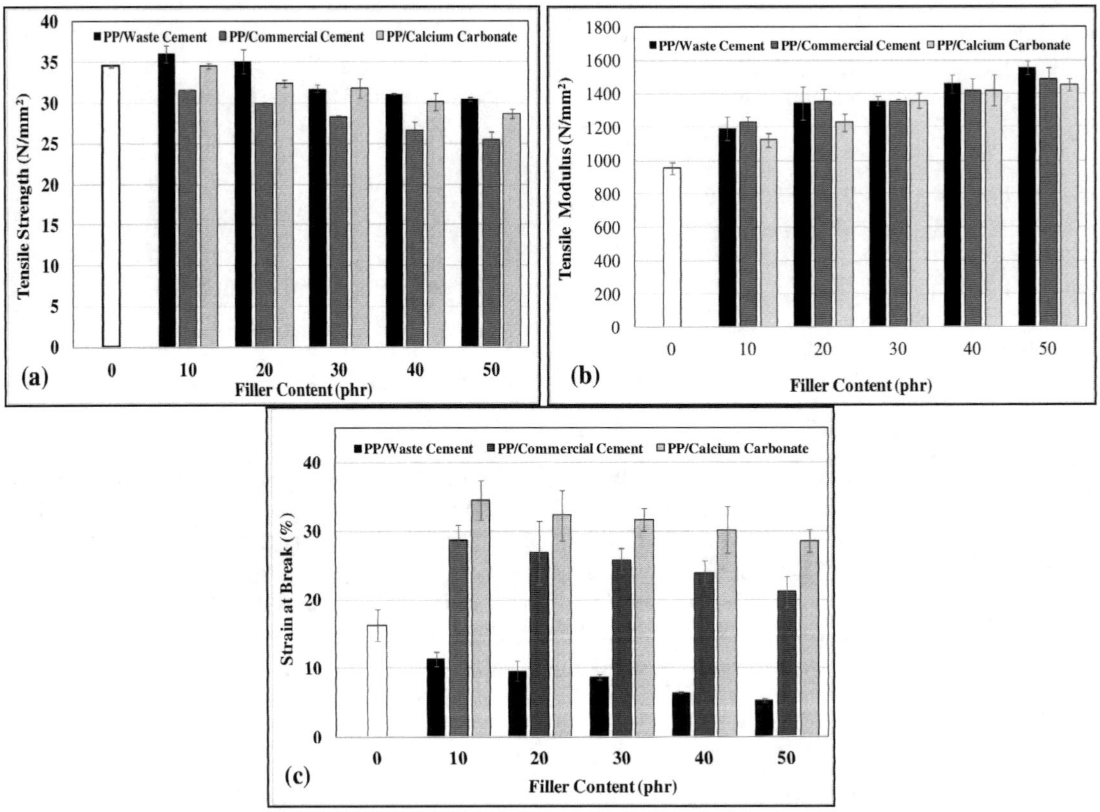

Figure 1. The tensile properties of polypropylene filled with various filler types at different filler loading; (a) tensile strength, (b) tensile modulus and (d) strain at break.

Figure 2 shows the flexural properties of polypropylene filled with various filler types at different fillers loadings. It is found that the flexural strength decreased slightly with increasing filler loading. The slight decrease of flexural strength illustrates the weak interaction between the fillers and the polypropylene matrix. Meanwhile, flexural modulus increased with increasing filler loading. This indicates that the fillers improved the flexural properties of polypropylene. Comparisons between different fillers show that the waste cement improved the flexural properties of polymer composites more than calcium carbonate, which normally used in industry.

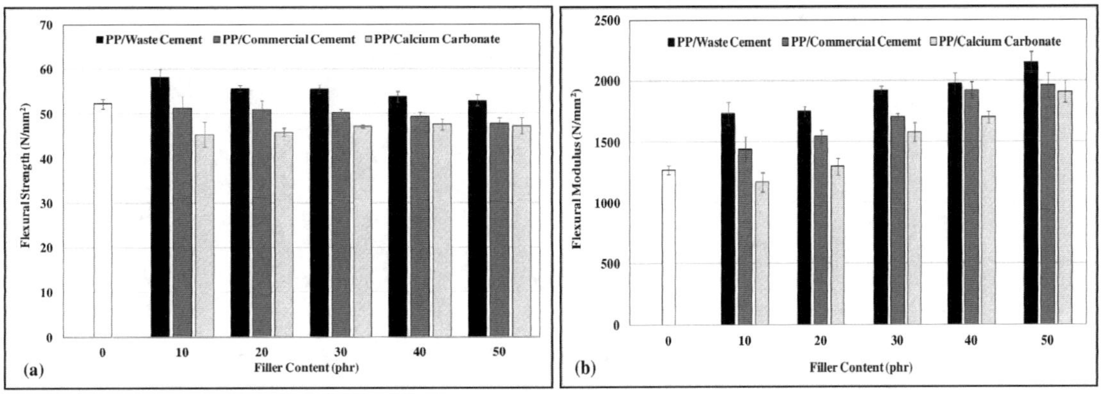

Figure 2. Flexural properties of polypropylene filled with various filler types at different filler loading; (a) flexural strength and (b) flexural modulus.

Figure 3 shows impact and hardness polypropylene filled with various filler types at different filler loadings. It is found that the impact strength and hardness of polypropylene filled with waste cement increased with increasing waste cement loading. For polypropylene filled with calcium carbonate and commercial cement, it was found that the impact strength slightly decreased with increasing filler loading, while the hardness value increased with increasing filler loading. Interestingly, at the same filler content, the trend of impact strength and hardness value of polypropylene filled with waste cement was higher than calcium carbonate.

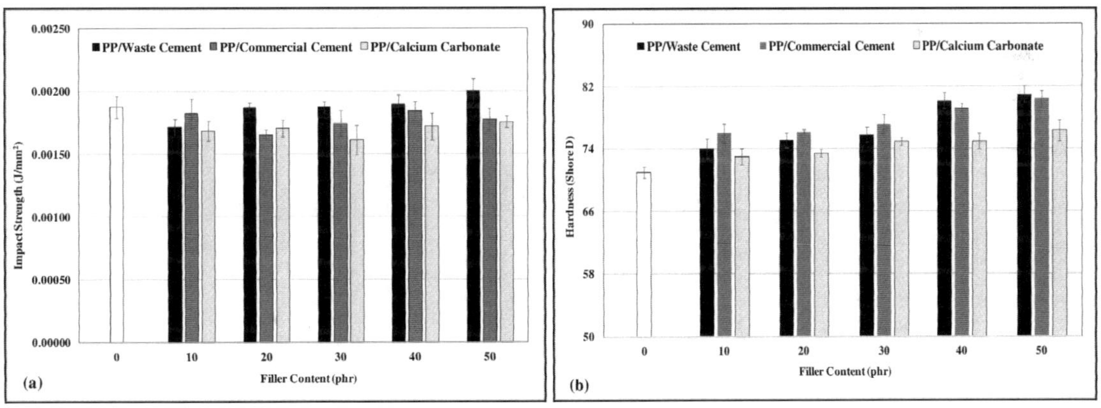

Figure 3. (a) Impact and (b) hardness properties of polypropylene filled with various filler types at different filler loadings.

This study shows the potential use of waste cement as a filler for preparing polymer composites. Since it improved the mechanical properties of polymer composites, as seen from the SEM micrographs in figure 4, the fractured surface of polymer composites showed a good distribution of waste cement in the polypropylene matrix. This indicates that waste cement improved the properties of polypropylene composites.

IOP Conf. Series: Materials Science and Engineering **811** (2020) 012018 doi:10.1088/1757-899X/811/1/012018

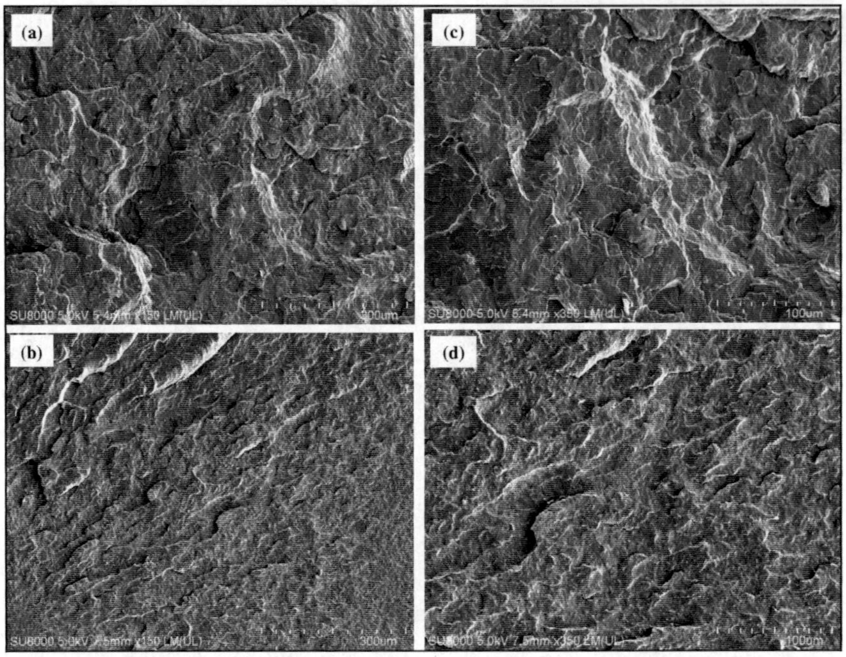

Figure 4. SEM micrographs of polypropylene filled with waste cement;
(a) 10 phr (x150), (b) 10 phr (x350), (c) 50 phr (x150) and (d) 50 phr (x350).

4. Conclusions
In this research, waste cement was used as a filler for polypropylene and compared with commercial cement and calcium carbonate. The mechanical and morphological properties were examined. From the results, waste cement can improve the properties of polypropylene similar to calcium carbonate, which are normally used as a filler in polymer industries. This observation was supported by SEM micrographs. Therefore, there is potential for waste cement to be used as a filler for prepared polymer composites, which may be useful for waste management from construction industries and to reduce costs.

Acknowledgment
A partial support was received from the Division of Polymer Engineering Technology, Department of Mechanical Engineering Technology, College of Industrial Technology, King Mongkut's University of Technology North Bangkok.

References
[1] Rai U S, Singh R K 2004 Synthesis and mechanical characterization of polymer-matrix composites containing calcium carbonate/white cement filler, *Mater. Lett.* **58** : 235-40
[2] Kausar A 2017 Physical properties of hybrid polymer/clay composites, *Hybrid Polymer Composite Mater.* : 115-32.
[3] Ouchiar S, Stoclet G, Cabaret C, Georges E, Smith A, Martias C, Addad A, Gloaguen V 2015 Comparison of the influence of talc and kaolinite as inorganic fillers on morphology, structure and thermomechanical properties of polylactide based composites, *Appl. Clay Sci.* **116-117** : 231-41.
[4] Pukánszky B 2016 Mineral Filled Polymers, *Reference Module Mater. Sci. Mater. Eng.* : 1-6.
[5] Žlebek T, Hodul J, Drochytk R 2017 Experimental testing suitability of the waste glass into the polymer anchor materials based on epoxy resin, *Procedia Eng.* **195** : 220-7.
[6] Çınar M E, Kar F 2018 Characterization of composite produced from waste PET and marble dust, *Constr. Build. Mater.***163** : 734-41.

[7] Aman A, Yaacob M M, Alsaedi M A, Ibrahim K A 2013 Polymeric composite based on waste material for high voltage outdoor application, *Int. J. Electr. Power Energy Syst.* **45** : 346-52.

[8] Wang Z, Xie G, Wang X, Li G, Zhang Z 2006 Rheology enhancement of polycarbonate/calcium carbonate nanocomposites prepared by melt-compounding, *Mater. Lett.* **60** : 1035-8.

[9] Pinheiro I F, Ferreira F V, Souza D H S, Gouveia R F, Lona L M F, Morales A R, Mei L H I 2017 Mechanical, rheological and degradation properties of PBAT nanocomposites reinforced by functionalized cellulose nanocrystals, *Eur. Polym. J.* **97** : 356-65.

[10] Karsli N G, Aytac A 2013 Tensile and thermomechanical properties of short carbon fiber reinforced polyamide 6 composites, *Composites Part B* **51** : 270-5.

Morphology and Mechanical Properties of Polyamide 6 and Polybutylene Terephthalate Blends Compatibilized with Epoxidized Natural Rubber

P Sapsrithong[1], T Sritapunya[1], S Tuampoemsab[1], A Rattanapan[1], M Nithitanakul[2]

[1]Division of Polymer Engineering Technology, Department of Mechanical Engineering Technology, College of Industrial Technology, King Mongkut's University of Technology North Bangkok, Bangkok, 10800, Thailand
[2]The Petroleum and Petrochemical College, Chulalongkorn University, Bangkok, 10330, Thailand

pornsri.s@cit.kmutnb.ac.th

Abstract. Blends of polyamide 6 (PA6) and, polybutylene terephthalate (PBT) have been prepared in this work. They were compatibilized using epoxidized natural rubber (ENR-50) by varying amount of ENR-50 i.e., 0, 3 and 5 phr. The effect of blend compositions and compatibilization on the phase morphology and properties of the resulting blends were investigated using a scanning electron microscope, universal testing machine, impact testing machine, and hardness shore durometer. All blends were prepared and shaped by using twin-screw extruder at 260°C and injection molding machine, respectively. The results revealed that the phase morphology of the PA6/PBT blend was incompatible. The separation of PA6 and PBT was observed wherein PBT is dispersed in the form of spherical domains in the PA6 matrix. The mechanical properties exhibit loss in tensile and impact strength. This could be attributed to poor interfacial adhesion between the two polymers. SEM micrographs showed that the addition of ENR-50 improved the compatibility of PA6/PBT blends. With the addition of ENR-50 as a compatibilizer, the uniformity and the maximum reduction of dispersed phase sized were achieved. The water absorption of PA6/PBT blends decreased with an increasing amount of PBT. Additionally, the results indicated that, as the amount of ENR-50 increased, the mechanical properties, including tensile stress at peak, tensile modulus, and izod impact strength were improved.

1. Introduction

High-performance plastics are playing a more important role in the automotive industry because of their many advantages, including high strength, lightweight, minimal corrosion, flexibility, excellent appearance and, recyclability. Many types of plastic are being employed for the application in the automotive industry, such as polystyrene (PS), polypropylene (PP), polyethylene (PE), polyvinyl chloride (PVC), polyamide (PA), acrylonitrile butadiene styrene (ABS), polycarbonate (PC), acrylonitrile styrene acrylate (ASA) and polybutylene terephthalate (PBT), etc. They are frequently used when a low cost, high mechanical strength, good heat and chemical resistant, durable material is desired [1]. Unfortunately, a number of problems limit their extensive applications. So, the utilization of polymer blends is an increasingly important part of the plastic industry. There are many studies focus on polymer blend such as PS/PE, PA/PE, PP/PE, PA/PC, etc [2-5]. Additionally, PA6 and PBT are both high-performance engineering polymers with commercial significance. The advantages of PA6 and PBT are mentioned, i.e., high strength, good thermal and chemical resistance and excellent barrier property which versatile industrial applications. However, the limitation of PA6 was also remarked, it has high moisture absorption. In order to create new and useful materials, blends of polyamide (PA) and polybutylene terephthalate (PBT) were successfully determined by many researchers since both polymers are among the most frequently used [6-9]. Unfortunately, blend of the two different materials cannot be compatible with each other leading to phase separation and reduce mechanical properties. Therefore, a suitable interaction between the phases must be added to the blends in order to obtain finer morphology and reduce the interfacial tension between the two polymers. The formation of the polyester-polyamide copolymer can be occurred by ester-amide interchange reaction [7]. The compatibility of polyamide and polyester could improve the adhesion between two phases, but the mechanical properties and crystallinity will reduce. Numerous research works have been carried out in order to improve interfacial adhesion and properties of the polymer blend. The addition of suitable compatibilizer is one of the most effective methods to improve compatibility and also properties [10-12]. In 2001, Han and coworker have reported that the compatibility between PA6 and PBT was observed in the presence of ethylene glycidyl methacrylate (EGMA) as a compatibilizer [6]. They found that EGMA could be used as a compatibilizer leading to increasing mechanical properties of the blends. Huang and Chang studied the improvement of morphology and properties of PA/PBT blends with the use of epoxy resin as a compatibilizer, resulting in good mechanical properties were obtained [7]. The addition of acrylic modified polyolefin type ionomer to polyester/polyamide6 blends was also reported by Samios and Kalfoglou [13]. They found that a polymer blend with ionomer compatibilizer leads to improve tensile and impact properties. Solid-state epoxy resin has also been used by several researchers to compatibilize the polymer blends, including PA6/PBT, PC/poly(acrylonitrile-butadiene-styrene), PC/PBT/LCP, PP/PBT [5]. Solid epoxy resin is an attractive compatibilizer because of the formation of in situ block copolymer between the epoxide end-groups with end-groups of PA6, PBT or PC. Epoxidized natural rubber (ENR) is derived from natural rubber by introducing hydrophilic groups (with epoxide units randomly with 25 or 50 mol% epoxidation) along the isoprene backbone. The epoxidation of ENR leads to improve properties and also the miscibility with polar polymers because the oxirane ring increases the hydrophilicity. Recently, epoxidized natural rubber (50 mol% epoxidation: ENR-50) has been considerable interest in the research of polymer blend and composite due to the ability of the epoxide group of ENR that can be used to improve the compatibility of the system [14, 15]. The interaction can take place between the epoxide group of ENR and the polar group of the polymer. The effectiveness of ENR-50 was previously demonstrated in poly(lactic acid)/rice starch composites [14], linear-low-density polyethylene (LLDPE)/soya powder blends [15], poly(lactic acid)/natural rubber blends [16] and PA6/LDPE blends [17]. They reported that the improvement in morphology and mechanical properties was observed. The compatibility between the phase increases because of the incorporation of ENR-50 into the system. The aim of this work is to investigate the compatibilizing effect of epoxidized natural rubber (ENR-50) on the morphology, and mechanical properties of PA6/PBT blends.

2. Experimental

2.1. Materials

Polyamide 6 (PA6 grade 1013 NW8), an injection molding grade polymer, was supplied by UBE Industries, Ltd. Polybutylene terephthalate (PBT: Ultradur® B2550) supplied by BASF (Thai) Limited. Epoxidized natural rubber (50 mol% epoxidation: grade ENR-50) compatibilizer was supplied by San-Thap International Co., Ltd. All materials were used as received.

2.2. Compounding and Specimen Preparation Characterizations

PA6 and PBT were externally mixed using a tumble mixer for 10 minutes, followed by drying them in a hot air oven at 60°C for 24 h before use, to remove any absorbed moisture. Blends of PA6 and PBT were prepared in 100/0, 90/10, 80/20 and 0/100 % by weight. The amount of ENR-50 used as a compatibilizer was 0, 3 and 5 phr for PA6/PBT: 90/10 and 80/20 % by weight. The materials were blended in a twin-screw extruder at 260°C using a screw speed of 40 rpm. The extruded pellets were dried in an oven at 60°C for at least 24 h. Test specimens were prepared using a Boy 22M injection molding machine at 260°C.

2.3. Characterizations

The fracture micrographs, as well as the dispersed structure of the fractured samples, were studied by using FE-SEM (Hitachi S-4800, Japan) at a voltage of 15.0 kV. The samples were fractured under liquid nitrogen and then coated with platinum on a sputter coater, under vacuum, to make them electrically conductive. The water absorption measurement was conducted per ASTM D570. Samples were dried, weighed after that immersed in distilled water. After 24 h., specimens were removed from the water and wiped with a dry cloth and weighed again. A Universal Testing Machine was used to measure the tensile properties of the obtained materials in accordance with ASTM D638. Izod impact strength was measured using a CEAST Impact tester according to ASTM D256 test procedure method with a 2.7 J pendulum. The hardness was tested by shore hardness durometer (Shore D) in accordance with ASTM D2240.

3. Results and Discussion

Phase morphology of the blends between PA6 and PBT was determined using a field emission scanning electron microscope. Figure 1 presents SEM micrographs taken from the cryogenically fractured surface of 80/20 PA6/PBT blends with different ENR-50 content from 0-5 phr. Morphology of the uncompatibilized blends exhibited two-phase morphology with phase separation between PA6 and PBT and contained numerous voids and rough surfaces. The presence of PBT dispersed phase, consisting of predominantly spherical droplets embedded in a PA6 matrix, was clearly observed from the micrographs. Moreover, big gaps between two phases were clearly observed implying relatively poor interaction. The average diameter of the PBT dispersed phase is about 4.6 μm. However, most of the PBT dispersed phase still kept intact within the PA6 matrix. This could be due to some of the interaction between the polar group of both PA6 and PBT. When the 3 and 5 phr of ENR-50 was used in PA6/PBT blend, the dispersed phase size was reduced. This phenomenon can be explained by the presence of ENR-50, which acted as a compatibilizer leading to enhance the interfacial bonding between the PA6 matrix and PBT. A more stable morphology with a smooth surface and a smaller droplet size can be observed with increasing ENR-50 content in blends. The number average dispersed phase size ranged between 1.2 and 1.8 μm, which were much smaller than in the uncompatibilized blends. This observation is an agreement with work reported by Yew *et al.* [14] that the presence of the polar group of ENR-50 promoted the compatibility between matrix and filler.

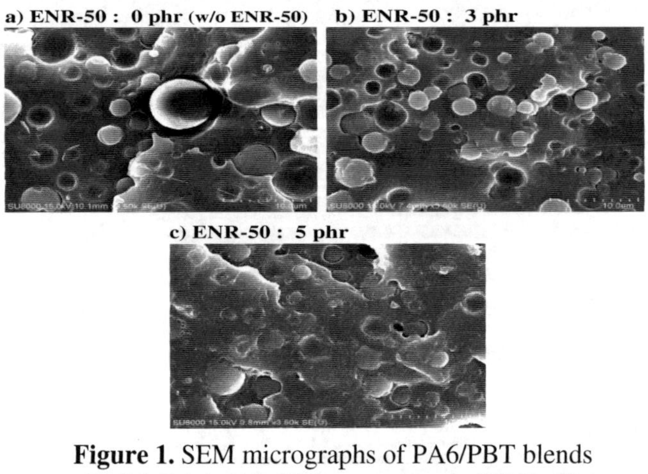

Figure 1. SEM micrographs of PA6/PBT blends (80/20 %wt) with different amount of ENR-50.

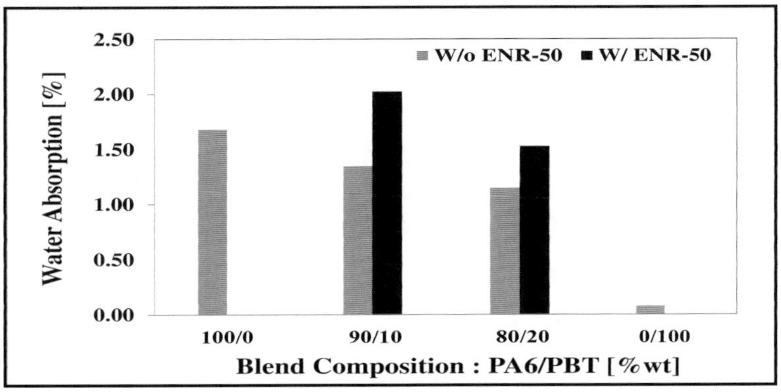

Figure 2. Percentage of water absorption for PA6/PBT/ENR-50 blends.

Figure 2 illustrated the percentage of water absorption of PA6/PBT blends both with (w/ ENR-50) and without ENR-50 (w/o ENR-50) as a compatibilizer. It is well known that polyamide 6 has a drawback that is moisture sensitive. According to Gadekar *et al.* the increase in the amount of water in PA6 was mainly due to the amide groups, thus blending PA6 with non-sensitive materials can overcome this problem [3]. Usually, the value of the percentage of water absorption of PBT was lower than that of pure PA6. So, the addition of PBT decreased the amount of water absorbed by PA6/PBT blends (from 1.6% to 1.1%). The percentage of water absorption of PA6/PBT blends increased with the incorporation of ENR-50. This may be attributed to the polarity of ENR-50 that will enhance bonding formation with water molecules. The effect of blend compositions and ENR-50 content on the tensile properties of PA6/PBT blends is shown in figure 3. For PA6/PBT blends without compatibilizer, a decrease in tensile stress and young's modulus were observed. This could be due to the poor interfacial adhesion between PA6 and PBT. However, the tensile properties between 90/10 and 80/20 wt% of PA6/PBT blends were slightly different. The addition of ENR-50 (50 mol% epoxidation) into PA6/PBT blends improved both tensile stress and young's modulus. This may be attributed to the compatibilization effect of ENR-50 that could form interaction with the polar group of PA6 and PBT, which results in better stress transfer between two phases. This behavior can be explained by the observation from SEM micrographs of the blends, which clearly indicated that the finer morphology with the smaller size of the dispersed phase was observed.

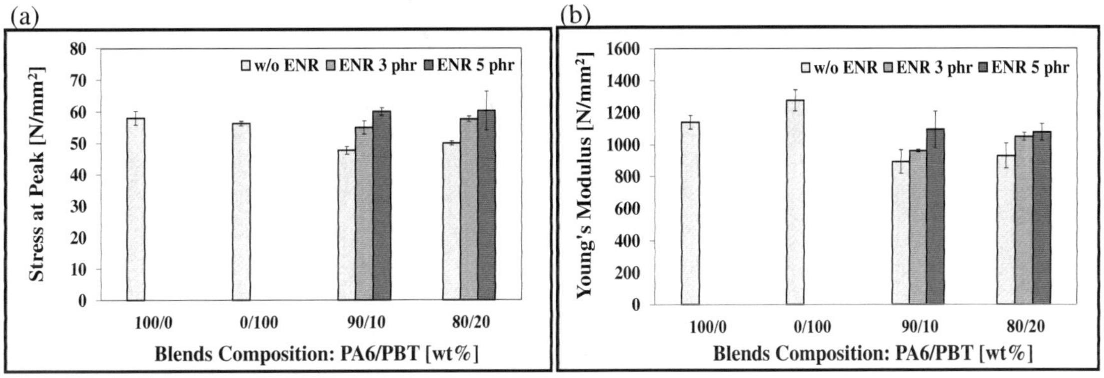

Figure 3. Tensile Stress and Young's Modulus of PA6/PBT/ENR-50 blends.

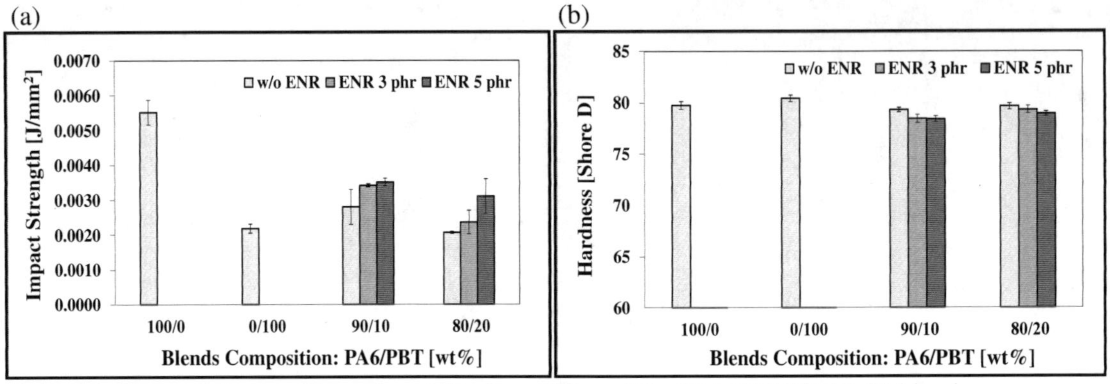

Figure 4. Impact Strength and Hardness (Shore D) of PA6/PBT/ENR-50 blends.

The notched impact property of blending between PA6 and PBT is illustrated in figure 4 (a), for PA6/PBT blends without ENR-50, a drastic decrease in notched impact strength with an increasing amount of PBT was reported, which is attributed to phase separation between matrix and dispersed phase. The incorporation of ENR-50 as a compatibilizer was found to improve the impact strength and further increase with an increasing amount of ENR-50. This improvement can also be explained by the reduction of interfacial tension between phases, which allow absorbed energy to transfer from one phase to another and elastomeric behavior of ENR-50 [14, 15]. Additionally, according to figure 4 (b), it could be observed that the hardness of PA6/PBT blends is slightly different as compared with pure PA6 and PBT. The effect of compatibilizer loading on hardness was also investigated. In PA6/PBT blends containing compatibilizer, the values of hardness were lower than that of pure PA6 and PBT for all compositions. The hardness of PA6/LDPE/ENR-50 was slightly decreased as the ENR-50 content increased due to the addition of the elastomeric component.

4. Summary

Blends of two significant engineering plastics were successfully prepared and investigated in this study in order to create new and suitable materials for the automotive industry. The effect of blend composition and compatibilizer content on phase morphology, water absorption and mechanical properties of PA6/PBT blends were reported. The percentage of water absorption of PA6/PBT blends was found to decrease with increasing PBT content. Mechanical properties, including tensile stress and young's modulus of blends with different amounts of PBT are slightly different and still lower than pure materials due to poor adhesion between the phases. Results of the study also indicated that the epoxidized natural rubber with 50 mol% epoxidation (ENR-50) was capable for using as a compatibilizer for blends of PA6 with PBT. The improvement in phase morphology and mechanical properties of PA6/PBT blends was observed. The number average size of dispersed phase of the compatibilized blends decreased from 4.6 to 1.2 μm. The reduction of dispersed phase size could be attributed to a strong interaction between the polar group of ENR-50, PA6, and PBT, which reduced the interfacial tension between phases. From the results of mechanical properties, it was found that tensile and impact properties increased with the increase in ENR-50 content.

Acknowledgments

Authors gratefully acknowledge the scholarship provided by College of Industrial Technology, King Mongkut's University of Technology North Bangkok, and Bangkok, Thailand.

References

[1] Patil A, Patel A, Purohit R 2017 An overview of Polymeric Materials for Automotive Applications, *Mater. Today: Proceed.* **4** : 3807-15.

[2] Dobrovszky K, Ronkay F 2016 Investigation of compatibilization effects of SEBS-g-MA on polystyrene/polyethylene blend with a novel separation method in melted state, *Polym. Bull.* **73** : 2719-39.

[3] Gadekar R, Kulkarni A, Jog J P 1998 Blends of Nylon with Polyethylene: Effect of Compatibilization on Mechanical and Dynamic Mechanical Properties, *J. Appl. Polym. Sci.* **69** : 161-8.

[4] Tai C M, Li R K Y, Ng C N 2000 Impact behaviour of polypropylene/polyethylene blends, *Polym. Test.* **19** : 143-54.

[5] Tjong S C, Meng Y Z 2004 Structural-mechanical relationship of epoxy compatibilized polyamide 6/polycarbonate blends, *Mater. Res. Bull.* **39** : 1791-801.

[6] Han M S, Lim B H, Jung H C, Hyun J C, Kim S R, Kim W N 2001 Reactive blends of poly(butylene terephthalate)/polyamide-6 with ethylene glycidyl methacrylate, *Korea-Aust. Rheol. J.* **13** : 169-77.

[7] Huang C C, Chang F C 1997 Reactive compatibilization of polymer blends of poly(butylene terephthalate) (PBT) and polyamide-6,6 (PA66): 1. Rheological and thermal properties, *Polymer* **38** : 2135-41.

[8] Huang C C, Chang F C 1997 Reactive compatibilization of polymer blends of poly(butylene terephthalate) and polyamide 6,6: 2. Morphological and mechanical properties, *Polymer* **38** : 4287-93.

[9] Wang P, Meng K, Cheng H, Hong S, Hao J, Han C C, Haeger H 2009 Reactive compatibilization of polyamide-12/poly(butylene terephthalate) blends with hyperbranched PEI-g-PA12: Morphology and thermal properties, *Polymer* **50** : 2154-60.

[10] Mengual A, Juárez D, Balart R, Ferrándiz S 2017 PE-g-MA, PP-g-MA and SEBS-g-MA compatibilizers used in material blends, *Procedia Manuf.* **13** : 321-6.

[11] Chiu H-T, Hsiao Y-K 2006 Compatibilization of Poly(ethylene terephthalate)/Polypropylene Blends with Maleic Anhydride Grafted Polyethylene-Octene Elastomer, *J. Polym. Res.* **13** : 153-60.

[12] Chiou K-C, Chang F-C 2000 Reactive Compatibilization of Polyamide-6 (PA6)/Polybutylene Terephthalate (PBT) Blends by a Multifunctional Epoxy Resin, *J. Polym. Sci. Part B Polym. Phys.* **38** : 23-33.

[13] Samios C K, Kalfoglou N K 1999 Compatibilization of poly(ethylene terephthalate)/polyamide-6 alloys: Mechanical, thermal and morphological characterization, *Polymer* **40** : 4811-9.

[14] Yew G H, Mohd Yusof A M, Mohd Ishak Z A, Ishiaku U S 2005 Water absorption and enzymatic degradation of poly(lactic acid)/rice starch composites, *Polym. Degrad Stab.* **90(3)** : 488-500.

[15] Sam S T, Ismail H, Ahmad Z 2010 Effect of Epoxidized Natural Rubber on the Processing Behavior, Tensile Properties, Morphology, and Thermal Properties of Linear-Low-Density Polyethylene/Soya Powder Blends, *J. Vinyl. Addit. Technol.* **16(4)** : 238-45.

[16] Nematollahi M, Jalali-Arani A, Modarress H 2019 High-performance bio-based poly(lactic acid)/natural rubber/epoxidized natural rubber blends: effect of epoxidized natural rubber on microstructure, toughness and static and dynamic mechanical properties, *Polym. Int.* **68** : 439-46.

[17] Tuampoemsab S, Riyajan S, Sritapunya T, Pakeyangkoon P 2014 Effect of Epoxide group in Thermoplastic Elastomer on the Properties of Polyamide6 and Low-density Polyethylene Blends, *Adv. Mater. Res.* **979** : 143-6.

IOP Conf. Series: Materials Science and Engineering **811** (2020) 012020 doi:10.1088/1757-899X/811/1/012020

Identification Plastic Properties of Spot Welded Joints Using the Instrumented Indentation Technique

I N Budiarsa[1], I N G Antara[1], I M G Karohika[1], I W Widhiada[1], N L Watiniasih[2]

[1]Mechanical Engineering, Udayana University, Bali. Indonesia
[2]Faculty of Marine and Fisheries, Udayana University, Bali. Indonesia

nyoman.budiarsa@unud.ac.id

Abstract. The parameters of elastic-plastic material and fracture of materials can be readily determined when standard specimens are available, however, for a spot welded joint, standard testing is not applicable to characterize the HAZ and nugget due to their complex structure and small size. This has opened up the possibility to characterize material properties based on the Indentation method to inversely characterize the parameters of the constitutive material laws for the nugget, HAZ and the base metals. In a mixed numerical-experimental approach, the load-deformation data of the material is used as input data to a numerical finite element (FE) model that simulate the geometry and boundary conditions of the experiment. The numerical approach based on the Finite Element (FE) model has been developed and validated. The established formulation is used for reverse (inverse) prediction of the nature of constitutive material (ie yield stress (σ_y), strain hardening coefficient (n)) for the welded joint zone namely the nugget, HAZ and parent metals (base). Then able to predict the effect of the nugget size and the thickness of the sheet metal on the strength of the spot welded joint with dissimilar material.

1. Introduction

The application of dissimilar metal welding (such as: steel and lighter material) on the vehicle body is an increasing effort to reduce the weight of the vehicle's body as well as increasing protection against corrosion and increasing crashworthiness [1, 2]. In the assembly industry many applications are used using welding with different materials [3]. Most research is carried out on the influence of the thickness and length of the welding point with an emphasis on physical and mechanical properties with variations in welding time. [4]. Another problem with diameter size of nuggets [5]. Different weld joints are still being researched and developed intensively, besides because the weld results are interactions between electrical, thermal, metallurgical and mechanical phenomena [6]. Also because of the complex nature of the structure at the joint deformation.

One of the main problems is to characterize material properties. The parameters of elastic-plastic material and material fracture parameters can be easily determined when standard tests are available. However, for spot welded joints, standard testing is not applicable to characterize HAZ and nuggets due to their complex structure and small size. Therefore, requires a non-standard method to be able to predict its characteristics more accurately. The most active research currently being carried out is the prediction of the dimensions and strength of the weld joint with finite element modelling to simulate the welding process [4, 7]. Another active area of research is Instrumented indentation of their complex structures [4]. Previous studies have shown the value of hardness is closely related to the correspondence between representative stress (σ_r) and representative strain (ε_r) which is the average value of the plastic strain produced in the indentation process [8-10]. The numerical approach through the finite element model developed has been successfully used in analysing representative strains and stresses based on Instrumented indentation with sharp indenter (Vickers) using a fixed indenter angle. In this case the resulting relationship between the parameters of material properties with force - indentation depth (ph curve) is finally used to estimate the hardness of material parameters. Simulations with validated models have been carried out over various ranges of material properties ($\boldsymbol{\sigma}_y$: 100 - 900 MPa, and n: 0.0 - 0.3) the relationship between the yield stress ($\boldsymbol{\sigma}_y$), the strain hardening coefficient (n), and the known hardness (HV) value. This will lead to the predictive results of their plastic properties ($\boldsymbol{\sigma}_y$ and n) in spot welded joint because of their complex structures, can be predicted more accurately through Instrumented indentation

2. Materials and Experimental

Specimens for spot welded with dissimilar materials consisting of two materials with different thicknesses (stainless steel and mild steel) were used in this study. The material specifications used are grade 304 stainless steel with a width of 25 mm, thickness of 0.8 mm. Other specimens are Mild steel with dimensions of width 25 mm and thickness of 1.44 mm. The welding process by applying Spot welding to the joining of different combinations of materials is prepared and tested. The hardness is carried out at the welded joints and the parent metal samples are sectioned and mounted prior to testing with a diameter elliptical Φ5 mm and a length of 90 mm. The Vickers hardness test was carried out using the Duramin-1 Struers Vickers hardness test machine using the direct loading method with various loads from 19.61 N to 490.3 mN. Indenter has the right pyramidal shape with a square base and angle 136° to face the opposite side [11].

Welding joints of dissimilar material sheet metals are prepared and tested. Tensile shear test on welded joint is carried out using the Lloyd LR 30K Universal material testing machine with a maximum capacity of 30 kN with accurate readings of up to 0.5% of force. Test specimens Tensile shear is carried out using an initial load of about 50 N, clamped with two gaskets to avoid bending during testing. Tensile testing was carried out at a loading rate of 5 mm/min based on ASTM E 8-04.

3. Identification of Materials parameters of Spot Welded Zones

The plastic behavior is normally described by the constitutive material equations. In many cases, the three parameter power law hardening rule (equation 1) is used for steels:

$$\sigma = \sigma_0 + K\varepsilon \qquad (1)$$

Where the parameter (σ_0) is the yield stress, K is the strength coefficient and 'n' is the strain hardening coefficient. These material parameters influence both the yielding strength and work hardening behavior of the spot welded joint. The ratio L/Lo is the extension ratio, denoted as λ. Using these relations, it is easy to develop relations between true (t) and engineering (e) measures of tensile stress and strain.

$$\sigma_t = \sigma_e\,(1 + \varepsilon_e) = \sigma_e \,.\, \lambda \qquad (2)$$

Based on Hooke's law and Von Mises criteria, then the true strain (ε) is generally expressed as:

$$\varepsilon = \begin{cases} \sigma/E & \text{for } \sigma \le \sigma_y \\ \sigma_y/E \left(\frac{\sigma}{\sigma_y}\right)^{1/n} & \text{for } \sigma > \sigma_y \end{cases} \qquad (3)$$

where E is Young's modulus, n is the strain hardening coefficient. In the plastic region strain can be described as yield strain ($\boldsymbol{\sigma}_y$) and true plastic strain. Due to the complex structure of the welds to find out the behavior of the nature of the failure and fracture of the weld spot joints more use the Gurson Model approach. This model is widely applied in ductile fracture mechanics, where material fractures are thought to be the result of void growth in the volume of material. The Gurson model most commonly used in failure behavior analysis is the Gurson Tvergaard Needleman (GTN) model [12] the model approach assumes that the plastic properties are produced from a porous material, where the yield surface is a function of emptiness formulated as follows

$$\Phi = \frac{3S_{jk}S_{jk}}{2\sigma_{ys}^2} + 2\,q_1\,\mathrm{f}\cosh\left(\frac{3\,q_2\,\sigma_m}{2\,\sigma_{ys}}\right) - (1 + q_3\mathrm{f}^2) = 0 \tag{4}$$

where σ_y is the yield stress of the material, $\boldsymbol{\sigma_m}$ is the average stress, f is the volume fraction of voids. f=0 means that the material is fully dense and conditions Gurson yield reduces von Mises; f = 1 means that the material is full void and has no stresses. $\mathbf{S_{jk}}$ is a component of the deviator (j,k=1,2,3), is defined as $\mathbf{S_{jk}=\sigma_{jk}-\sigma_m\ \delta_{jk}}$ and $\boldsymbol{\delta_{jk}}$ is Kronecher delta $\boldsymbol{\delta_{jk}} = 1$ if j=k and $\boldsymbol{\delta_{jk}} = 0$ if j≠k. while the parameter values $\mathbf{q_1}$, $\mathbf{q_2}$, $\mathbf{q_3}$ are material constants, the values most found in most of the literature areas. $\mathbf{q_1} = 1.5$, $\mathbf{q_2} = 1$, $\mathbf{q_3} = 2.25$. In Elasto's plastic power law if the representative strain ($\boldsymbol{\varepsilon_r}$) is a plastic strain point of reference, the Representative stress strain ($\boldsymbol{\varepsilon_r}$), represents the average plastic strain defined by Tabor (1948). Techniques for estimating points on the curve true stress-true strain is developed, in this case representative stress ($\boldsymbol{\sigma_r}$) can be expressed as:

$$\sigma_r = \sigma_y \left(1 + \frac{E}{\sigma_y}\,\varepsilon_r\right)^n \tag{5}$$

Where E=Young's modulus indenter, and v = Poisson ratio. By combining the effects of the elasticity of an elastic indenter and elasto plastic solid can be written:

$$P = P\,(h, E^*, \boldsymbol{\sigma_y}, n) \text{ than } P = P\,(h, E^*, \boldsymbol{\sigma_y}, \boldsymbol{\sigma_r}) \tag{6}$$

Where

$$E^* = \left[\frac{1-v^2}{E} + \frac{1-{v_i}^2}{E_i}\right]^{-1} \tag{7}$$

In this work, the main material group to be investigated is steel, so the E value is fixed at 200 GPa rather than using true E* value (~187 GPa with $E_{\mathbf{indenter}}$ = 1220 GPa and $E_{\mathbf{steel}}$ = 200 GPa) to avoid uncertainty in the value of E* from different sources. So equation (6) can be simplified as $\mathbf{P} = \mathbf{P(h, E, \boldsymbol{\sigma_y}, n)}$ and incorporating equation (5), Applying the Π theorem in dimensional analysis, equation (6) becomes

$$P = \sigma_t\,h^2 \prod 1 \left(\frac{E}{\sigma_r}, n\right) \tag{8}$$

Where ∏1 is a dimensionless function. similarly, applying the theorem to equation (7), loading curvature C may alternatively be expressed as [13].

$$\mathbf{C} = \frac{\mathbf{P}}{\mathbf{h^2}} = \boldsymbol{\sigma_y} \prod \mathbf{1} \left(\frac{\mathbf{E^*}}{\boldsymbol{\sigma_y}}, \frac{\boldsymbol{\sigma_r}}{\boldsymbol{\sigma_y}}\right) \tag{9}$$

In equation (6), normalization is required with respect to the yielding stress ($\boldsymbol{\sigma_y}$) or the representative plastic stress ($\boldsymbol{\sigma_r}$). Through simulation formulated the relationship between $\mathbf{C_v}$ Vs normalized relationship between material properties and normalized properties of materials of Vs $\mathbf{C_s}$ (strain hardening exponents (n) and yield stress ($\boldsymbol{\sigma_y}$)), P-h curves for Vickers and Spherical indentations has the following relationship. Where ∏ 1 is a dimensionless function, the dimensionless given in equation 9 and the normalization was taken with respect to E^* instead of $\boldsymbol{\sigma_y}$ or $\boldsymbol{\sigma_r}$. Curve fitting has been performed by iterating the relationship between loading curvature indentation and properties material ($\boldsymbol{\sigma_y}$, n) also the relationship between normalized $\mathbf{C_s}$ Vs properties material (strain hardening coefficient (n) and Yielding stress) as following equations.

$$\mathbf{C_v} = 384.08.\,e^{3.0617\,n} - 8.22.56\,(\boldsymbol{\sigma_y})^{0.7282} \tag{10}$$

Where P is load and (h) indentation depth at each load curve. $\mathbf{C_v}$ is the Vickers indentation curvature coefficient and the ball indenter designated as $\mathbf{C_s}$ and the Curvature is a function of the relationship between the yield stress (σ_y) and strain hardening coefficient (n). This will provide a potential relationship allowing the prediction of material parameters of the test continuous indentation

4. Prediction Plastic Properties Spot welds

The method for determining the properties of plastic materials through indentation instrumentation was developed with an inverse prediction using FE modeling with input hardness values of materials are known to identify constitutive material properties (σ_y, n). In the first phase, finite element models that are developed systematically with simulation space boundaries cover a variety of potential material properties. In the next stage, the P-h curve established is used in the spatial boundary simulation. A comparative approach has been developed to predict material sets based on actually indented curvature. Through force input data (F) vs depth (h) related to indentation hardness value. Simulations with validated models that have been carried out on various material properties (σ_y: 100-900 MPa, and n: 0.0-0.5) then the relationship between the yield stress (σ_y), strain hardening coefficient (n), hardness value (HV) can be known (figure 1)

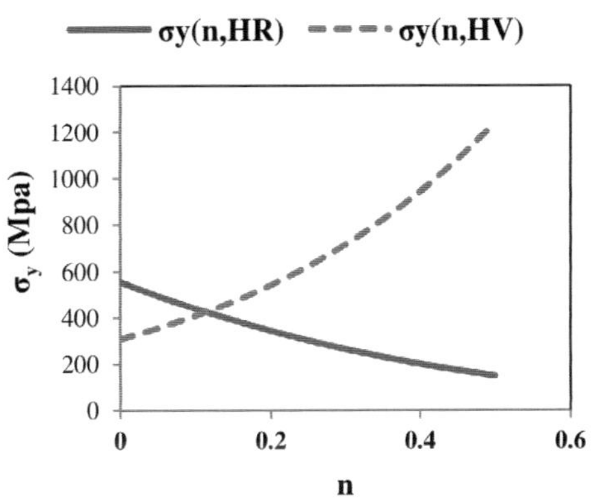

Figure 1. Typical intersection curve σy (n, HRB) and σy (n, HV) in the FE Modeling approach for the prediction of parameters constitutive material (σy, n) for the parent metal (base).

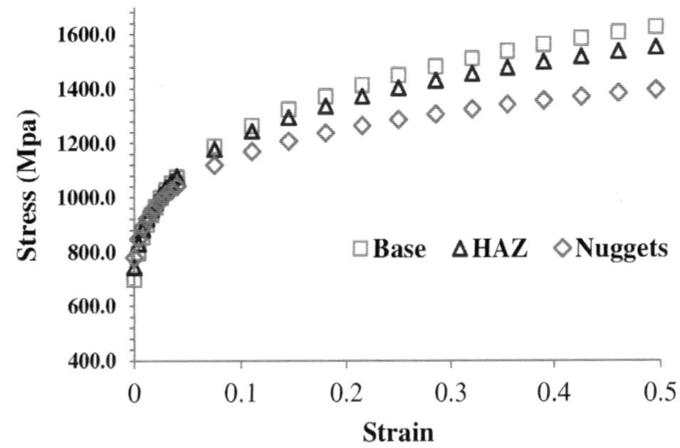

Figure 2. Typical prediction of the plasticity for Nugget, HAZ and the parent metal (base metal) as an application of FE Modeling constitutive parameters in the prediction of material (σy, n).

As shown in figure 2 shows the relationship between modeling results with experimental data. This means that the material laws predicted (σ_y, n) by the Finite element-based instrumentation of indentation elements for different material weld zones are accurate. The slight differences between numerical and experimental results on the fracture behavior suggest that detailed fracture for each material zone has to be obtained rather than using parameters from the base material, which requires further investigation.

5. Conclusion

The method for determining the properties of plastic materials through an inverse prediction approach uses finite elements modelling with known input hardness values to identify constitutive material properties (σ_y, n). has been developed and validated. The results show a good agreement with the experimental data. This means that the material law (σ_y, n) predicted by the Finite element-based instrumentation of the indentation for different material weld zones is accurate.

An evaluation of the prediction results based on experimental data shows an accurate similarity with the numerical approach of continuous curve indentation. The developed approach has been successfully used to characterize the plasticity of different zones at spot welding. (nugget; σ_y = 780 MPa, n = 0.12, HAZ: σ_y = 745 MPa, n = 0.15, Base: σ_y = 700 MPa, n = 0.17). These plastic material parameters used in finite element modelling for tensile shear deformation spot welding and welded joints showed good correlation with experimental results. Validated FE models are then used to predict the effect of nugget size and thickness of the sheet metal on the strength of the weld joint points with different materials (dissimilar material).

References

[1] Litman T 2013 Comprehensive Evaluation of Energy Conservation and Emission Reduction Policies, *Transport. Res. A*, **47** 153-66.

[2] Anaraki A G, et al. 2002 Experimental and numerical analysis of low cycle fatigue of spot welded joints under peel-tension loading, *Mech. Computational*, **221** 786-804.

[3] Irfan D, et al. 2012 Mechanical Evaluation of Joining Methodologies in Multi Material Car Body, *Int. J. Adv. Eng. Technol.* **5(1)** 259-68.

[4] Rahman M M, et al. 2008 An Investigation into the effects of spot diameter and sheets thickness on fatigue life of spot welded structure based on FEA, *Res. J. Appl. Sci.* **3(1)** 10-5.

[5] Aslanlar S 2006 The effect of nucleus size on mechanical properties in electrical resistance spot welding of sheets used in automotive industry, *Mater. Des.* **27** 125-31.

[6] Nachimani C 2013 An Experimental Investigation on Spot Weld Growth on Dissimilar Joints of 304L Austenitic Stainless Steel and Medium Carbon Steel (Part1), *Int. J. Adv. Appl. Sci. (IJAAS)* **2(1)** 25-32.

[7] Hou Z, et al. 2007 Finite element analysis for the mechanical features of RSW process, *J. mater. Process. Technol.* **185(1-3)** 160-5.

[8] Budiarsa I N, Antara I N G, Karohika I M G 2019 Indentation Size Effect of the Vickers Indentation to Improve the Accuracy of Inverse. Materials Properties Modelling Based on Hardness Value, *IOP Conf. Series: Earth Environ. Sci.* **248** 012009.

[9] Budiarsa I N, Ary Subagia I D G, Widhiada I W, Suardana N P G 2015 Characterization of Material Parameters by Reverse Finite Element Modelling Based on Dual Indenters Vickers and Spherical Indentation, *Procedia Manufact.* **2** 124-9.

[10] Bucaille J L, et al. 2003 Determination of plastic properties of metals by instrumented indentation using instrumented sharp indentation, *Acta. Mater.* **49** 3899-918.

[11] Budiarsa N, Norbury A, Su X X, Bradley G, Ren X J 2013 Analysis of Indentation Size Effect of Vickers Hardness Tests of Steels, *Adv. Mater. Res.* **652-654** 1307-10.

[12] Cricrì G, et al. 2013 A consistent use of the Gurson-Tvergaard-Needleman damage model for the R-curve calculation, **7(24)** 161-74.

[13] Dao M, Chollacoop N, Van Vliet K J, Venkatesh T A, Suresh S 2001 Computational modelling of the forward and reverse problems in instrumented sharp indentation, *Acta. Mater.* **49(19)** 3899-918.

The Study on Chitosan Coating on Poly(Lactic Acid) Film Packaging to Extend Vegetable and Fruit Life

J Kumbunleu[1], A Rattanapun[1], P Sapsrithong[1], S Tuampoemsab[1], T Sritapunya[1]

[1]Division of Polymer Engineering Technology, Department of Mechanical Engineering Technology, College of Industrial Technology, King Mongkut's University of Technology North Bangkok, Bangkok, Thailand

thritima.s@cit.kmutnb.ac.th

Abstract. The objective of this work is the study of possibility of extending the life of vegetables and fruits packed with polylactic acid (PLA) film coated with squid chitosan with various contents (0, 1, 3 and 5 phr), acted as an antimicrobial agent and the study of mechanical properties of these films. In the first step, PLA was compounded with 20 percents of glycerol by using a two-roll mill. After that, all compounds were fabricated by blown film extrusion. In next step, the chitosan solutions were prepared in 1 wt% hydrochloric acid solution to coat on one side of PLA film. The last step, the prepared films were examined tensile properties and storage time of 4 types of vegetables and fruits including broccolis, Leb Mue Nang bananas, red grapes and queen tomatoes by packing. The results showed that the tensile strength of PLA films coated with chitosan was closed to that of neat PLA/glycerol film. For spoilage test, it was found that all vegetables and fruits were more shriveled and wilted with more chitosan content through respiration process due to the leakage of chotisan coated PLA film and weight loss of vegetables and fruits., especially for PLA film coated with 5 phr chitosan since more remained acetic acid solvent accelerated more hydrolysis rate of PLA film.

1. Introduction

One of environmental concerns for all countries in the world is a reduction of marine animals caused from litter of oil-based plastic which cannot be biodegraded, especially food packaging plastic. After many marine and wild animals were killed by plastic waste in Thailand, Thai government issued policies to reduce plastic waste problems as they are aware of the increasing volume of plastic waste and its negative environmental effects [1]. Several local plastic resin producers in Thailand e.g. PTTGC, Total Corbion PLA initiate plan to build polylactic acid (PLA) plant for composable packaging that is a solution that uniquely fits Thailand's waste management infrastructure [2]. However, adoption rate of PLA for food packaging in Thailand by packaging manufacturers and brand owners is gradually decreasing because of its limited applicability and function/performance-to-cost ratio. For example, PLA has limited applicability to fabricate packaging film for short shelf-life products due to its brittleness and high water vapor permeability. Thus, blending of low-cost biodegradable polymers or additives to develop functional attribute and performance to PLA but still maintaining its compostability with reasonable cost is demanding to promote bioplastic adoption and replace conventional plastic for food packaging [3, 4]. Rong Zhang *et.al.* [5] researched the blending of poly (butylene adipate-co-terephthalate) (PBAT) in PLA films to preserve quality of passion fruits during shelf-life. Their results were found that PLA/PBAT film can retard the metabolism of passion fruits, reduce weight loss, shrinkage index, firmness, total sugar, total acid, ascorbic acid content, ethanol content, sensory evaluation, and aroma components compared with PE film. However, in contrast to PE, both PLA and PBT were biodegradable polymers.

Furthermore, chitosan is one of the bio-derived and biodegradable additives that is widely used as functional additive for plastic packaging. Due to its antimicrobial properties and permeability-reducing properties [6], it has potential to extend shelf-life of vegetable and fruit products. Chitosan is derivative of chitin, which is amino polysaccharide and basically found in shell of crab, lobster, shrimp, insect and squid [7]. In 2013, Irkin R and Guldas M [8] have studied antimicrobial action of chitosan on fungal in red grape and honey melon samples. They found that chitosan coating on red grape and honey melon successfully inhibits fungi (*Fusarium oxysporum*) growth. Another study, Sanga S and Hongsriphan N [9] prepared antibacterial food packaging films from 90/10 wt% of PLA/PBS by dip coating of chitosan solution on films. Corona treatment was applied on film surface before chitosan coating to improve the adsorption of chitosan. The obtained films revealed higher antibacterial activity against *E.coli* and *S.aureus.* with higher chitosan concentration. However, chitosan prepared by low-cost process has not been investigated.

This current research work aimed to investigate the applicability of α-chitosan from squid pen in PLA packaging film to extend shelf-life of vegetable and fruit products and also aimed to characterize the tensile properties of PLA biofilm with chitosan bio-additive to become an alternative to a single-use plastic packaging.

2. Experimental

2.1. Materials

Polylactic acid pellet (PLA), (Ingeo^{TM} 4043D, MFI of 6 g/10 min at 210°C/ 2.16 kg.) was supplied by, NatureWork LLC, USA). High molecular weight α-chitosan powder from squid pen (food grade, particle size 18 mesh, viscosity > 1,000 Cps, degree of deacetylation >95%) was supplied by Bona Fides Marketing co., ltd. (Bangkok, Thailand). Glacial acetic acid (99.5% purity) was purchased from SNP General Trading co., ltd. (Bangkok, Thailand). And glycerol (99.5% purity) was purchased from Lab Valley LP. (Bangkok, Thailand). For the spoilage study, broccolis, Leb Mue Nang bananas, red grapes, and queen tomatoes were purchased in a local supermarket. All chemicals were used without further purification.

2.2. Compounding and film processing

Before compounding, PLA pellets and glycerol were premixed by manually stirring in beaker with a PLA/glycerol ratio of 80/20wt%. Then, they were dried at 60°C in an air-circulating oven for 24 h in order to remove moisture. The mixture of PLA and glycerol were compounded by two-roll mill at 180°C for 15 min. PLA compound was cooled at room temperature and granulated by cutter grinder.
The film-forming process was carried out by blown film extruder with a screw speed of 60 rpm, nip roll speed of 4.0 m/min, winding speed of 4.2 m/min, internal bubble cooling speed of 1600 rpm and temperature profile ranging from 160°C to 180°C. Film thickness was measured using a micrometer.

2.3. Chitosan coating on PLA films

PLA film was cut in glass mold size (20 cm x 30 cm) and then weighted to determine the required amount of chitosan. Chitosan powder with various amounts (1, 3 and 5 phr of sized PLA film) was dissolved in a 30 ml of 1wt% acetic acid with continuous stirring at room temperature until solution was clear. Chitosan solution was poured onto PLA film mounted on glass mold. Chitosan coated PLA film was left at room temperature for 3 days to evaporate acetic acid and was then removed from glass mold for further testing. Thus, the single side chitosan coated PLA film was obtained.

2.4. Characterization and Testing

2.4.1. Tensile testing. Prepared films of pure PLA (with glycerol), PLA/1 phr chitosan, PLA/3 phr chitosan and PLA/5 phr chitosan were determined tensile properties in accordance to ASTM D882 using a Universal Testing Machine (Testometric, Model M500-25AT) in both machine direction (MD) and transverse direction (TD). The samples were tensioned with a crosshead speed of 50 mm/min using load cell of 100 N.

2.4.2. Spoilage testing. The broccolis, Leb Mue Nang bananas, red grapes, and queen tomatoes were separated from clusters and they were packed in sealed PLA film coated with different chitosan concentrations along with chitosan coated side of film was turned inward to expose to vegetables and fruits. Then, they were put on laboratory bench at room temperature for a week. The photos of vegetables and fruits were taken in everyday to observe texture and appearance change and all samples (including vegetable or fruit and PLA film) were weighted to analyze weight loss corresponding to spoilage rate.

3. Results and Discussions

3.1. Tensile properties of PLA film coated with different concentrations of chitosan
Tensile strength, elongation at break and Young's modulus of both MD and TD of PLA film adding 20wt% of glycerol together with 1, 3 and 5 phr of chitosan by coating were investigated and shown in figure 1. It was found that the tensile properties of uncoated PLA film were very low that might due to the presence of immiscible glycerol in PLA matrix. Compared with coated PLA films, PLA film coated with 1 phr of chitosan presented quite low tensile properties especially for tensile strength and Young's modulus in MD. This is due to roughness on film surface which was occurred from poor dispersion of low chitosan content resulting in stress concentration that caused easily breaking of films [10]. While, all tensile properties of PLA films coating chitosan of either 3 or 5 phr were not significantly different from that of uncoated film in both film directions. Also the Young's modulus was slightly increased with chitosan increasing. This was attributed to the rigid cellulosic-like structure of chitosan. Therefore, it did not impact on practical application of the coated film samples compared with uncoated one.

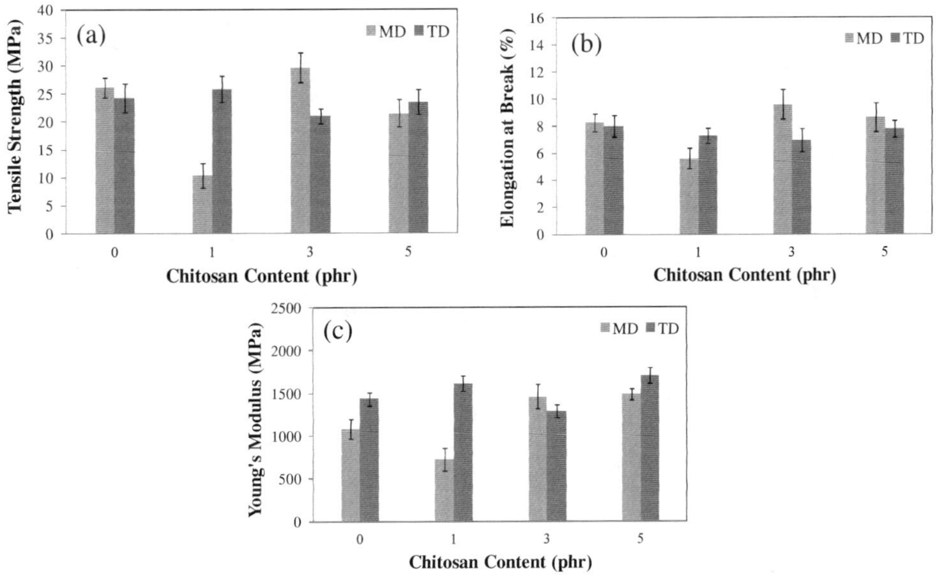

Figure 1. (a) Tensile strength, (b) elongation at break and (c) Young's modulus of PLA film coated with different chitosan contents in machine direction (MD) and transverse direction (TD).

3.2. Spoilage testing of vegetables and fruits in chitosan coated PLA films
The loss of appearance qualities such as aroma, texture, taste, or color is generally used to describe 'spoilage of food'. Therefore, the visual appearance including color and texture were carried out to simply considerate the spoilage of studied vegetables and fruits. The visual appearance of broccolis,

Leb Mue Nang bananas, red grapes, and queen tomatoes packed in PLA films packaging of 10 cm x 15 cm in size are presented in figure 2, 3. The figures showed that color of some parts of banana peels became black, broccolis became yellowish, as well as grapes and tomatoes peels were darker when they were left at room temperature for longer time with higher chitosan contents. In addition, a head of broccoli was more shriveled as well as external textures of bananas, red grapes and queen tomatoes were more wilted during 7 days due to water loss causing condensation of water on film surface coated with chitosan. Compared with uncoated PLA film, the film packages were clear and not found water drop on film surface caused by high water vapor permeability of PLA film. In generally, factors of vegetables and fruits spoilage are the physiological age, the storage temperature, the storage atmosphere (humidity) and microorganism presence. This work intended to improve vapor barrier property of PLA and reduce microorganism growth using chitosan coating and the result also revealed that water drop was more occurred when vegatables and fruites were packed in PLA film for longer time with higher chitosan contents. It meant that chitosan coating can protect water permeation but it cannot reduce water loss of vegetables and fruits. Therefore, too high humidity trapped in packaging can aid to fungi growth. However, mold which is a one of fungus species were not appeared on vegetable and fruit samples. These results indicated that chitosan can against microbial attack.

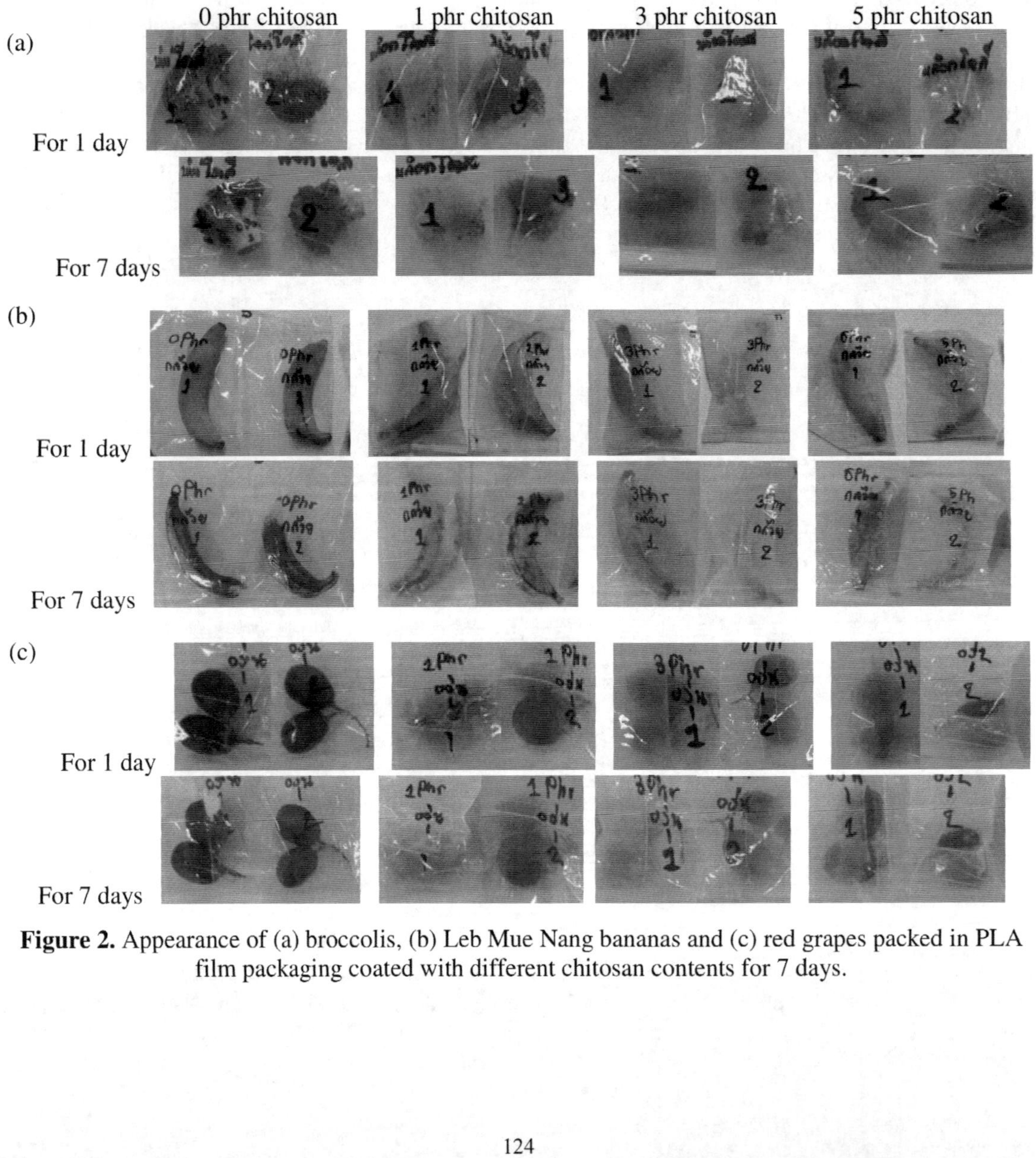

Figure 2. Appearance of (a) broccolis, (b) Leb Mue Nang bananas and (c) red grapes packed in PLA film packaging coated with different chitosan contents for 7 days.

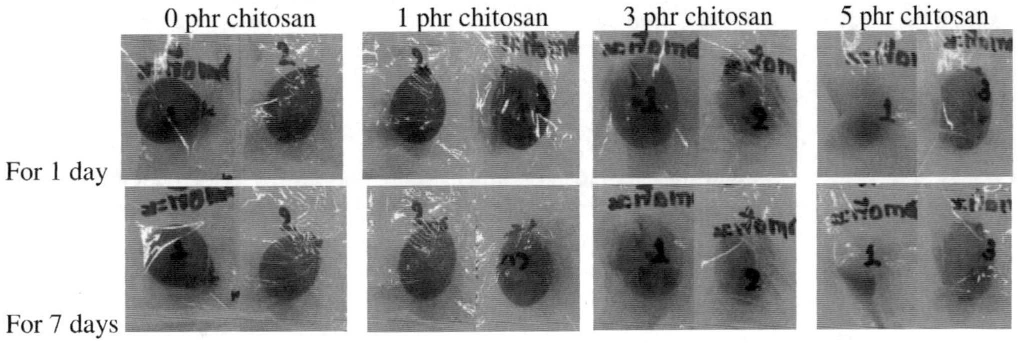

Figure 3. Appearance of queen tomatoes packed in PLA film packaging coated with different chitosan contents for 7 days.

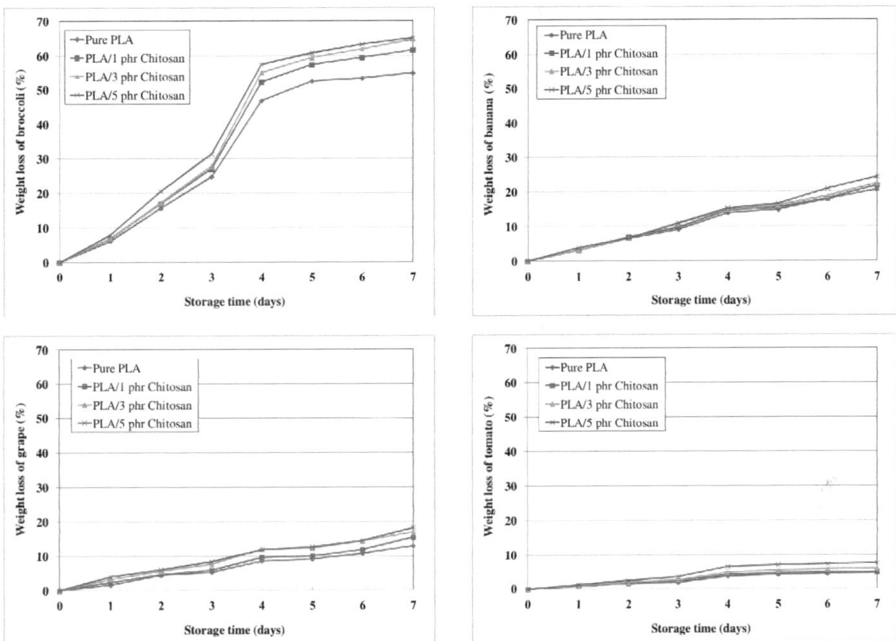

Figure 4. Weight loss of (a) broccolis, (b) Leb Mue Nang bananas, (c) red grapes and (d) queen tomatoes packed in PLA film packaging coated with different chitosan contents.

In addition, this work determined the weight loss of vegetables and fruits to evaluate an acceptable quality for marketing and to evidence the metabolism of harvested plants. These results were shown in figure 4, they presented that an increase chitosan increased the weight loss of all vegetables and fruits and the broccoli exhibited the highest rate of weight loss following by bananas, grapes and tomatoes, respectively. Normally, the weight loss was related to respiration and water loss of harvest plants and their rate depended on type of plant, which generally faster in leafy vegetables [11]. As above description, we found water drop were trapped in only chitosan coated PLA film indicating that water were lost from vegetables and fruits but it was not lost from film packaging whereas their weight loss was enhanced with chitosan increasing. This is suggested that the weight loss of vegetables and fruits was caused by respiration which is process that plants take in oxygen to break their carbohydrate into carbon dioxide and water. So, the presence of water drop in chitosan coated PLA film is the product of plant respiration which cannot permeate this film while a small molecule of carbon dioxide can do. This result was in contrast to the uncoated PLA film, the prepared chitosan coated PLA film also had a good water vapor barrier property but it had poor gas barrier property. Therefore, it was possible that acetic acid solvent was remained after chitosan coating and it induced the leakage of carbon dioxide since acidic sensitivity of PLA caused some hydrolytic degradation [12].

4. Conclusion

Chitosan coated PLA film was prepared by cast coating of squid chitosan on a single side of 80%PLA/20% glycerol films which was an economical simple process. For spoilage test of vegetables and fruits packed in chitosan coated PLA film, it can be concluded that chitosan coating on PLA film can satisfyingly inhibit mold or microorganism growth on vegetables and fruits even high humidity. In addition, both 3 and 5 phr of chitosan did not significantly change the tensile properties of PLA film. However, chitosan coating using casting method also left acetic acid solvent on PLA film, resulting in some hydrolytic degradation of PLA. This result intended to increase respiration rate of vegetables and fruits taking air or oxygen that permeated chitosan coating PLA films. Besides, both water and carbon dioxide were produced from this metabolism which the water was trapped into these films but the carbon dioxide was leaked trough the film leading to shriveling and weight loss of vegetables and fruits. Although both vegetables and fruits were not spoiled by microorganism, they were shriveled and wilting that is unacceptable for sale. In conclusion, chitosan also was a suitable antimicrobial agent for bioplastic package, but alternative coating technique could be considered.

Acknowledgments
The authors would like to thank College of Industrial Technology, King Mongkut's University of Technology North Bangkok, Bangkok, Thailand for all supports.

References

[1] Milne G 2019 Plastics a growing concern an asean perspective *Ipsos Business Consulting* British Chamber of commerce Thailand.
[2] Schiavo A 2019 Biodegradation, biocomposites, and the future of bioplastics in southeast asia *13th Bioplastics Market Summit (Bangkok)*.
[3] Homklin R, Hongsriphan N 2013 Mechanical and thermal properties of PLA/PBS co-continuous blends adding nucleating agent, *Energy Procedia* **34** 871-9.
[4] Chen R, Zou W, Wu C, Jia S, Huang Z, Zhang G, Yang Z, Qu J 2014 Poly (lactic acid)/ poly (butylene succinate)/calcium sulfate whiskers biodegradable blends prepared by vane extruder: Analysis of mechanical properties, morphology, and crystallization behaviour *Polym. Test* **34** 1-9.
[5] Zhang R, Lan W, Ding J, Ahmed S, Qin W, He L, Liu Y 2019 Effect of PLA/PBAT antibacterial film on storage quality of passion fruit during the shelf-life, *Molecules* **24** pii: E3378.
[6] Reuck K D, Sivakumar D, Korsten L 2009 Effect of integrated application of chitosan coating and modified atmosphere packaging on overall quality retention in litchi cultivars, *J. Sci. Food Agr.* **89** 915-20.
[7] Bose A, Wong T W 2018 Oral colon cancer targeting by chitosan nanocomposites, *Appl. Nanocomposite Mater. Drug Delivery* (Amsterdam: Elsevier) Chapter 18 409-29.
[8] Irkin R, Guldas M 2014 Chitosan coating of red table grapes and fresh-cut honey melons to inhibit fusarium oxysporum growth, *J. Food Process Pres.* **38** 1948-56.
[9] Sanga S, Hongsriphan N 2015 Chitosan coating on biodegradable film modified surfaces by corona treatment, *12th Eco-Energy Mater. Sci. Eng. Symp. (Krabi)* 99-103.
[10] Shin G H, Lee Y H, Lee J S, Kim Y S, Choi W S, Park H J 2002 Preparation of Plastic and Biopolymer Multilayer Films by Plasma Source Ion Implantation, *J. Agric. Food Chem.* **50** 4608-14.
[11] Kays S J, Paull R E 2004 *Postharvest Biology* (Athens: Exxon Press).
[12] Casalini T, Rossi F, Castrovinci A, Perale G 2019 A perspective on polylactic acid-based polymers use for nanoparticles synthesis and applications, *Front Bioeng. Biotechnol.* **7** 259.

IOP Conf. Series: Materials Science and Engineering **811** (2020) 012022 doi:10.1088/1757-899X/811/1/012022

Unvulcanized Rubber Tape Made from Natural Rubber and Synthetic Polyisoprene Modified with Rosin as a Tackifier

S Tuampoemsab[1], A Rattanapan[1], P Sapsrithong[1], T Sritapunya[1]

[1]Division of Polymer Engineering Technology, Department of Mechanical Engineering Technology, College of Industrial Technology, King Mongkut's University of Technology North Bangkok, 1518 Pracharat 1 Road, Wongsawang, Bangsue, Bangkok 10800, Thailand

surakit.t@cit.kmutnb.ac.th

Abstract. This presented work was to study mechanical and tack properties of unvulcanized rubber tapes made from STR 20 and IR 2200 with various amount of carbon black and rosin as reinforcing agent and tackifier, respectively. Mixing process of solid rubbers and all chemicals was carried out by two roll mill machine. All rubber compounds were subjected to increasing the stickiness by aging under hot air at 150°C for 3 hours. From the experiment, green strength of the sticky rubber tape mixed with carbon black and rosin showed higher than that of the commercial rubber tape. The formulation exhibited the highest tensile stress at 400% strain is STR 20 mixed with 15 and 5 phr of carbon black and rosin, respectively. Mooney viscosity of the rubber tape made from STR 20 and IR 2200 was also reduced with the higher content of rosin. Although Mooney viscosity of the adhesive rubber tape normally decreased after heating under hot air, its tack property investigeted by rolling ball test on the surface predominantly became increased. Moreover, Mooney viscosity of the rubber tape made from IR 2200 would be remarkably enhanced by blending with only 20% of STR 20.

1. Introduction

Unvulcanized rubber insulation tape is well known as several names such as cork insulation tape, rubberized cork tape, sticky rubber tape, adhesive rubber tape, etc. However, their main purposes are regularly designed for use on heating, refrigeration, air conditioning and piping applications. Hot pipes, valves, tees, fittings and other gasket materials are normally employed by the tape. This tape can also be efficiently used on cool and hot pipes to prevent the condensation, provides insulation as well as reduce vibration [1, 2]. One of special properties of the adhesive rubber tape is a wide range of a service temperature [3, 4]. There are a lot of types and brands of the rubber insulation tape not only depending on applications but also using conditions. For example, cork insulation tape is a combination of cork, rubber, tackifier and other polymers to improve the adhesion performance [5]. It is well recognized that Thailand is the number one producer of raw natural rubber in the world for a few decades. There is a lot of rubber products that have been daily manufactured in the country. Although unvulcanized rubber tape has normally been launched in the industrial market many brands, fewer papers and researches have been reported, excepting for many patents. Thus, the authors tried to domestically make a value-added for a raw rubber. Standard Thai rubber (STR) can be separated into many grades according to prices, usage and production processes. STR 20, a solid natural rubber, was selected to develop the unvulcanized rubber tape comparing to synthetic polyisoprene IR 2200 in this work.

2. Experimental

2.1. Preparation of unvulcanized rubber tapes
STR 20 and IR 2200, representatives of natural and synthetic polyisoprene rubbers, were masticated and mixed with both of carbon black and rosin on a two-roll mill (Labtech Engineering LMR 1100) at 60°C by using a rolling nip of 0.3 mm. Mixing time of all rubber compounds were about 30-60 min depending on the amount of fillers. All rubber tapes were subjected to increasing the stickiness by heating under force air flow oven (Memmert model UF450) at 150°C for 3 hours.

2.2. Investigation of green strength
Tensile specimen was prepared by hot press machine (Labtech Engineering LP20-B) at 100°C for 15 min with a flat mold before cutting to a dumbbell shape according to ASTM D412 (type C). Green strength of the rubber tape was carried out by a Testometric model M500-25 AT with 100 N of load cell under crosshead speed of 300 mm/min. Three samples were measured and the average value was calculated. Tensile stress at 400% strain (M400) was determined as a representative data to report because of a limitation of maximum strain of all tensile specimens.

2.3. Determination of Mooney viscosity
Measurement of Mooney viscosity (ML1+4) was carried out by an EKTRON Mooney viscometer model EKT-2001M. Approximately 20-25 g of rubber sample was divided into two portions to cover up and down of a large rotor. Pre-heating and testing temperatures was set at 100.0 ± 1.0 °C while the shear strain rate was applied at 2 sec^{-1}. The rubber specimen was preheated for 1 min and then continuously tested for 4 min before decay of torque when the rotor was stopped.

2.4. Analysis of tack property
The tack property of all unvulcanized rubber tapes after heat aging was determined by using a rolling ball tester. Steel ball with a diameter of 4.0 mm was released from the starting point of an inclined plane with an angle of 30°. The ball was allowed to accelerate down and rolled onto a rubber tape specimen. The stickiness of rubber tape is obtained by means of measuring the distance that the ball moves across the tape before stopping. Three times of measurements with other lanes were carried out to get an average value.

3. Results and Discussion
The first part is an attempt to characterize both of mechanical and tack properties of the commercial rubber tape so as to choose two types of solid raw rubber continually making an unvulcanized rubber tape as a model. Table 1 presents Mooney viscosity and M400 of sticky rubber tapes made from STR 20 and IR 2200 with various amount of carbon black. It was revealed that Mooney viscosity of the commercial rubber tape exhibited very low at 5.84, while its stress at 400% strain (M400) and the tack properties could not be investigated. This is because a specimen preparing from the commercial rubber tape was too weak. Mooney viscosity of the unvulcanized STR 20 tape was found to be increased with the increasing of carbon black loading both before and after accelerated heat aging but not for the IR 2200. However, all types of the sticky rubber tape showed decreasing in Mooney viscosity after heat aging. That means unvulcanized rubber tapes might be thermal-oxidative degraded under heating in the oven at 150°C for 3 h. It is well recognized that carbon black is one of the best reinforcing agent to improve the hardness and strength of the compounded rubber [6]. Thus, STR 20 and IR 2200 mixed with various contents of the carbon black up to 45 phr were studied. The results showed that carbon black is an effective additive by serving the M400 of all unvulcanized rubber tapes. Figure 1 shows rolling distance of a steel ball on the surface of sticky rubber tapes made from STR 20 and IR 2200 with various amount of carbon black. All unmodified rubber tapes made from both STR 20 and IR 2200 displays the highest rolling distance at 35.0 and 22.5 mm, respectively. This implies that stickiness on surface of the rubber tape is the lowest comparing to their mixed with carbon black.

Table 1. Mooney viscosity and stress at 400% strain (M400) of sticky rubber tapes made from STR 20 and IR 2200 with various amount of carbon black.

Type of Rubber	Carbon Black)phr)	Mooney Viscosity (ML 1+4)		M400 (MPa)
		Before heat aging	After heat aging	
STR 20	0	54.42 ± 4.84	23.34 ± 2.35	0.11 ± 0.01
	15	55.19 ± 4.82	20.80 ± 2.12	0.17 ± 0.01
	30	73.09 ± 4.09	31.00 ± 0.71	0.43 ± 0.04
	45	67.55 ± 3.18	39.11 ± 1.71	0.84 ± 0.07
IR 2200	0	60. 77±0 .66	33.82 ± 4.55	0.10 ± 0.02
	15	53. 16±3 .87	22.76 ± 3.20	0.08 ± 0.00
	30	63. 26±4 .33	21.28 ± 1.39	0.17 ± 0.02
	45	58. 03±1 .32	28.79 ± 0.30	0.27 ± 0.04
Commercial Rubber Tape	-	5.84 ±1 .19	N/A	N/A

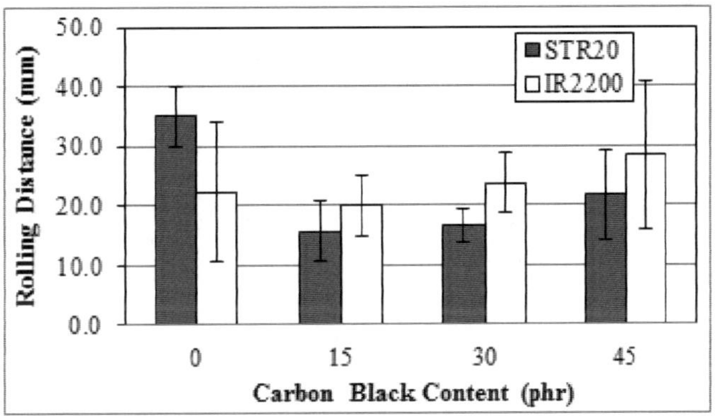

Figure 1. Rolling distances of a steel ball on sticky rubber tapes made from STR 20 and IR 2200 with various amount of carbon black.

The presence of only 15 phr of carbon black might improve the stickiness of the sticky rubber tape. However, the higher amount of carbon black loading demonstrated higher in rolling distance or lower in stickiness of the rubber surface. It might be explained by the dilution effect of the rubber phase. Moreover, the results of Mooney viscosity and rolling distance of all rubber tape prepared by adding the carbon black were significantly consistent. Rosin, brittle resin obtained from pines, was chosen as a tackifier in this work to improve the tackiness of sticky rubber tapes [7]. Mooney viscosity and M400 of compounded rubber tapes made from STR 20 and IR 2200 mixed with 15 phr of carbon black and various amount of rosin up to 10 phr was demonstrated in table 2. Mooney viscosity of the rubber tapes modified with 2.5 phr of rosin exhibited the highest at 32.37 and 24.05 for STR 20 and IR 2200, respectively. Hence, rolling distance (see figure 2) and M400 of the rubber tapes were seemingly to be lower when the rosin content was increased. This is actually due to the effectiveness of the rosin.

Table 2. Mooney viscosity and stress at 400% strain (M400) of sticky rubber tapes made from STR 20 and IR 2200 mixed with 15 phr of carbon black and various amount of rosin.

Type of Rubber	Rosin)phr)	Mooney Viscosity (ML 1+4)	M400 (MPa)
STR 20 / Carbon Black 15 phr	0	20.80 ± 1.70	0.17 ± 0.01
	2.5	32.37 ± 3.08	0.15 ± 0.01
	5.0	26.68 ± 1.91	0.14 ± 0.01
	10.0	24.62 ± 0.81	0.14 ± 0.00
IR 2200 / Carbon Black 15 phr	0	22.76 ± 3.35	0.08 ± 0.00
	2.5	24.05 ± 2.90	0.05 ± 0.00
	5.0	15.26 ± 2.18	0.03 ± 0.01
	10.0	5.31 ± 0.59	N/A

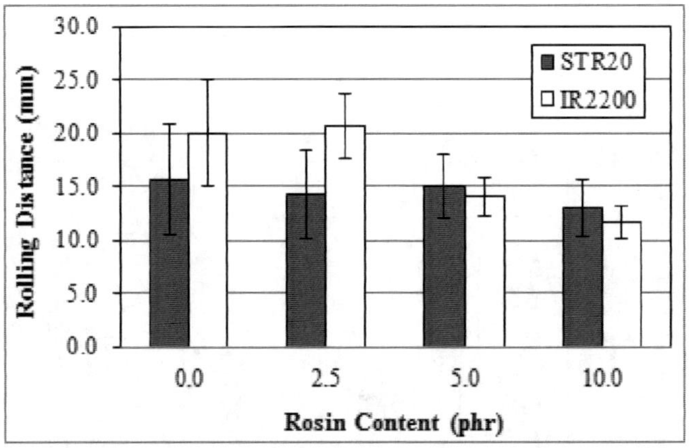

Figure 2. Rolling distances of a steel ball on sticky rubber tapes made from STR 20 and IR 2200 mixed with 15 phr of carbon black and various amount of rosin.

From the previous results, mechanical property of sticky rubber tape made from natural rubber or STR 20 shows outstanding while tack property of the tape made from synthetic rubber or IR 2200 presents predominantly. In order to possess the optimized properties of the sticky rubber tape, effect of STR 20/IR 2200 blends was finally studied. Table 3 shows Mooney viscosity and M400 of sticky rubber tapes made from STR 20/IR 2200 blends modified with fix amount of carbon black and rosin at 15 and 5 phr, respectively. It was found that IR 2200 blended with natural rubber in the range of 20 to 80% clearly resulted in enhancing of the Mooney viscosity or maximum increasing by about 146%. This is probably because STR 20 generally composes of natural branching structure, which was generated or initiated by non-rubber component [8]. Among all blending formulations, the ratio of STR 20/IR 2200 at 20/80 was found to be the highest in stickiness demonstrated by the lowest in rolling distance as shown in figure 3.

Table 3. Mooney viscosity and stress at 400% strain (M400) of sticky rubber tapes made from STR 20/IR 2200 blends mixed with 15 phr of carbon black and 5 phr of rosin.

Ratio of STR 20/IR 2200	Mooney Viscosity (ML 1+4)	M400 (MPa)
0/100	15.26 ± 1.51	0.03 ± 0.01
20/80	33.90 ± 2.79	0.13 ± 0.00
40/60	36.05 ± 1.97	0.11 ± 0.02
50/50	33.77 ± 2.33	0.09 ± 0.00
60/40	37.56 ± 2.04	0.10 ± 0.01
80/20	32.31 ± 2.56	0.06 ± 0.01
100/0	26.68 ± 3.28	0.14 ± 0.01

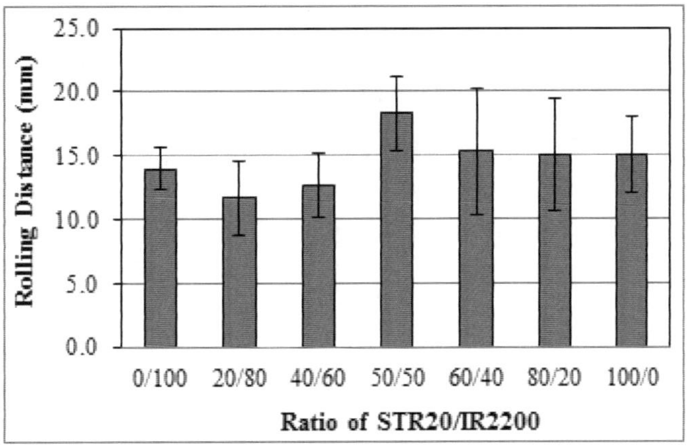

Figure 3. Rolling distances of a steel ball on sticky rubber tapes made from STR 20/IR 2200 blends mixed with 15 phr of carbon black and 5 phr of rosin.

For the overall experimental results, the unvulcanized rubber tapes were preliminary tried to apply on a steel pipe comparing with the commercial rubber tape. Figures 4 and 5 are the appearances of sticky rubber tapes made from IR 2200 and STR 20 mixed with 15 phr of carbon black and 5 phr of rosin comparing to commercial rubber tape after cooling at 5°C for 5 days and after sunlight exposure for 5 days, respectively. After testing under cooling condition the rolled unvulcanized rubber tapes on a steel pipe (see figure 4), it was found that both of IR 2200 and STR 20 seem to be completely tighten on a steel pipe but not for the commercial tape.

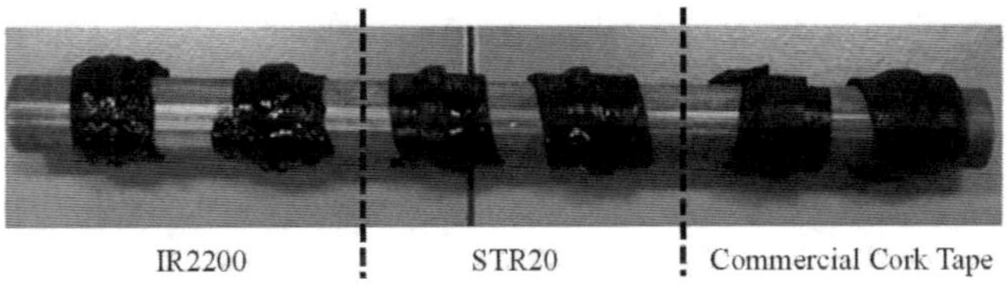

Figure 4. Appearance of sticky rubber tapes made from IR 2200 and STR 20 mixed with 15 phr of carbon black and 5 phr of rosin comparing to commercial rubber tape after cooling at 5°C for 5 days.

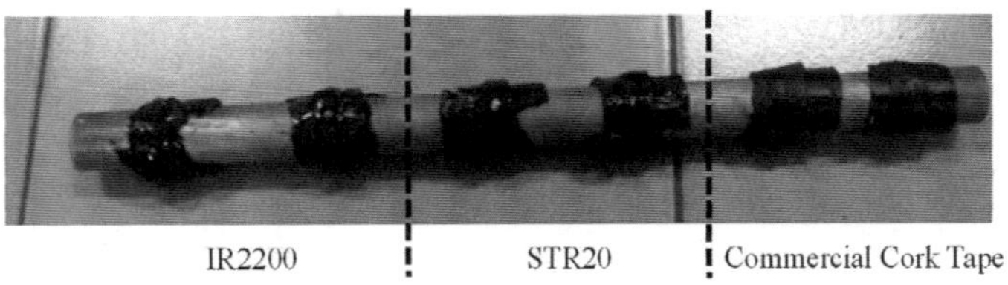

Figure 5. Appearance of sticky rubber tapes made from IR 2200 and STR 20 mixed with 15 phr of carbon black and 5 phr of rosin comparing to commercial rubber tape after sunlight exposure for 5 days.

On the other hand, after exposure by the sunlight both unvulcanized rubber tapes made from IR 2200 and STR 20 appear a bit of disintegration as demonstrated in figure 5. This might be explained by the fact that all recipes of the sticky rubber tapes compounded in this work were not consisted of stabilizers or antioxidants. Although many properties with various amounts of chemicals and types of the rubber have been already reported in this paper, other significant characterizations such as peel strength and chemical resistance of the unvulcanized rubber tape should be studied further.

4. Summary

The unvulcanized rubber tapes have been successfully produced both from STR 20 and IR 2200. Preliminary characterizations such as Mooney viscosity, M400 and tack properties of all sticky rubbers tape have been clarified as well as blending ratio between STR 20 and IR 2200 to improve the mechanical properties. Mooney viscosity of the unvulcanized rubber tapes can significantly reduce by heat aging the compounded rubbers at 100°C for 15 min. The best recipe for the unblended rubber exhibited that the highest M400 is STR 20 mixed with 15 and 5 phr of carbon black and rosin, respectively. After characterization of the blending ratio of STR 20/IR 2200, it might be concluded that 20% addition of the natural rubber can enhance the Mooney viscosity of the rubber tape.

Acknowledgements

The authors would like to acknowledge College of Industrial Technology, King Mongkut's University of Technology North Bangkok for providing space, materials and equipment to do this research.

References

[1] Khan I, Poh B T 2011 Natural rubber-based pressure-sensitive adhesives: A review, *J. Polym. Environ.* **19** : 793-811.

[2] Zaikin A E, Sofina S Y, Stoyanov O V 2011 Tape polymer materials for anti-corrosion insulation of pipelines, *Polym. Sci. Ser. D* **4** : 209-20.

[3] Ansarifar M A, Fuller K N G, Lake G J 1993 Adhesion of unvulcanized elastomers, *Int. J. Adhes. Adhes.* **13** : 105-10.

[4] Neetha J, Rani J 1998 Rubber-to-steel bonding studies using a rubber compound strip adhesive system, *J. Adhes Sci. Technol.* **12** : 59-69.

[5] Poh B T, Chee C L 2007 Effect of coumarone-indene resin on adhesion property of SMR 20-based pressure-sensitive adhesives, *Int. J. Polym. Mater. Po.* **56** : 247-55.

[6] Edge M, Allen N S, Gonzalez-Sanchez R, Liauw C M, Read S J, Whitehouse R B 1999 The influence of cure and carbon black on the high temperature oxidation of natural rubber I. Correlation of physic-chemical changes, *Polym. Degrad. Stabil.* **64** : 197-205.

[7] Poh B T, Yong A T 2008 Effect of molecular weight of rubber on tack and peel strength of SMR L-based pressure-sensitive adhesives using gum rosin and petroresin as tackifiers, *J. Macromol. Sci. A.* **46** : 97-103.

[8] Tanaka Y 2001 Structure characterization of natural polyisoprenes: Solve the mystery of natural rubber based on structure study, *Rubber. Chem. Technol.* **74** : 355-75.

IOP Conf. Series: Materials Science and Engineering **811** (2020) 012023 doi:10.1088/1757-899X/811/1/012023

Traffic Safety Benefits of Using Highway Lighting Technology: (Case Study: 4-Lanes Expressways in the Rural Areas)

Muamer Abuzwidah[1, *], Mohsin Balwan[2], Mohamed Abdel-Aty[3], Ghazi Al-Kateeb[4]

[1]Department of Civil and Environmental Eng., University of Sharjah, Sharjah, U.A.E
[2]Traffic Engineering Department., Sharjah Roads & Transport Authority, Sharjah, U.A.E
[3]Department of Civil, Environmental, and Construction Eng., University of Central Florida, U.S.A
[4]Department of Civil and Environmental Eng., University of Sharjah, Sharjah, U.A.E

mabuzwidah@sharjah.ac.ae, m.balwan@srta.gov.ae, m.aty@ucf.edu, galkhateeb@sharjah.ac.ae

Abstract. Highway Lighting (HL) is designed and implemented on the transportation system for expected societal, safety and security reasons. The use of Technologies such as LED lighting and the solar power can encourage many transportation agencies to implement the Highway Lighting on their roads, especially on the rural roads where there are difficulties to connect the power, also it known that the typical Highway Lighting is need a lot of energy. The main goal of this paper is to evaluate the safety effectiveness of the implementation of highway lighting on the Night-Time Crashes on 4-Lanes expressways in the rural areas. An advanced Cross-Sectional method was used to evaluate the effect of adding highway lighting on the safety performance of expressways as a major part of the transportation system. The data was carefully selected and only expressways with 4-lanes were chosen because there were enough treated and reference segments for them. In the Cross-sectional models, a total number of 22 treated segments and 155 untreated segment located on the 4-lanes expressways in Florida State were used. The results showed that adding HL has a positive effect on Crash reduction for the night-time Crashes (all crash types and all severity levels) by an approximately 30 percent. The night-time fatal and serious injury crashes were reduced by 27 percent while the Non-Injury crashes were reduced by 21 percent. Overall, implementing HL technologies is highly encouraged especially on the rural roads to improve safety and security. The Crash Modification Factors developed would help transportation officials to benefit from the extensive research in adding highway lighting to the road network, especially on expressways by providing quantitative information on crash analysis and evaluation for decision making in planning, design, operation, and maintenance.

1. Introduction

In recent decades, urbanization has increased rapidly, resulting in an unprecedented growth in transportation crashes that cause losses of life and property. Preventing traffic crashes in order to improve public safety, transportation system performance, and safe routing has become crucial. It is highly useful to predict potential future crash situations (how, when, or where), not only for individual travelers and transportation administrators but also for public safety stakeholders etc. [1, 2].

Highway Lighting (HL) is designed, fabricated and installed for expected societal and security reasons at night. Determination of the value of adding lighting to transportation system is hard to quantify, because its value rests not simply upon its tangible implementation and operation costs but on its expected benefits, which are inherently difficult to estimate [3].

The literature showed that highway lighting can improve safety on the transportation system by an average 20 percent [4, 5]. However, this statement must be carefully qualified in light of the potential biases that are inherent to study the effect of the treatment. First, and foremost, these estimates may be biased because of other safety measures closely associated with the implementation of lighting [6]. Moreover, these statistics give no indication of where and when lighting might or might not affect transportation safety. But, it appears for example that lighting has little benefit in areas where there is limited chance of vehicle-vehicle or vehicle-pedestrian conflict [7]. Previous research has been done by National Cooperative Highway Research Program (NCHRP) to assess the possible role of lighting in transportation safety. Some samples of highway lighting presence and crash data were assembled and analyzed to evaluate the impacts of highway lighting on traffic safety in general [3, 9].

Overall, an extensive data collection that includes all 4-lanes expressways in Florida was collected in this study including, traffic data, geometric and crash data. It's worth to mention that in Florida, there are more than 750 miles of expressways (toll roads). These roads vary in classification (urban, sub-urban, or rural) and the highway lighting condition [10]. To the best of our knowledge, there are no specific study that evaluated the safety impacts of adding highway lighting on the 4-Lanes expressways. So, the main goal of this paper is to evaluate the traffic safety effectiveness of the implementation of the highway lighting on the transportation system performance.

2. Data Prepration

The detailed objectives were achieved for this study by the following main procedures; 1) Data Collection including all 4-Lanes expressways that have HL treatment on the transportation system in Florida. 2) Estimating advanced statistical crash risk evaluation models. 3) Analyzing crash injury severity for these Highways, and 4) Developing crash risk evaluation models.

An advanced Cross-Sectional method was used to evaluate the effect of adding highway lighting on the transportation safety performance of the 4-Lanes expressways. The Cross-sectional method requires the development of crash prediction models known as Safety Performance Functions (SPFs) for the calculation of Crash Modification Factors (CMFs). The data was carefully selected and only expressways with 4-lanes were chosen as a major part of the transportation system.

The models were developed using crash data for both treated and untreated sites for the same time period. The literature recommended that, Cross-Sectional method should include more samples than the Before-After study, (more than a hundred sites) [11]. Sufficient sample size is particularly important when many variables are included in the SPF. This ensures large variations in crash frequency and variables, and helps better understand their inter-relationships. The treated and untreated sites must have comparable geometric characteristics and traffic volume [12].

In this study, a total number of 177 segments (22 treated segments and 155 untreated segment) located on 4-Lanes expressways in the State of Florida were used. A set of SPFs using Negative Binomial (NB) models were developed to estimate CMFs for the HL treatment at expressways. Negative binomial distribution, also known as Pascal distribution, is a constituent of probability distribution that is used alongside random discrete variables [10]. Compared to the NB model, the Poisson-lognormal model provides more flexibility; however, it does have certain limitations, such as its multifaceted approximation of parameters because of the fact that there is no closed form in the Poisson-lognormal distribution.

3. Analysis and Results

Cross-sectional method was used in this study; it should be noted that the CMF for certain treatments can only be estimated using the Cross-sectional method, but not Before-After method. This is because it is difficult to isolate the effect of the treatment from the effects of the other treatments applied at the same time using the Before-After method.

IOP Conf. Series: Materials Science and Engineering **811** (2020) 012023 doi:10.1088/1757-899X/811/1/012023

The method is used in the following conditions [12]: 1) the date of the treatment installation is unknown, 2) the data for the period before treatment installation are not available, and 3) the effects of other factors on crash frequency must be controlled for creating a Crash Modification Function (CMFunction).

The Cross-sectional method requires the development of crash prediction models (i.e. SPFs) for calculation of CMFs. The models are developed using the crash data for both treated and untreated sites for the same time period (3-5 years). According to the Highway Safety Manual [12], 10~20 treated and 10~20 untreated sites are recommended. However, the Cross-sectional method requires much more samples than the Before-After study, say 100~1000 sites.

The SPFs describe night-time crash frequency as a function of explanatory variables including the presence of adding lighting, AADT and length of highway segments as follows:

$$F_i = \exp(\alpha + \beta_1 * Adding\_Lighting_i + \beta_2 * Length_i) * AADT_i^{\beta_3} \tag{1}$$

where,

- F_i = Night-Time crash frequency on a road segment i;
- Lighting $_i$ = presence of adding lighting on a road segment i (= 1 if the Lighting of a segment i is implemented, = 0 if the Lighting of a segment i is not implemented);
- $Length_i$ = length of a road segment i (mile);
- AADTi = average annual daily traffic on a road segment i (veh/day);
- α = constant;
- β = coefficients for variables.

The results of SPFs for four-lane expressways by severity (injury/non-injury) are shown in table 1. All the factors are statistically significant at a 95 percent confidence level.

Table 1. Adding Lighting Specific SPFs for 4-lanes Expressways.

Night-Time Crashes										
Acci d-ent Type	Sev* levels	Intercept		Log(AADT)		Adding Lighting		Length		ODP*(K)
		Est*	P-Value	Est*	P-Value	Est*	P-Value	Est*	P-Value	
All types	All	-15.216	<.0072	1.669	<.0058	-0.310	0.0441	0.664	<.0001	0.982
	Non-Injury	-12.181	<.0002	1.495	<.0001	-0.230	0.0361	0.724	<.0001	0.954
	F+I	-14.312	<.0372	1.469	<.0108	-0.350	0.0081	0.554	<.0001	0.857
	Serious injury	-13.764	<.0079	1.376	<.0037	-0.320	<.0325	0.986	0.0035	1.320

Sev*=Severity, **Est*=** Estimate, **ODP*=** Over Dispersion Parameter and F+I = fatal and Injury crashes.

After developing the SPFs, the Crash Modification Factors were calculated using the following equation:

$$CMF = \exp(\beta_1 * (1 - 0)) = \exp(\beta_1) \tag{2}$$

The above model can be applied to for prediction of the total crash frequency or frequency of a specific crash type or severity. The standard error (SE) of the CMF is calculated as follows [13]:

$$SE = \frac{\exp(\beta_k * (x_{kt} - x_{kb}) + SE_{\beta_k}) - \exp(\beta_k * (x_{kt} - x_{kb}) - SE_{\beta_k})}{2} \tag{3}$$

where,

SE = standard error of the CMF;

$SE_{\beta k}$ = standard error of the coefficient of the variable (*Adding_Lighting*).

Table 2. Recommended CMFs for Adding Lighting to 4-lanes Expressways.

Night-Time Crashes				
Cross-Sectional method				
Crash Type	Severity Levels	CMF	Standard Error	Safety Effectiveness of the treatment (1- CMF) x 100=%
All types	fatal and Injury crashes	0.73	0.09	(1- 0.73)*100 = 27%
	Non- Injury	0.79	0.10	(1-0.79)*100 = 21%
	Serious injury	0.73	0.09	(1-0.73)*100= 27%
	All Severity	0.70	0.10	(1-0.70)*100 = (30%)

It is clear that the treatment has positively affect the night time crashes on the transportation network (4-lane expressways) by reducing all crash types and severity levels as shown in table 2. The fatal and Injury crashes (F+I) were reduced by 27% while the Non-Injury crashes were reduced by 21%. Similar, the Serious injury crashes were reduced by 27% while the all crashes all severity were reduced by 30%.

4. Conclusion

Highway Lighting (HL) is designed and implemented on the transportation system for expected societal, safety and security reasons. The use of Technologies such as LED lighting and the solar power can encourage many transportation agencies to implement the Highway Lighting on their roads, especially on the rural roads where there are difficulties to connect the power, also it known that the typical Highway Lighting is need a lot of energy.

The main goal of this paper is to evaluate the safety effectiveness of the implementation of highway lighting on the Night-Time Crashes on 4-Lanes expressways in the rural areas. An extensive data collection that includes all 4-lanes expressways in Florida was collected in this study including, traffic data, geometric and crash data. It's worth to mention that in Florida, there are more than 750 miles of expressways (toll roads). These roads vary in classification (urban, sub-urban, or rural) and the highway lighting condition. To the best of our knowledge, there are no specific study that evaluated the safety impacts of adding highway lighting on the 4-Lanes expressways.

The results of this study showed that adding HL has a positive effect on crash reduction for the night-time crashes (all crash types and all severity levels) by an approximately 30 percent. Also, the night-time fatal and Serious injury crashes were reduced by 27 percent while the Non-Injury crashes were reduced by 21 percent.

Overall, implementing HL technologies is highly encouraged especially on the rural roads to improve safety and security. The Crash Modification Factors developed would help transportation officials to benefit from the extensive research in adding highway lighting to the road network, especially on expressways by providing quantitative information on crash analysis and evaluation for decision making in planning, design, operation, and maintenance.

References

[1] Abuzwidah M 2011 Evaluation and Modeling of the Safety of Open Road Tolling System, Master Thesis. Orlando, University of Central Florida.

[2] Ahmed M, Abdel-Aty M, Qi S, Abuzwidah M 2014 Synthesis of state-of the-art in visibility detection systems' applications and research, *J. transport. safety security*, **6(3)** 183-206.

[3] Bullough J, Rea M, Zhou Y 2009 Analysis of Visual Performance Benefits from Roadway Lighting (NCHRP Project 5-1 report), submitted to National Cooperative Highway Research Program.

[4] Donnell E, Shankar V, Porter R 2009 Analysis of Safety Effects for the Presence of Roadway Lighting (NCHRP Project 5-19 report), submitted to National Cooperative Highway Research Program.

[5] Fisher A 1977 Road Lighting as an Accident Countermeasure, *Australian Road Res.* **7(4)** 3-15.
[6] Bullough J D, Hickcox K, Klein T R, Narendran N 2011 Effects of flicker characteristics from solid-state lighting on detection, acceptability and comfort, *Light. Res. Technol.* **43** 337-48.
[7] Beyer F, Ker K 2009 Street Lighting for Preventing Road Traffic Injuries, *Cochrane Database Syst. Rev.* CD004728.
[8] Donnell E, Shankar V, Porter R 2009 Analysis of Safety Effects for the Presence of Roadway Lighting (NCHRP Project 5-19 report), submitted to National Cooperative Highway Research Program.
[9] Bullough J, Rea M, Zhou Y 2009 Analysis of Visual Performance Benefits from Roadway Lighting (NCHRP Project 5-1 report), submitted to National Cooperative Highway Research Program.
[10] Abuzwidah M, Abdel-Aty M 2018 Crash Risk Analysis of Different Designs of Toll Plazas, *J. Safe. Sci.* **107** 77-84.
[11] Carter, J. A., Agol, E., Chaplin, W. J., et al. 2012 A Pair of Planets with Neighboring Orbits and Dissimilar Densities, *Sci*, **337** (6094) 556. .
[12] AASHTO, 2010. The Highway Safety Manual, American Association of State Highway Transportation Professionals, Washington, D.C., http://www.highwaysafetymanual.org.

IOP Conf. Series: Materials Science and Engineering **811** (2020) 012024 doi:10.1088/1757-899X/811/1/012024

Chemical Quality of Fresh New Zealand White Rabbit Meat in Batu, Indonesia

A Fadlilah[1], D Rosyidi[2*], A Susilo[2]

[1]Postgraduate student of Animal Science Faculty, Brawijaya University, Indonesia
[2]Lecturer of Animal Science Faculty, Brawijaya University, Indonesia

djalal_tht@ub.ac.id

Abstract. Batu was one of the cities in the province of East Java, Indonesia that one of the leading tourist cities in Indonesia. Rabbit satay was one of the Batu city famous foods. Many tourists visited Batu City for consume it. Rabbits were obtained from rabbit breeders in the Batu city. The environment and the pattern of rearing rabbits had an impact on the quality of the meat and specific characteristic of the meat production. This study aimed to determine the chemical quality of the fresh New Zealand White (NZW) rabbit meat in the Batu City using the proximate analysis. This Research used 9 males of NZW rabbits, each of them was taken 100gr of loin flesh for proximate analysis. Rabbits were kept intensively. The results showed that fresh NZW Rabbit Meat had a carbohydrate content was 0.43%, Protein content was 19.99%, Moisture was 75.84%, Ash content was 1.43% and Energy gross was 110.47 Kcal/100g.

1. Introduction

Batu is one of cities in the East Java Province, Indonesia. It located at the high altitude of 700-2.000 meter above sea level (masl) with average temperature around 11-19°C. Batu City is known as one of a tourism city in Indonesia because its natural beauty. The surrounding region around Batu City has been known as mountainous areas with cool air that is comfortable for holiday. The total area of Batu City is about 202.30 km² and most of Batu's topographical conditions are dominated with highland areas, and located on the slopes of two large mountains, Arjuno-Welirang and Butak-Kawi Panderman. The soil is contained a lot of minerals from volcanic eruptions so that it has a high fertility rate.

Rabbit satay is one of the famous foods in Batu City. The high demand of rabbit satay by tourists could increase the consumption of rabbit meat in Batu City. The quality of rabbit meat, such as a high protein with low fat and cholesterol content, makes it healthier meat for consumption. The rabbit meat was produced by local breeders from Batu City. The local breeders usually bred New Zealand White (NZW) rabbit.

The New Zealand White rabbit is meat-producing type because of it sturdy and compact meat. The New Zealand White rabbit was known as commercial meat producer. It had fast growth rate, good carcass quality, high fertility and good maternity [1]. The first mating age was 8-9 month with average litter size was 8-9 kits and weaned was 37 days. The NZW rabbit had pregnancy time was 30 days, calving interval was 27 days [2]. The breeder use pattern of rearing rabbits in Batu City generally an intensively pattern, which the rabbits were kept inside the cage and given forage originating from the remnants of agricultural products. The temperature of Batu City is around 11-19°C and the amount of forage is sufficient for the needs of rabbits causes rabbit livestock in Batu can develop rapidly.

Nowdays, several factors can be used to determine the quality of fresh meat are its nutritional content meat and safety meat. Some factors that affect the quality of rabbit meat were environmental factors, rearing technique, feeding effects, preslaughter conditions, stunning conditions, genetics, age, body weight, feed, cooling conditions and the technology used after cutting [3]. Based on the condition of Batu City and rearing technique of Batu rabbits' breeders, it has an impact on the quality of rabbit meat production, so this article will discuss the quality of the meat in terms of the chemical quality of fresh NZW rabbit meat of in Batu City, Indonesia and the results of this research can be used as one of the bases in the development of processing animal products from rabbit meat.

2. Materials and Methods

2.1. Collection and Preparation Sample

This research used meat from 9 tails of Male NZW Rabbit of "Adi Jaya" Batu Breeder Group. This research was on August, 2019. The rabbits were rearing intensively that kept inside cage and fed the forage and complete feed twice a day. Water was supplied ad-libidtum in the respective water (automatic Niple) kept inside the cage. This research was carried out under cold condition having ambient temperature range from 11°C (minimum) to 19°C (Maximum). The rabbits were slaughtered manually according to Islamic procedure and aged 3-4 months, then deskinned and eviscerated on spot, then taken 100gr fresh meat on the loin flesh of each rabbit tail, kept separately in polyethylene bags and transported with a minimum of delay to the laboratory of Meat, Faculty of Animal Science, Brawijaya University.

2.2. Chemical Analysis

Chemical analysis used in this research was a proximate analysis, consisting of analysis of proteins, fats, water and ash. The all of methods of proximate analysis used in this study are according to the procedure of Association of Analytical Communities [4]. The Carbohydrate content was analysis using calculations: 100 - (protein + fat + ash content) [5]. Gross energy was analysis using calculation: Gross energy (Kcal/100g) = (protein% x 4) + (fat% x 9) + (carbohydrate% x 4) [6].

3. Results

Based on the results of the research on the chemical quality of the New Zealand White rabbit meat in Batu were obtained the following data on Table 1.

Table 1. Proximate Composition of Fresh NZW Rabbit meat, Batu Indonesia.

Composition	Value
Gross Energy	110.47 Kcal/100g
Carbohydrate	0.43 %
Protein	19.99 %
Fat	2.31 %
Moisture	75.84 %
Ash	1.43 %

The carbohydrate content of Fresh NZW Rabbit meat, Batu was 0.43%. The carbohydrate content in rabbit meat was relatively low when compared to carbohydrate content in red meat. This research results show that it was higher when compared with the results of the carbohydrate levels in the fore quarter and hind quarter rabbit meat was 0.33% [7]. Gross Energy calculation result for NZW rabbit meat was 110.47 Kcal/100g. Based on Food and Agriculture Organization (FAO) the gross energy content of rabbit meat was around 120 Kcal/100g [8], so the result of the study was compared with the FAO standard, the research results was lower than the standard. The gross energy content of local rabbit meat and Rex rabbits that reared in Bogor (one of city in Indonesia) was 136.0-171.6 Kcal/g. The Male rabbit behaviour had a greater eating behaviour than females' rabbits, so it would further affect the energy content of the meat [17]. Genetic differences affect the energy content of rabbit meat. The main energy value was more closely related to protein content than fat content, while carbohydrate content can be neglected [9].

Proximate analysis measured crude protein levels were actually not only protein fractions but also all nitrogen-containing compounds. The protein content was one of the most important indicators in food. High-quality food ingredients had high protein content. The protein content of rabbit meat, when compared with the meat of other species was quite high. The protein content of rabbit meat from this research was 19.99%, it was higher than protein content of the fore quarter's rabbit meat was 18.6% and lower than the protein content of the hindquarters at 20.59% [10] and it was higher when compared with the protein content of local rabbit meat in Bogor Indonesia was 18.36-18.78% [17]. Rabbit meat protein content of NZW was 18.63-18.86% [11]. The protein content of rabbit meat increased with increasing age of the rabbit [11] and has no significant effect on diet and genotype [9] and sex [12].

The fat content of fresh rabbit meat of NZW Batu City was 2.31%, the result showed that rabbit meat was classified as low-fat meat. Fat content of rabbit meat of NZW in Batu City was lower when compared with the results of 8.8%in the forelegs rabbit meat and 3.4% in the hind legs rabbit meat [9], higher when compared with the results of the hind quarters's rabbit was 1.78% [7]. The value of fat content was 1.69-1.81 for Belgian Giant rabbits [13]. The NZW fat content in the Longissimus Dorsi muscle section was 5.82% [11]. Fat content of meat was influenced by breed, age, species, muscle location and feed [18]. The meat fat content of Rex rabbits that reared in Bogor Indonesia was 2.20 - 5.01%, and local rabbits was 0.89-1.34%. The meat fat content of Rex rabbits was higher because it was one of the fur producing rabbits whose behaviour was tame, so that the energy consumed was more deposited in fat than for activity. In other hand Rex rabbit was subtropical rabbits, so the meat fat content was higher than local rabbits whose live in tropical environments. It was one of the effect of the environment [17]. The fat content will increase with increasing age of rabbits. Cutting types, diets, and class/family show significant differences in fat content in rabbit meat [11], while sex and genotype did not have a significant effect on rabbit meat fat content [9, 14].

The moisture level in rabbit meat of NZW from this research was 75.84%. It was lower than the results of the Longissimus Dorsi muscle was 76.19% [14], and higher when compared to the results of the study of lion muscle of NZW rabbit meat showed a value of 74.1% [15]. The moisture of rabbit meat was ranged from 63.6-75.93% both female and male rabbits [13]. Water availability was greatly influenced the endurance of the meat as food, especially in the context of microbiological damage. The water measured by the direct heating method in testing the moisture content was water that evaporates because its boiling point was exceeded and lost during heating (±105° C) included both surface water and water on the inside of the meat. The type of breed and the interaction between type of breed and gender significantly was significant effect on Rex rabbit meat and Bogor-local rabbit meat. The average moisture of Rex female rabbits was 70.57±2.51 was smaller than the average moisture of male and female local rabbits locally was 73-75% [17]. The moisture of fresh meat was more influenced by the level of dehydration of livestock before slaughtered [19]. The condition of livestock at the time of slaughtering could cause the differences in the chemical composition of meat produced. Carcass weight was a reflection of the condition of livestock. The carcass weight was influenced by interactions between breed and feed which showed that the efficient utilization of energy, protein and possibly mineral feed was relatively different among breed and feed treatments, but it was not always reflected in differences in the chemical composition of meat. The relationship between moisture and fat content had a significant negative correlation with meat fat content and moisture [20]. The moisture meat of Rex rabbits had the smaller water content and higher fat content than local rabbit, while the local rabbit meat had higher moisture and the smaller fat content [17]. The moisture level of the rabbit meat would decrease significantly with increasing age of rabbit and did not have a significant effect on rabbit species and sex [9].

The ash content of fresh NZW rabbit meat in Batu was 1.43%. The ash content of fresh NZW rabbit meat in Batu was higher when compared with the results of the four quarters's rabbit meat was 1.21% and hind quarters's rabbit meat was 1.17% [7], while lower when compared the ash content of male rabbit meat was 2.27% [16]. The results of NZW rabbit meat was 1.15% ash content in the rabbit longissimus muscles [11], and the rabbit loin meat of NZW 1.35% [15]. The ash content of Rex rabbit meat and Bogor-Local rabbit meat was not influenced by the type of breed, gender and interaction between the them. The average ash content for all rabbits was 1.06 - 1.19%. This difference can be influenced by the mineral content of feed [17]. The meat ash content of New Zealand, Californian, Chinchilla and Rex rabbits had 4.44, 4.52, 4.43 and 4.49% [21]. Age, class/family, sex, diet or carcass section did not show a significant effect on ash content in rabbit meat [9, 15], whereas an increasing in age significantly affects the ash content in rabbit meat of Soviet Chinchilla [11].

4. Conclusion

In summary the results of this research was fresh meat of New Zealand Rabbit in Batu city, Indonesia had a chemical quality that were almost the same as the meat chemical qualities of the rabbit was kept in other countries. Based on the chemical qualities of the fresh NZW rabbit meat in Batu, it could be concluded that the fresh rabbit meat is healthier over that other meats and can be recommended for consumption as functional food with the high protein content (19.99%) and low fat content (2.31%). The results of this research are preliminary research for make the new product as the the development of livestock products processing from rabbit meat.

Acknowledgments

This research is funded by Lembaga Pengelola Dana Pendidikan (LPDP) Scholarship, Ministry of Finance of Indonesia.

References

[1] Cheeke P R, Patto N M, Lukefahr S D, Mcnitt J L 1987 *Rabbit Production.* 6th Ed. The Interstate Printers and Publisher, Inc., Danville, Illinois
[2] Fauzi A Z, Sri M, Nur C 2015 Performance Reproduction Of Rabbit In Bumiaji Sub-District Batu City.
[3] Dalle Zotte A 2000 Main factors influencing the rabbit carcass and meat quality, In: *Proceedings of the 7th World Rabbit Congress (Valencia, Spain)* 1-32.
[4] AOAC 2012 Official Methods of Analysis of AOAC International, 19th Ed., Latimer G. W. (edit). Gaithersburg, Md Publisher.
[5] Jurgens M H, Bregendahl K 2007 Animal Feeding and Nutrition, 10thEd Kendall/Hunt Publishing Company, Iowa, USA.
[6] Merrill A L, Watt B K 1973 Energy Value of Foods: Basis and Derivation, Agriculture Handbook No. 74, ARS United States Department of Agriculture, Washington DC.
[7] Abd-allah S M S, Doa M A 2018 Nutritional Value and Quality Profile of Fresh Rabbit Meat in Assiut City, Egypt, *Int. J. Res. Agri. Food Sci.* **4** 1-15.
[8] FAO 1997 *The Rabbit - Husbandry, Health and Production.* Lebas, F. Coudert, P. De Rochambeau H. and. Thébault R.G. (edits).FAO Animal Production and Health Series No. 21 (new revised version), Food and Agriculture Organization of the United Nations, Rome (Italy).
[9] Belichovska D, Belichovska K, Pejkovski Z, Uzunoska Z 2017 Effect of genotype on physico-chemical characteristics of rabbit meat, *Meat Tech.* **58(1)** 10-5.
[10] Dalle Zotte A, Cullere M, Sartori A, Szendrő Zs, Kovàcs M, Giaccone V, Dal Bosco A 2014 Dietary Spirulina (Arthrospira platensis) and Thyme (Thymus vulgaris supplementation to growing rabbits: Effects on raw and cooked meat quality, nutrient true retention and oxidative stability, *Meat Sci.* **98(2)** 94-103.
[11] Haque A, Rahman M, Bora J 2016 effect of breed, weaning age and feeding regime on chemical composition of rabbit meat, *Int. J. Vet. Sci. Animal Husbandry* **1(1)** 12-3.
[12] Ghosh N, Mandal L 2008 Carcass and meat quality traits of rabbits (Oryctolagus cuniculus), *Livestock res. Rural Develop.* **20(9)**.
[13] Frunza Gabriela T, Lazar R, Biosteanu P C 2014 Research regarding nutritional quality of rabbit meat (Belgian giant breed), *University of Agriculture Science and Veterinary Medicine Iasi, Scientific Papers – Animal Science Series,* **62** 45-8.
[14] Simonova M P, Chrastinova L, Mojto J, Laukova A, Szaboova R, Rafay J 2010 Quality of rabbit meat and phyto-additives, *Czech J. Food Sci.* **28** 161-7.

[15] Mattioli S, Cardinali R, Balzano M, Pacetti D, Castellini C, Dal Bosco A, Frega N G 2017 Influence of dietary supplementation with prebiotic, oregano extract, and vitamin E on fatty acid profile and oxidative status of rabbit meat, *J. Food Quality*, **5** 9-17.

[16] Fadare A O, Doka Z B, Odimayo O M, Fesobi M E 2017 Effect Of Genotype On The Proximate Composition and Sensory Properties Of Rabbit Meat, *FUTA J. Res. Sci.* **13(2)** 266-73.

[17] Brahmantiyo B, Setiawan M A, Yamin M 2014 Physical and Chemical Properties of Rex and Local Rabbit's (Oryctolagus cuniculus) Meat, *J. Peternakan Indonesia.* **16(1)** 1-7.

[18] Judge M D, Aberle E D, Forrest J C, Hedrick H B, Merkel R S 1989 *Principles of Meat Science*. 2nd. Ed. Kendall / Hunt Publishing Co. Dubuque, Iowa.

[19] Lawrie R A 2003 *Meat Sci*. 5thEd. Perganon Press, Oxford.

[20] Soeparno 1992 Meat Scince and Technology. Gadjah Mada University Press, Yogyakarta.

[21] Hernandez J A, Rubio Lozano M S, Carregai R D 2001 *Effect of Breed and Sex On Rabbit carcass Yield and Meat Quality*. Meat Science Labolatory, Facultad de Medicina Veterinaria y Zootecnia Universidad Nacional Autonoma de Mexico, Mexico.

IOP Conf. Series: Materials Science and Engineering **811** (2020) 012025 doi:10.1088/1757-899X/811/1/012025

Evaluation of Chemical Composition and Lipid Components in Maize Straw based Complete Feed Diets Supplemented by Conden Tannin and Myristic Acid

M Muchlas[1], S Chuzaemi[2*], Mashudi[2*]

[1]Postgraduate student of Animal Science Faculty, Brawijaya University, Indonesia
[2]Lecturer of Animal Science Faculty, Brawijaya University, Indonesia

schuzaemi@ub.ac.id

Abstract. Although maize straw as an agricultural by products could be used as an alternative feed ingredient for ruminants, its low nutritional value and palatability require an early processing. Supplementation of feed additive that functions as a methane inhibitor. The study was to examine the chemical composition in maize straw based complete feed diets supplemented by conden tannin and myristic acid. The experimental method used a group-randomized trials by ANOVA with 4 dietary treatments and 3 replications. The treatments are T1: Maize Straw Based Complete Rations (Control Diet), T2: Control diet +Mimosa powder (MP) 1,5% /kgDM + Myristic acid (MA)2% /kgDM, T3: Control diet + MP 1,5% /kg DM + MA 3% /kg DM, and T4: Control diet + MP 1,5% /kg DM + MA 4% /kg DM. Proximate and van soest analysis was analysed in the Feed and Nutrition Laboratory, animal Science faculty, Brawijaya University, Malang. The result showed that the maize straw complete feed without supplemented has not different with a chemical compositions Maize straw with 3 level of Conden tannin and myristic acid supplementation. Besides higher EE components contained on T4 which supplemented by CT 1,5 %/kg DM and MA 40 g/kg DM. It can be concluded that CT and MA in Maize Straw based complete feed diets can be used as rumen modifier without effected on chemical component.

1. Introduction

The major constraint in improving ruminant production in Indonesia is discontinuous availability of forage feed throughout the year, particularly in a dry season when the forage sources are limited. Utilization of agricultural by-products is usually practiced to overcome this problem. Agroindustry residues resulting from the processing of plant products, such as meals, seeds, pulps and peels, are generated in large quantities and have the potential to be used to feed animals of zootechnical interest [16]. Alternative sources of energy for livestock, of low commercial value and that do not impair animal performance, such as agroindustry co-products, may represent an increase in competitiveness in the zootechnical activity. However, the level of substitution of such foods has to be assessed, since each ingredient has intrinsic positive and/or negative characteristics. Feed is an absolute necessity that should always be considered in the maintenance of ruminants i.e. cattle, buffaloes, goats and sheep. However, availability feed always be a constraint, especially in the dry season, feed fresh forage difficult to obtain, there are only the remains of plants such as straw. One of them is the maize straw that has excellent potential as a food source, it's just low quality. To improving the quality and benefits of maize straw will require technology that is easy and simple to do farmer.

IOP Conf. Series: Materials Science and Engineering **811** (2020) 012025 doi:10.1088/1757-899X/811/1/012025

Previous studies showed that the use of maize straw as a source of crude fibre can be combined with other feed ingredients with high protein and energy content so that a good nutritional content is obtained or can be called a complete feed. The provision of complete feed is expected to be able to meet the nutritional needs of livestock and can improve business efficiency [5]. Although it is palatable, the fibre content of corn cob silage was high [23], which is potential in producing methane during fermentation in the rumen. Methane emission by ruminants has negative effects on the animal and causes greenhouse effect to the environment. Methane which is an end product of fermentation process of feed is formed through methanogenesis. Methanogenesis cause 2–12% energy loss of the digested energy depending on diet type [10]. Methane emission from enteric fermentation contributes 25% to the total global greenhouse gases from agriculture [15].

These conditions make researchers interested in conducting research to increase the nutritional value of maize-based complete feed as ruminant animal feed and add natural compounds found in plants as feed additives to increase livestock productivity [24], including in reducing methane production [20]. One effort to reduce methane gas production in ruminants is by manipulating feed without disturbing livestock productivity. One method for the process of reducing methane gas in livestock is the addition of compounds into feed such as tannins, saponins, fat administration, organic acids, and inhibitors [8]. Reduction methane emissions have been observed through adding feed additive. Ruminant nutritionists have studied essential oils [3] and dietary lipids [2] as rumen manipulators to improve ruminal fermentation characteristics as well as potential anti-methanogenic feed additives.

Palm oil (*Elaeis guineensis Jacq.*) is one of the world's most produced and consumed oils. This cheap, production-efficient and highly stable oil is used in a wide variety of food, cosmetic and hygiene products, and can be used as a source for bio-fuel or biodiesel. Indonesia is the largest producer and exporter of palm oil worldwide. Ten types of fatty acids from crude palm oil or CPO were observed consisting of laurate, myristate, palmitate, palmitoleate, stearic, oleic, linoleic, linolenic, arachidate, and gadoleic [18]. One sterified lauric acid (C12) was reported to have a particularly high potential in suppressing ruminal methanogenesis, followed by myristic acid (C14) [4]. The addition of myristic acid showed a stronger growth-inhibiting effect on the investigated microbes[6]. the mitigation potential of C14:0 was suggested to be limited when not combined with C12:0 [20]. The addition of saturated fatty acid by 7.5% can reduce the number of protozoa by 23.95% and reduce the amount of methane gas production by 18.51%[19]. Tannins or polyphenols are compounds that have the potential to reduce methane emissions among natural compounds found in plants [9]. Tannins are naturally occurring plant polyphenols and their main characteristic is that they bind and precipitate proteins [17]. The natural state that occurs when oil is added to the feed is that the oil will coat the body of the protozoa so that the protozoa cannot carry out metabolic activities that cause the protozoa to die in the rumen.

2. Materials and Methods

2.1. Collection and Preparation Sample

Maize straw ware procured from research farm laboratory of Faculty of Animal Science, Brawijaya University. This area is located in East Java island Indonesia at an altitude of 506 m above sea level with an average annual rainfall and temperature of 369 mm and 22,7 – 25,1°C, respectively. Myristic acid (99% purity) and Conden tannin ware bought from CV. Sumber Berlian Kimia, Jakarta (Indonesia). Agriculture and agro-industrial by-products for composition of total mixed rations were purchased from a feedstuff supplier in Malang city. The samples were oven dried at 50°C for 48 h then ground to pass through a 2.0 mm sieve. The dry samples were further ground to pass through a 1.0 mm sieve for the in vitro gas production experiments and chemical analysis. DM of feed component was determined by drying at 100 °C, organic matter by ashing at 550°C for 4h and crude protein by the Kjeldahl technique [1]. The acid detergent fibre (ADF) and neutral detergent fibre (NDF) content were determined according to [22]. All chemical analyses were carried out in triplicate. The experimental method used a group-randomized trials by ANOVA with 4 dietary treatments and 3 replications. The experimental diets were:

- T1: Maize Straw Based Complete Rations (Control Diet)
- T2: Control diet +Mimosa powder (MP) 1,5% /kgDM + Myristic acid (MA)2% /kgDM
- T3: Control diet + MP 1,5% /kg DM + MA 3% /kg DM
- T4: Control diet + MP 1,5% /kg DM + MA 4% /kg DM

2.2. Chemical Analysis

Chemical analysis used in this research was a proximate analysis, consisting of analysis of proteins, fats, water and ash. The all of methods of proximate analysis used in this study are according to the procedure of Association of Analytical Communities [1]. The Carbohydrate content was analysis using calculations: 100 - (protein + fat + ash content). Gross energy was analysis using calculation: Gross energy (Kcal/100g) = (protein% x 4) + (fat% x 9) + (carbohydrate% x 4) [11].

3. Results

Based on the results of the research on the chemical maize straw based complete feed diets supplemented by conden tannin and myristic acid were obtained the following data on table 1.

Table 1. Chemical composition of the feeds used during the experiment.

No	Component	DM	OM*	CP*	CF*	EE*	NFE*
1	Maize straw (*Zea mays*)	60.31	89.83	5.13	36.43	0.63	47.64
3	Coffee waste (*Coffea arabica*)	94.14	89.42	10.11	34.00	1.49	43.82
4	Rice bran (Oryza sativa L)	90.63	87.40	10.15	16.20	13.00	48.05
5	Tapioca by-Product (*Manihot esculenta*)	92.59	82.87	1.76	25.39	0.44	55.28
6	Soybean Meal (*Glycine max*)	93.53	91.62	47.53	4.04	2.57	37.48
7	Palm kernel meal (*Elaeis guineensis* Jacq.)	95.39	94.97	14.24	20.91	10.01	49.81
8	Copra Meal (*Cocos nucifera L.*)	95.69	92.23	22.12	21.78	2.45	45.88
9	Urea	99.88	99.93	244.60	-	-	
10	Molasse	78.47	84.56	4.54	-	-	
Chemical composition of chemical treated and untreated - maize straw							
	T1	93.44	90.09	16.18	24.00	2.94	46.97
	T2	93.82	90.38	15.59	24.53	4.07	46.19
	T3	93.53	90.57	13.79	24.60	5.44	46.74
	T4	93.78	90.55	14.09	24.90	7.12	44.44

Keterangan: - Analysis results at the Laboratory of Nutrition and Animal Feed Faculty of Animal Science, Brawijaya University (2019)
-*100% DM Basis

Data in table 1. showed that DM of treatment feed content of ranged from 93,44% to 93,82%, whilst DM content of complete feed component ranged from 60,31% to 99,88%. Regarding the crude protein content of maize straw, the relatively low protein content can be corrected by a combination of protein and mineral feed ingredients. However, the selection of protein sources must pay attention to the characteristics of the protein source to be used in addition to considering the price of feed ingredients to be easily affordable [14]. The use of maize straw as ruminant animal feed requires supplementation of energy and protein feed sources, because of its low quality [21]. Nutrient supplementation both energy and protein together is intended for optimization of microbial growth so that the utilization of fibrous feed can be optimized.

The crude protein content (CP) in the complete treatment of maize straw-based feed can be seen in the tables in P1, P2, P3, and P4 respectively 16.18%, 15.59%, 13.79%, and 14.09%. Crude protein in complete feed is adjusted by NRC [13], which is the feed for beef cattle fattening in 13%. Maize straw has a low PK which can be seen in the table that is equal to 5.13% so that in this study corn grain is treated as complete feed. The processing of maize straw as a complete feed is intended as a feed that is sufficient nutrition for certain animals in a certain physiological level. Complete feed is formed or mixed to be given as sole food and is capable of caring for the basic life or production (or both) of livestock without the addition of other substances except water [23].

In this study, maize straw is a source of fibre feed ingredients with additional other feed ingredients as a supplement to the lack of nutrients. The largest group of methane gas-producing feedstuffs is the forage of fibre sources, then followed by the forage of protein sources and the constituent feed ingredients. Selection of feed ingredients with various types can also be done in the activity of formulating complete feed in an effort to reduce methane emissions from livestock [22]. The crude fat content of CF in each treatment also showed a quite noticeable difference (table 1) with the highest value at P4 7.12% and the lowest at P1 2.94%. This difference in crude fat content correlates with the addition of myristic acid levels in the treatment. Myristic acid (myristic acid) or tetra decanoic acid is a saturated fatty acid composed of 14 atoms of C. In the study of Soliva Supplementation of Mystic acid in dairy cattle feed can increase fat in milk [20].

4. Conclusion

Based on the results of the study, it can be concluded that the higher fat components contained on T4 which supplemented by CT 1,5 %/kg DM and MA 40 g/kg DM. It can be concluded that CT and MA in Maize Straw based complete feed diets can be used as rumen modifier without effected on chemical component.

Acknowledgments

I am enormously grateful to LPDP (Lembaga Pengelola dana Pendidikan-Indonesia/Endowment fund for Education) for continuous encouragement on my study and funding to participate in this conference. I am thankful for Siti Mrs Chuzaemi and Mr Mashudi for kindly advise throughout my study.

References

[1] AOAC 2012 *Official Methods of Analysis of AOAC International.* 19th Ed., Latimer G. W. (edit). Gaithersburg, Md Publisher.

[2] Beauchemin K A, S M McGinn and H V Petit 2007 Methane abatement strategies for cattle: lipid supplementation of diets Canadian Journal Anim. Sci.

[3] Calsamiglia S, M Busquet, P W Cardozo, L Castillejos and A Ferret 2007 Invited review: Essential oils as modifiers of rumen microbial fermentation Journal Dairy Science.

[4] Dohme F, Machmüller A, Wasserfallen A, Kreuzer M. Comparative efficiency of various fats rich in medium-chain fatty acids to suppress ruminal methanogenesis as measured with RUSITEC. Canadian Journal of Animal Science. 2000;80(3):473–482.

[5] Gustiani E, dan Permadi K 2015 Study on Productivity of PO Cattle Provided by Complete Feed Containing Fermented Corn-Cobs in Majalengka District, West Java.Jurnal Peternakan Indonesia, Februari 2015 Vol 17 (1) ISSN 1907-1760

[6] Henderson, C 1973 The effect of fatty acids on pure cultures of rumen bacteria. Journal of Agricultural Science 81, 107–112.

[7] Hess BW, Moss GE, Rule DC 2008 A decade of developments in the area of fat supplementation research with beef cattle and sheep. J Anim Sci. 86:E188-E204.

[8] Hook S E, Wright A D G, McBride B W 2010 Methanogens: methane producers of the rumen and mitigation strategies, *Archaea* 1-11.

[9] Jayanegara A, Sofyan A 2008 Determination of biological tannin activity in some forages in vitro using the 'Hohenheim Gas Test' with polyethylene glycol as a determinant, *Med. Pet.*31**(1)** 44-52.

[10] Johnson K A, Johnson D E 1995 Methane emissions from cattle, *Journal Animal Sci.* **73(8)** 2483-92.

[11] Jurgens M H, Bregendahl K 2007 *Animal Feeding and Nutrition.* 10thEd Kendall/Hunt Publishing Company, Iowa, USA.

[12] Makkar H P S, Francis G, Becker K 2007 Bioactivity of phytochemicals in some lesser known plants and their effects and potential applications in livestock and aquaculture production systems, *Animal* **1** 1371-91.

[13] National Research Council 1996 Nutrient requirements of beef cattle. Fifth Revised Edition. National Academy of Science. Washington DC.

[14] Natsir A 2012 Efficient Utilization of Fibre for Ruminants. Masagena Press.

[15] Olivier J G J, van Aardenne J A, Dentener F, Ganzeveld L, Peters J A H W 2005 Recent trends in global greenhouse gas emissions: Regional trends and spatial distribution of key sources, In A. van Amstel (Ed.) Non-CO2 Greenhouse Gases (NCGG-4) 325-30.

[16] Prado I N, Moreira F B 2002 Supplementation of moringa in the pasture and alternative feeds used in cattle farming PR: Eduem.

[17] Sahoo A, Singh B and Bhat T K 2010: Effect of tannins on in vitro ruminal protein degradability of various tree forages. Livestock Research for Rural Development. Volume 22, Article 119. Retrieved December 7, 2019, from http://www.lrrd.org/lrrd22/7/saho22119.htm.

[18] Siregar H A, Rahmadi H Y, Wening S, dan Suprianto E 2018 Composition of Fatty Acid and Carotene of Oil Palm Elaeis Oleifera, Interspecific Hybrid, and the First Pseudo-Backcross in North Sumatra INDONESIA J Pen Kelapa Sawit 26(2): 91-101.

[19] Sitoresmi P D, Mira L Y, dan Hartadi H 2009 Effects of Addition of Coconut Oil, Sunflower Seed Oil, and Palm Oil to the Decrease of Methane Production in the Rumen *In vitro*. Buletin Peternakan **33(2)** 96-105.

[20] Soliva CR, Hindrichsen IK, Meile, L, Kreuzer M, Machmüller A 2003 Effects of mixtures of lauric and myristic acid on rumen methanogens and methanogenesis in vitro Lett Appl. Microbiol 37 35–39.

[21] Suhartanto B, Widyobroto B P, dan R Utomo 2003 Production of complete feed (complete feed) and supplementation of undegraded protein to increase production and quality of beef. Research institutions of Universitas Gadjah Mada. Yogyakarta.

[22] Van Soest P.J., Robertson J.D. and Lewis B.A. 1991 Methods for dietary fibre, neutral detergent fibre and non-starch polysaccharides in relation to animal nutrition. J. Dairy Sci. 74, 3583-3597.

[23] Widiawati Y, Puastuti W, dan Yulistiani D 2017 Profile Of Methanic Gas From Ruminansia Raw Feed Materials. Proceedings of the Livestock Technology and Agribusiness Seminar V: Livestock Technology and Agribusiness to Support Food Security, Faculty of Animal Husbandry Universitas Jenderal Soedirman.

[24] Yulianti D L, Permata Ika Hidayati P I, dan Shodiq A 2018 Complete Feed Formulation (Complete Feed) Based on Agricultural Waste as Goat Animal Feed in Kromengan District, Malang Regency. JPM (Journal of Community Empowerment) **3(1)**.

IOP Conf. Series: Materials Science and Engineering **811** (2020) 012026 doi:10.1088/1757-899X/811/1/012026

A Semi-Stack Approach for Accurate Network Traffic Classification Using Multi-View Stacking

Adil Fahada,*

[a] Department of Computer Science, College of Computer Science and IT, Albaha University, Albaha, Saudi Arabia

Email: afalharthi@bu.edu.sa

Abstract. The network traffic classification is used as a basis of network management related works, from security monitoring and Intrusion Detection Systems (IDS) to Quality of Service (QoS). Many works have been carried out on this theme that proposed various approaches. Most of the proposed approaches utilize predefined class label provided by an expert to perform the network traffic classification. Nevertheless, the difficulty to acquire consistent, adequate and up-to-date ground truth for classifying network flows effectively is still as serious problem. This work addresses such an issue via multi-view stacking to fuse the metadata from heterogeneous types of semi-supervised classifiers for accurate classification thru the following steps. Firstly, the original data is represented as multi-view using dimensionality reduction methods, with the aim to have strong discrimination capability. Secondly, integrate different semi-supervised learning algorithms from an ensemble learning perspective, to deliver a better stability and quality classification output. Finally, propose N-fold cross validation on metadata for training the meta-classifier, to produce the final class decision and predict the unlabelled traffic data. Experiment results on four existing network traffic datasets provide the best average accuracy mark is above 96% on all datasets and the best stability performance is up to 99.50%.

1. Introduction

Network classification is the task of classifying a set of network flows into classes based on two categories such as the used protocols (e.g. UDP, TCP) or applications (e.g. Web, Email, etc.). Traffic classification methods have been extensively developed lately as essential tools [1-5, 7] to support network management activities. In turn, the tools assist the Internet Service Provider (ISPs) from providing better QoS, to security control. Promising classification methods using statistical characteristics of Internet Protocol (IP) traffic flows and machine learning algorithms have been proposed and developed by the researchers and the networking industries over the past decade to address the limitations of the traditional network classification methods, including port-based methods and deep packet inspection methods [2, 6, 10, 12, 17]. However, accuracy and efficiency problems are still remain unsolved. The reasons are the proposed methods used force assignment where the classification of the unlabelled traffic flows should be compulsory assigned to fixed predefined classes of protocols or applications, and disregard to learn the appearance of new traffic patterns and emerging applications (e.g. zero-day attacks). Thus, researchers need an automatic procedure for accurate and efficient labelling and building up-to-date ground truth. Such an automatic procedure is an essential to train and test various network classification approaches based on the different machine learning algorithms, and also to replace the manual labelling procedure which is tedious and costly.

To do so, this paper proposes a novel semi-supervised approach, namely Semi-Stack. The proposed approach fuses the metadata from heterogeneous of ensemble semi-supervised classifiers as a training set along with N-fold cross validation approach to train the meta-classifier and obtaining the final class label. Experiments using four existing network traffic datasets [1, 9, 15, 16] verify the significant improvements of the proposed approach's classification capability.

The paper is organized as follows: Section 2 discussed the related work in of network traffic classification. Section 3 introduces our proposed Semi-Stack approach. Section 4 presents the experimental settings, and discussion of the results. Finally, the conclusion and outlines of future work are presented in Section 5.

2. Related Work

Machine Learning-based classification techniques are categorized into supervised and unsupervised learning.

- Supervised learning: traffic flow classes are given before the learning [2, 3, 8, 11]. The process of learning technique consists of two main phases: training phase and testing phase. The former phase is about learn and analyse the given training dataset, then generate a predictive classification model. Through the training process, the predictive model builds a capability to generate a prediction for any new instances by examining the characteristic values of unknown flows as based on prior training data. In the testing phase, the obtained model from the training phase is used to classify new unseen traffic flow instances. Supervised learning creates knowledge structures that support the task of classifying new instances into pre-defined classes [13]. Thus, supervised learning concentrates on modelling the relationships of input and output to produce knowledge presented as a decision tree, classification rules, a flowchart, etc. and will be used to classify a new instance.
- Unsupervised learning: is used to make inferences from datasets without labelled traffic flows. Cluster analysis is the most usual unsupervised learning technique used for investigative data analysis to discover unseen patterns. McGregor et al. in [8] and Rotsos et al. [14] are ones of the earliest works to employ unsupervised machine learning technique to model the underlying structure and distribution of network traffic flows. The authors use Expectation Maximization (EM) method and consider Transport Layer attributes as traffic flow features. Zander et al. [19], extend the works by introducing different EM algorithm named as Auto Class. The authors investigate the best set of attributes to use. Then, using Bayesian clustering technique combined with standard EM algorithm will guarantee the convergence of the combined method to a local maximum. Next, the Auto Class performs the EM searches iteratively, starting from pseudo-random points in parameter space to search the global maximum. Therefore, it works considerably superior to the original EM method. The proposed idea in [3] has shown that the use of Transport Layer attributes only, enables cluster analysis to classify Internet traffic flows. Furthermore, the use DBSCAN and K-Mean clustering methods to evaluate the predicting performance has been presented by Erman et al. [5]. The authors verified that both DBSCAN and K-Mean provide better accuracy and faster processing time compared to the Auto Class clustering method presented in [19]. Overall, the unsupervised learning techniques are not as good as supervised techniques. Therefore, this work attempts to explore the benefits of both learning techniques to improve essentially stability and accuracy of traffic flow classifiers.

3. The Proposed Semi-Stack Approach

The proposed Semi-Stack approach considers the advantages combination of semi-supervised and meta-learning algorithms as a basis in its development.

3.1. The proposed Semi-Stack approach overview

A huge labelled traffic dataset is necessary for developing accurate network traffic classifiers based on machine learning (ML) methods. However, current ground truth creation for network classifiers requires a fantastic cost and workload. This work proposes a novel semi-supervised (Semi-Stack) approach for proper training and evaluation of network classifiers that reduces tiresome manual processes of labelling and at the same time reaches the goal at realistic cost. Figure 1 depicts the grand architecture of the proposed Semi-Stack approach. The Semi-Stack approach is structured by three layers: (1) the projection of the data to achieve diversity among traffic data, (2) ensemble Semi-Stack to increase the class prediction accuracy and the learning process efficiency thru unlabelled data exploitation, and (3) meta-learning to robust the automatic labelling process by generating the final prediction for the unlabelled traffic data.

3.2. Multi-view layer

This layer increases the performance of the labelling process by giving multiple representations of the initial data.

This work assumes that not all of the interesting patterns can be take out from a single view. On the other hand, a representation may have different data structures.

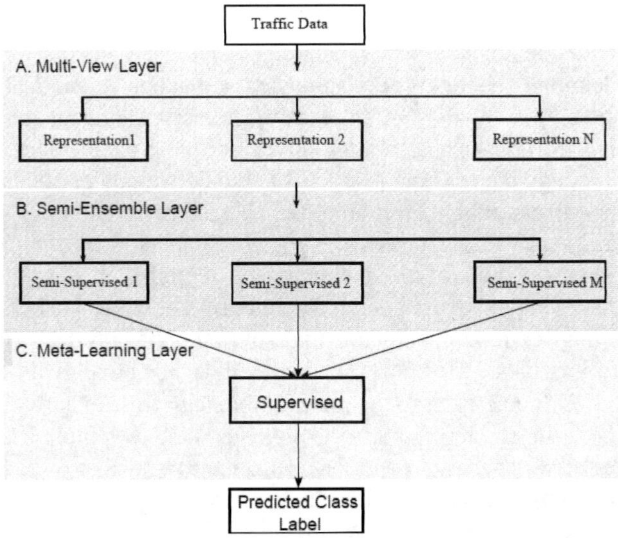

Figure 1. General Architecture of the Semi-Stack approach.

This layer process the network traffic data to reduce dimensionality and acts as representation methods. Three common famous methods are selected [20], they are:

- Random Projections (RP): projects the data into lower dimensional spaces using random projection matrices.
- Isomap: extends the multidimensional scaling metric and incorporates a weighted graph to estimates the geodesic manifold of intrinsic geometry of the nonlinear data.
- Kernel Principle Component Analysis (KPCA): generalizes the PCA. The PCA is a key statistical technique for extracting features and data modelling.

3.3. Semi-ensemble Layer

The proposed Semi-Ensemble Layer learns from different representations of the traffic flows generated previously by the former step (the multi-view layer), and then generates a pool of the predication values of the unlabelled data, referred to as meta-level data. Figure 2 shows a visualization of Semi-Ensemble Layer process.

IOP Conf. Series: Materials Science and Engineering **811** (2020) 012026 doi:10.1088/1757-899X/811/1/012026

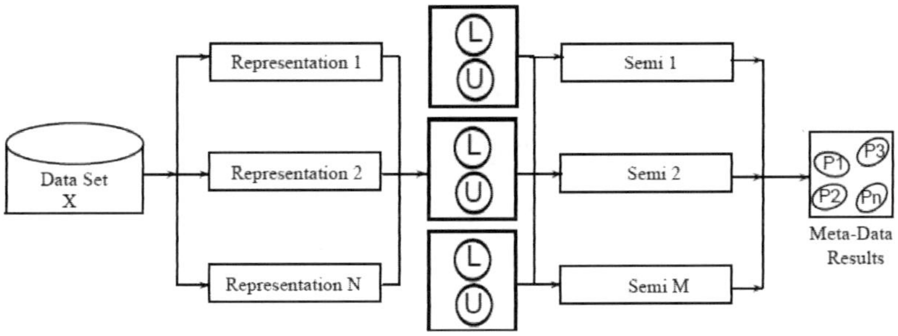

Figure 2. Overview of the Ensemble Semi-supervised Layer.

3.4. Meta-learning layer

In this work, the author uses a homogeneous base learner rather than a single base learning algorithm to decrease variance, bias and improve predictions in classifier model. A widely adopted approach to this objective is to apply meta-learning layer to benefit from its repetitive process and to compute descriptive models where the learning instances are not represented by original descriptive features, e.g. port number, IP address features, but by the output features (the meta-level data) generated previously from semi-ensemble layer. In particular, meta-learning layer has two steps. The first step is trained using the predictions of the semi-supervised classifiers (meta-level data) and the true class of the original traffic data. The second step generates the final prediction results by testing the meta-classifier model. Figure 3 shows the process of the meta-learning layer. Firstly, split the meta-level data into two sets: training set and testing set. The former set is used as initial labelled data to train the meta-classifier. The author uses the Decision Tree [3], an inherently interpretable algorithm. Once the training process is accomplished, the latter set, which comprises part of labelled data with unlabelled one, is applied on the Meta classifier model to obtain the final predictions. 10-fold cross validation is used to improve the quality of meta-classifier model by generating more meta-instances (traffic flows).

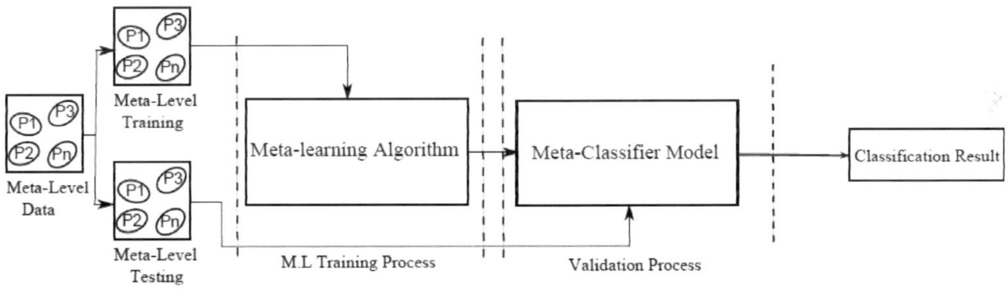

Figure 3. Overview of the Meta-Learning Layer.

3.5. Experimental procedure

Figure 4 shows the algorithm of experimental procedure of this work.

```
Algorithm 1: Experimental Procedure
Input:
M= 5;
SemiTech(ST)={Semi-Stack, PGM, ORTSC, BGCM, SemiBoost};
DATA= {DARPA, WTP, ··· , D_n};
Output:
PerfoMetrics={Accuracy, F-measure, Stability, Runtime};
foreach SemiTech_i ∈ [1, ST] do
    foreach D_i ∈ DATA do
        foreach times ∈ [1, M] do
            randomise instance-order for D_i;
            generate N bins from the randomised D_i;
            foreach fold ∈ [1, N] do
                Labelled_Data = bin[fold];
                Unlabelled_Data = D_i − Labelled_Data;
                Train'_Data = Labelled_Data + Subset from Unlabelled_Data;
                Test'_Data = Unlabelled_Data − Train'_Data;
                foreach ST_i ∈ SemiTech do
                    ST^i_Model = learning(Train'_Data, ST_i);
                    Results_i= testing(Test'_Data, ST^i_Model);
            OverallResult= OverallResult ∪ Results_i;
        PerfoMetrics=average (OverallResult);
```

Figure 4. Algorithm of experimental procedure.

4. Experimental Evaluation

4.1. Traffic datasets

Four publicly available traffic data sets are used for experimental evaluation of the proposed Semi-Stack approach, including Internet Traffic Data (ITD, defined in [11]); DARPA datasets [15]; wide dataset [1]; and ISP dataset [16].

This work focuses on TCP traffic flows. The advantage is it has the clear start-end information. Table 1 recapitulates attributes of each dataset, i.e.: the number of classes, the number of features, and the data allocation for instances in training and testing phases.

Table 1. Summary of dataset attributes.

Dataset	# of Classes	# of Features	Instances# (Training)	Instances# (Testing)
ITD	12	149	15750	5250
DARPA	2	41	7500	2500
wide	6	20	10030	3100
ISP	12	149	15750	5250

4.2. Experimental set up

The experiment uses the 10-fold cross-validation strategy to obtain robust results. The strategy is repeated 5 times to avoids bias toward the data ordering, which may degrade the performance of the Semi-Stack approach and the four comparison methods. All experiments were conducted on a computer with processor of 4-duo and Core(i7), 3.30 GHz Intel CPU, 8-GB RAM and 64-bit Windows system. The Semi-Stack approach is implemented using Java Platform and combined with Weka software [18].

4.3. Experimental results

Experiments are conducted to measure the performance criteria: overall accuracy, runtime, and stability.

1) Accuracy Performance: Table 2 indicates that all four baseline (comparison) methods achieve significant overall accuracy marks on all datasets (normal or attack). Specifically, Semi-Stack approach reaches the best average accuracy marks, by performing accuracy above 96% on all datasets. The rational is because the Semi-Stack approach is able to associate results of multiple labelling processes to gain accurate class label and remove unconfirmed results.

Table 2. Overall accuracy for each model and each dataset.

Dataset	Semi-Stack		SemTra [20]		PGM		BGCM		ORTSC	
	Mean	*St.D*	*Mean*	*St.D*	*Mean*	*St.D*	*Mean*	*St.D*	*Mean*	*St.D*
ITD	96.78	2.0	95.3	0.88	93.1	3.46	93.65	2.77	92.52	4.01
DARPA	97.08	0.34	95.8	0.25	93.24	3.37	93.13	3.61	93.88	2.55
wide	96.74	1.42	94.59	1.92	94.59	3.6	93.52	2.21	92.32	3.24
ISP	98.48	1.94	94.64	1.33	94.71	2.3	93.72	2.78	92.54	4.42

2) Runtime performance: The runtime measurement was repeated ten times for each method/approaches to give the average processing time with high confidence level. Table 3 shows the runtime measurement results. The ORTSC, PGM and BGCM are much faster than the Semi-Stack approach. The Semi-Stack is slightly faster than SemTra. Figure 5 shows that ORTSC and BGCM methods tend to scale better than PGM and Semi-Stack.

Table 3. Runtime performance in labelling process.

Dataset	Semi-Stack	SemTra	PGM	BGCM	ORTSC
ITD	64.85	79.75	1.24	1.87	0.39
DARPA	54.79	67.38	0.67	1.26	0.21
wide	30.48	37.49	0.86	1.77	0.28
ISP	64.11	78.85	1.25	1.90	0.39

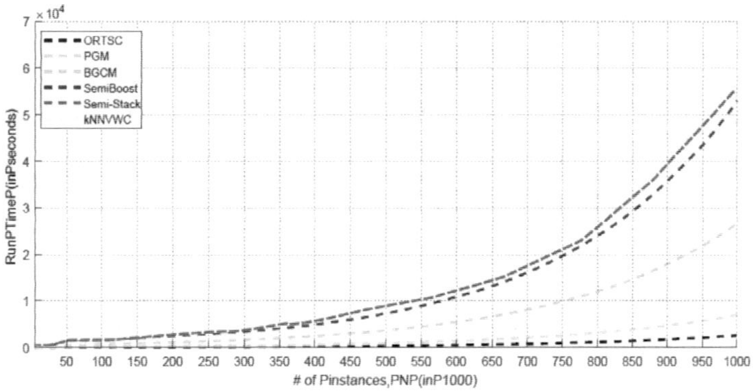

Figure 5. Experimental results on scalability.

3) Stability: The Semi-Stack approach also aims to increase the stability in producing accurate class labels for unlabeled flows for all possible runs. The number of trials was set to 100 in order to obtain higher confidence level of the results. The experiments use 30% of data in each class for testing and the remaining data was randomly selected to form testing. Table 4 exhibits the Semi-Stack approach achieved the best stability performance compared to the baseline methods with margin between 2.38% and 12.3%. The rational is that the Semi-Stack approach combines the advantages of several machine learning methods, which results in robustness of assigning class labels with low variance and bias of

the predictive model. Notably, the lowest stability belongs to the ORTSC method on all traffic data, this due to the sensitive and random initialization of K-means. The BGCM method was the second worse stability result, especially on ITD and DARPA datasets.

Table 4. Stability performance comparison.

Dataset	Semi-Stack	SemTra	PGM	BGCM	ORTSC
ITD	**97.44**	97.35	94.08	90.98	90.82
DARPA	**99.50**	98.98	93.77	91.48	91.55
Wide	98.11	**98.71**	94.90	92.73	92.93
ISP	**98.44**	97.91	94.97	91.55	90.82

5. Conclusion

This paper proposed a novel semi-supervised traffic classification approach, namely Semi-Stack. The proposed Semi-Stack approach combines the advantages of several machine learning methods, including: feature projection, ensemble learning and stacking, to enhance and improve the process of network traffic classification with low variance and bias of the predictive classifier model. Experiment results on four existing network traffic datasets verified the significant improvements of the proposed approach's classification capability. Even the Semi-Stack approach has the second worse runtime, however it reached the best average accuracy marks, by performing accuracy above 96% on all datasets. Besides, the Semi-Stack approach achieved the best stability performance up to 99.50% on DARPA dataset compared to the baseline methods with margin between 2.38% and 12.3%.

The author considers the use of parallel computing, i.e. Graphics Processing Units (GPUs) or multi-core processing (CPU) to minimize the execution time as one of future works.

References

[1] Mawi working group traffic archive. Available: http://mawi.wide.ad.jp/mawi/.

[2] Bernaille L, Teixeira R, Akodkenou I, Soule A, Salamatian K 2006 Traffic classification on the fly, *ACM SIGCOMM Comput. Communication Rev.* **36(2)** 23-6.

[3] Erman J, Mahanti A, Arlitt M 2006 Qrp05-4: Internet traffic identification using machine learning, *IEEE Global Telecommunications Conf.* 1-6.

[4] Erman J, Arlitt M, Mahanti A 2006 Traffic classification using clustering algorithms, *SIGCOMM workshop on Mining network data,* 281-6.

[5] Erman J, Mahanti A, Arlitt M, Cohen I, Williamson C 2007 Offline/real-time traffic classification using semi-supervised learning, *Perform. Evaluat.* **64(9)** 1194-213.

[6] Karagiannis A, Broido M, Faloutsos, et al. 2004 Transport layer identification of p2p traffic, *4th ACM-SIGCOMM conf. Internet measure.* 121-34.

[7] Karagiannis T, Papagiannaki K, Faloutsos M 2005 Blinc: multilevel traffic classification in the dark. *ACM-SIGCOMM Comput. Communication Rev.* **35** 229-40.

[8] McGregor A, Hall M, Lorier P, Brunskill J 2004 Flow clustering using machine learning techniques. *Passive Act. Network Measure., Springer,* 205-14.

[9] Moore A, Hall J, Kreibich C, Harris E, Pratt I 2003 Architecture of a network monitor. *Passive & Active Measurement Workshop2003 (PAM2003). Citeseer.*

[10] Moore A.W, Papagiannaki K 2005 Toward the accurate identification of network applications. *Passive Act. Network Measure. Springer,* 41-54.

[11] Moore A.W, Zuev D 2005 Internet traffic classification using Bayesian analysis techniques. *ACM-SIGMETRICS Perform. Evaluation Rev.* **33** 50-60.

[12] Nguyen T T, Armitage G A 2008 survey of techniques for internet traffic classification using machine learning. *IEEE Communications Surveys Tutor.* **10(4)** 56-76.

[13] Reich Y, Fenves S J 1991 The formation and use of abstract concepts in design concept formation: knowledge and experience in unsupervised learning. Citeseer.

[14] Rotsos C, Van Gael J, Moore A W, Ghahramani Z. 2010 Probabilistic graphical models for semi-supervised traffic classification. *Proceedings of the 6th International Wireless Communications and Mobile Computing Conference,* 752-7.

[15] Stolfo S J, Fan W, Lee W, Prodromidis A, Chan P K 2000 Cost-based modelling for fraud and intrusion detection: Results from the jam project. *IEEE-DARPA Information Survivability Conference and Exposition, (DISCEX'00).* **2** 130-44.

[16] Wang Y, Xiang Y, Zhang J, Yu S 2011 A novel semi-supervised approach for network traffic clustering, *5th IEEE Int. Conf. Network Syst. Security (NSS),* 169-75.

[17] Wang Y, Xiang Y, Zhang J, Zhou W, Wei G, Yang L T 2013 Internet traffic classification using constrained clustering. *IEEE Transact. Parallel Distributed Syst.* p.1.

[18] Witten I H, Frank E, Trigg L E, Hall M A, Holmes G, Cunningham S J 1999 Weka: Practical machine learning tools and techniques with java implementations.

[19] Zander S, Nguyen T, Armitage G 2005 Automated traffic classification and application identification using machine learning. *30th Anniversary IEEE Conf. Local Comput. Networks,* 250-7.

[20] Fahad A, Almalawi A, Tari Z, Alharthi K, Al Qahtani F S, Cheriete M 2019 SemTra: A semi-supervised approach for traffic flow labeling with minimal human efforts, Pattern Recognition, *Elsevier J.* **91** 1-12.

IOP Conf. Series: Materials Science and Engineering **811** (2020) 012027 doi:10.1088/1757-899X/811/1/012027

Improvement of Heat Sink Performance by Using Graphene Nanosheets Coated by Chemical Spray Method

A. A. Jaafar[1], S. A. W. Al-abassi[2*], H. A. Alhattab[2], M. A. Albaghdad[2], A. A. Mosa[2], H. K. Al-Musawi[2], L. M. Abo Gneem[2]

[1]Department of Computer Technical Engineering, College of Technical Engineering, The Islamic University, Najaf, Iraq
[2]Electronic and Communications Engineering, Department, University of Kufa, Najaf, Iraq

salam.waheed@uokufa.edu.iq

Abstract. Graphene nanosheets were fabricated on a metal (Al) substrate by using a chemical spray method. Annealing process has been applied to reducing the concentration of carbon and to increase the sheet hardness. Surface morphology measured by metallurgical microscope Device Corporation, X 1600. Thermal camera was used to measure the performance of coated heat sink. The temperature was measured from the two sides, one from heat sink side, and the other one is from power transistor side. The results showed that the temperature of the transistor was decreased by about 4.3 °C in the case of heat sink coating with 80 μm layer thickness while decreased by 6.9 °C in the case of the heat sink coating with 120 μm layer thickness.

1. Introduction

The rapid development within the field of semiconductor devices have resulted in a very vital increase in number of integrated chips per space units with needing to devices with very high-speed response, and the inception of imbedded systems, all that causing an increasing in generation of heat flux at chip level [1, 2].

For reaching this purpose many ideas have been proposed such as utilizing of micro heat sink [3-5], or coating of heat sink by high conductivity materials [6].

Some studies have been proposed multi approach to address significantly enhance the heat dissipation potential of conventional aluminium (Al) heat sinks by mechanically coated graphene nanosheets [7]. The study had significant enhancement of metal heat dissipation from mechanically exfoliated graphene nanosheets through thermal radiation effect. It is found that with the increase in coating coverage from 0 to 100%, the steady- state temperature is decreased by 7°C at a heat flux of 1.8 Wcm-1[8].

Thermally conductive graphene nanosheets (GNs) adorned with Cu nanoparticles or Nano clusters are manufactured through an efficient microwave-assisted (MA) by thermal reduction. As a result, this study proves that the robust designing of Cu/GN heterogeneous film offers an integrated skeleton for thermal diffusivity, benefiting practical applications such as Si-based integrated chip. In addition, where explored the possibility of integrating 1D CNT and 2D graphene into a 3D covalently bonded structure, i.e. a graphene-CNT hybrid material for thermal management application. It is found graphene-CNT hybrid material might be suitable to be used as both TIMs and heat spreader simultaneously. The graphene-CNT hybrid material was later investigated morphologically and thermally to observe its heat dissipation capability [9].

In this study, the heat transport of heat sink was improved by coating heat sink with graphene nanosheets fabricated by mechanical method. It is expected that the graphene nanosheets will improve the performance of heat sink because of high thermal conductivity of graphene.

2. Materials and Methods

2.1. Manufacturing of homogeneous mixture of ethanol and graphene

50 ml of ethanol (C2H6O) with density of 0.7893 g/cm3 [10], 0.5% natural graphene powders with density of 2.267 g/cm3 [11], to decide the percentage of graphene in solution the following equation (equation 1) has been used [12].

$$W_g = \frac{\frac{W_{ethnol}}{\rho_{ethnol}} \times \rho_{graphene} \times rate}{1 - rate} \quad (1)$$

Where:
Wg= weight of the graphene
Wethanol = amount of ethanol
ρethanol =density of ethanol
ρgraphene = density of graphene
Rate= amount of graphene to the amount of ethanol = 0.05

Amount of 60 μg of graphene was used and measured by a high sensitivity balance. Graphene and ethanol were mixed by using hotplate and stirrer for 20 minutes. For more homogeneity of solution, the mixture was putted in Ultrasonic mixer for 20 minutes with power-on time of 2S and power-off time of 2S. Obtained solution shown in figure 1 and can be used for chemical spray process.

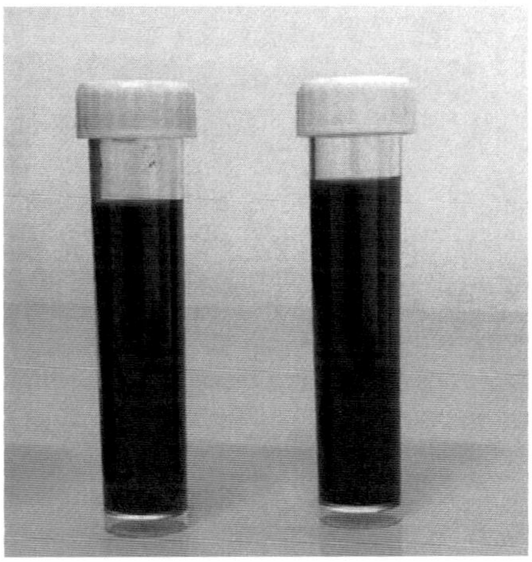

Figure 1. Prepare Liquid of graphene.

2.2. Coating system

Before coating process heat sink was cleaned by MTI ultrasonic cleaner device for 20 minutes.

Coating system is shown in figure 2 is a full auto controlled system designed by our lab to have a precious controlling in spray time and the distance between substrate and nozzle, additionally the substrate temperature also controlled by a heater. These features give to fabricated graphene nanosheets a homogenous appearance as shown in figure 3. Fabrication of nanosheets was under substrate temperature of 200 °C. Put the amount of solution prepared in the spray machine and coating the samples with a various spraying time to get a various thickness. Sample (a) was spired for 20 times each one for 2 second and between one and another 2 second, while sample (b) spired for 30 times each one for 2 second and between one and another 2 second. After this process, the samples were annealed under 120oC. Thickness was measured by use optical thin film measurement device, Lambda scientific Pty ltd. (LIMF-10). Surface morphology measured by metallurgical microscope Device Corporation, X 1600.

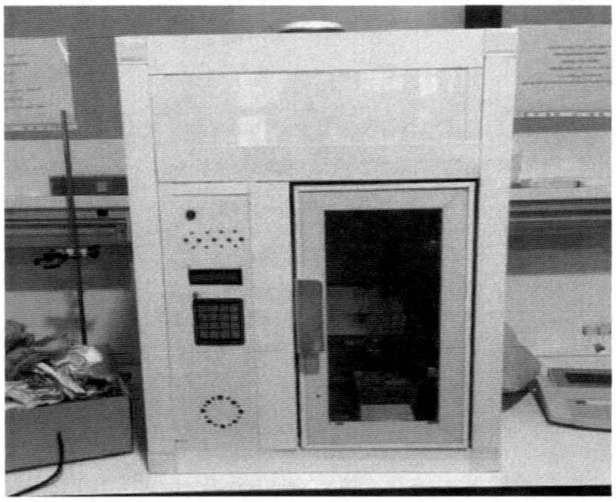

Figure 2. Chemical spray system.

Figure 3. Heat sinks after coating. From the left: uncoated, coated with 80 um sample (a) and coated with 120 um sample (b).

2.3. Testing of samples

To test the heat sinks that were fabricated power transistor type LM 7805 has been used. Three power transistors with three heat sinks were connected to electrical circuit. at same time the power switched on. Thermal camera FLIR ONE where used to measure the temperature before and after power on. One of these circuits use a heat sink that is not coating while the other circuits, one of them use heat sink coating thickness 80μm and the other use heat sink coating thickness 120μm. the measurement of heat sinks temperature was after applying a 2.6 watt for each circuit after 10 minutes, thermal camera was installed in front of it and take a picture to transistors from both sides.

Temperature of room was measured before measurement process which was 28°C.

3. Results and Discussion

Figure 4 shows the surface morphology of fabricated samples which indicating a rough surface in sample (a) and (b). This figure showing an island shape graphene surface due to substrate heating and annealing process which was make a main role to reducing of concentration of Carbone atoms on surface and shaped a graphene island [13].

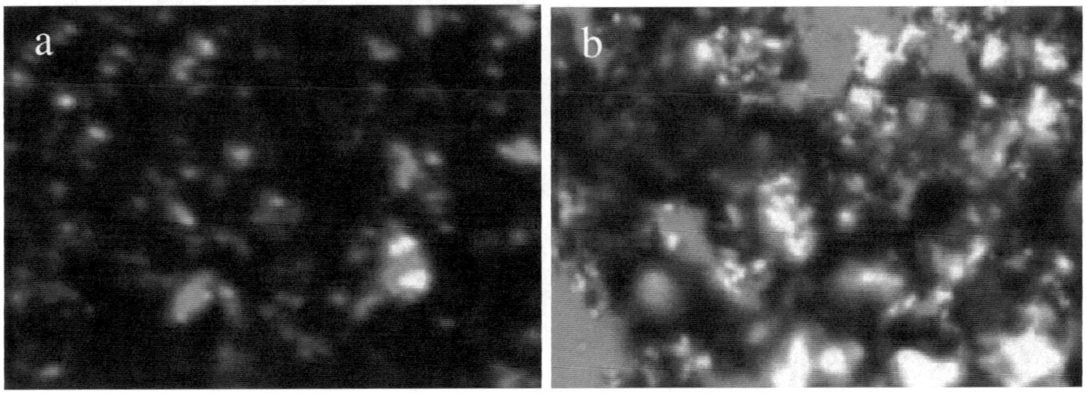

Figure 4. Graphene nanosheets with thickness of a) 80 μm, b) 120 μm.

Figure 5 shows the thermal camera picture for both samples and uncoated sample too. The temperature of the coated part, which may be almost double compared to the uncoated sample.

The temperature of the transistor connected with uncoated heat sink equal 94.3°C, while the transistor connected with heat sink coated thickness 80 μm (sample (a)) have been kept at a temperature equal 90°C, and the other transistor, which is connected with heat sink coated thickness 120 μm (sample(b)) have temperature equal 87.4°C.

Figure 5. Temperature distribution on the transistor body.
From the right: uncoated, coated with 80 um, and coated with 120 um.

From the results, the improvement in heat sinks that were coated by Graphene, occurred due to two reasons, first one is the roughness and surface morphology, which was increased causing increasing in surface area which came from island shape make the surface area larger than uncoated heat sink [14].

The second reason of this the improvement is due to a high conductivity of graphene [15]. Generally, the thermal energy is dissipated through convection and radiation, as shown in the equations below (Rea and West 1976) [16].

$$Q\_convection=h\times A\times(T-T\_room) \quad (2)$$

$$Q\_rad=\in\times\delta\times A\times(T\text{^}4-T\_room\text{^}4) \quad (3)$$

h = the Planck's constant (6.62607004 × 10-34 m2 kg / s).

δ = Stefan– Boltzmann constant (5.67 × 10-8 W m-2 K-4). ∈ the emissivity of bare and coated samples.

A = the surface area. T represents the steady-state temperature of bare and coated samples.

Troom =signifies the ambient temperature.

The obtained results shown that the temperature of the transistor decreased by about 4.3 °C degrees in the case of heat sink coating with 80μm layer thickness, while decreased by about 6.9 °C in the case of heat sink coating with 120μm layer thickness.

4. Conclusions

In this study, improved system to heat transport enchantment of power transistor using coating of graphene nanosheets by using a chemical spray method. This work abled to fabricate graphene nanosheets by chemical spray method on a metal substrate (AL). The improvement in coated samples came from increasing of surface area which was a result of island shape of coated graphene nanosheets, beside the high conductivity of graphene. Low cost and simple method showed a significant improvement in performance of heat sink.

References

[1] Murshed S S, De Castro C N 2017 A critical review of traditional and emerging techniques and fluids for electronics cooling *Renew. Sustain. Energy Rev.* **78** 821-33.

[2] Ramalingam V, Anandan S S 2008 Thermal management of electronics: A review of literature *Thermal Sci.* **12(2)** 5-26.

[3] Naqiuddin N H, Saw L H, Yew M C, Yusof F, Poon H M, Cai Z, San Thiam H 2018 Numerical investigation for optimizing segmented micro-channel heat sink by Taguchi-Grey Method *Appl. Energy* **222** 437-50.

[4] Hadad Y, Ramakrishnan B, Pejman R, Rangarajan S, Chiarot P R, Pattamatta A, Sammakia B 2019 Three-objective shape optimization and parametric study of a micro-channel heat sink with discrete non-uniform heat flux boundary conditions *Appl. Therm. Eng.* **150** 720-37.

[5] Alhattab H A, Albaghdadi M A, Hashim R S, Ali A H 2016 Design of micro heat sink for power transistor by using CFD *Al-Sadeq International Conference on Multidisciplinary in IT and Communication Science and Applications (AIC-MITCSA)* 1-5 IEEE.

[6] Bistritzer R, MacDonald A H 2009 Electronic cooling in graphene *Phys. Rev. Lett.* **102(20)** 206410.

[7] Hu J, Xu J, Zhu C, Li Q, Ullah Z, Liu F, Li W, Guo Y, Zhao X, Liu L 2017 Significant enhancement of metal heat dissipation from mechanically exfoliated graphene nanosheets through thermal radiation effect *AIP Adv.* **7(5)** 055315.

[8] Hsieh C T, Chen Y F, Lee C E, Chiang Y M, Hsieh K Y, Wu H S 2017 Heat transport enhancement of heat sinks using Cu-coated graphene composites *Mater. Chem. Phys.* **197** 105-12.

[9] Yan Z, Liu G, Khan J M, Balandin A A 2012 Graphene quilts for thermal management of high-power GaN transistors *Nat. Commu.* **3** 827.

[10] Goldemberg J 2007 Ethanol for a sustainable energy future *Sci.* **315(5813)** 808-10.

[11] Geim A K 2009 Graphene: status and prospects *Sci.* **324(5934)** 1530-4.

[12] Dhiaa A H 2016 Investigation of thermal transport properties enhancement of nanoparticles suspensions in the application of nanofluids *Ph.D. thesis, University of Baghdad.*

[13] McCarty K F, Feibelman P J, Loginova E, Bartelt N C 2009 Kinetics and thermodynamics of carbon segregation and graphene growth on Ru (0 0 0 1) *Carbon* **47(7)** 1806-13.

[14] Khan W A, Culham J R, Yovanovich M M 2006 The role of fin geometry in heat sink performance *J. Electr. Packag.* **128(4)** 324-30.

[15] Aliofkhazraei M (Ed) 2013 Advances in Graphene Science *BoD–Books on Demand.*

[16] Rea S, West S 1976 Thermal radiation from finned heat sinks *IEEE Transactions on Parts, Hybrids, and Packaging* **12(2)** 115-7.

Evaluation of Repair Quality for Concrete Structure by Non-Destructive Technique

Yung-Chiang Lin[1], Keng-Tsang Hsu[2], Chia-Chi Cheng[3], Yi-ching Lin[4]

[1]Postdoctoral Research Fellow, Chaoyang University of Technology, Taiwan, R.O.C.
[2]Assistant Professor, Chaoyang University of Technology, Taiwan, R.O.C.
[3]Professor, Chaoyang University of Technology, Taiwan, R.O.C.
[4]Professor, National Chung Hsing University, Taiwan, R.O.C.

yungchaing.lin@gmail.com

Abstract. The objective of the proposed research project is to develop a non-destructive test technique based on stress wave for quantitative evaluation of the repairing status of concrete. Concrete structures frequently exist various types of cracks due to an inferior environment and excessive loading. There are several crack repairing techniques used in concrete construction. Epoxy-injection is popular maintenance method in crack repairing of the damaged concrete structure. In this paper, a technique based on the stress wave was used to evaluate the repair quality of concrete crack. Numerical simulation is also under research and results show that the stress wave propagation is capable of identifying the infill of epoxy-injection in the crack of repaired structures. The new technique under development has a high potential for evaluation of the quality of the repaired cracks.

1. Introduction

Concrete structures frequently exist various types of cracks due to an inferior environment and excessive loading. To assure the safety and the carrying capacity, a structural reinforcement work of it is requested. The cracks of concrete structure are an important indicator for the structure health diagnosis. The repairing method which injects Epoxy resin into the crack of concrete members is relatively simple and cost-effective, increases the strength of existing concrete members, blocks moisture penetration of reinforced concrete members and reduces the chance of corrosion.

2. Time-of-Flight Diffraction Technique

To measure the depth of surface-opening cracks in concrete [1], two receivers are used and located on the opposite sides of the crack. Figure 1 shows the instrument arrangement. The first receiver is located on the impact side and has a distance H0 to the impact point. The second receiver is located on the opposite side of the crack. Figure (b) and (c) show typical waveforms recorded at the first and second receivers, respectively. After impact, the R-wave arrival at the first receiver will trigger the monitoring system. If the arrival time, t_1, and the speed, C_R, of R-wave are known, the time of impact initiation can be calculated as (t_1-H_0/C_R). The value of t_1 is negative because a pre-trigger mode is used in recording signals to avoid losing important information. The region behind the crack cannot be penetrated by waves generated by impact until diffraction at the crack tip occurs. Thus, the second receiver initially responds to the arrival of the diffracted P-wave. The arrival time, t_2, of the diffracted P-wave can be obtained by identifying the initial disturbance in the waveform recorded at the second receiver. Thus, the shortest travel time for P-wave from the impact point to the second receiver can be obtained as equation (1). Knowing the P-wave speed, C_P, in concrete, the shortest travel length of P-wave from the impact point to the second receiver can be calculated as $C_P \times \Delta_t$. If the distances from the crack to the impact point and the second receiver are H_1 and H_2 respectively, the depth, d, of the vertical surface-opening crack can be determined by the equation (2)

$$\Delta_t = t_2 - (t_1 - H_0/C_R) = t_2 - t_1 + H_0/C_R \tag{1}$$

$$\sqrt{\left[\frac{(C_p \times \Delta_t)^2 + {H_1}^2 - {H_2}^2}{2 \times C_p \times \Delta_t}\right]^2 - {H_1}^2} \tag{2}$$

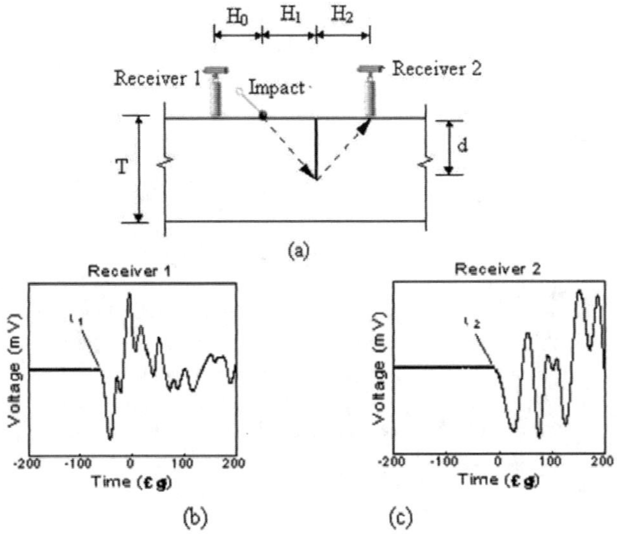

Figure 1. Test scheme: (a) Instrument arrangement; (b) Waveform 1; (c) Waveform 2.

3. Numerical Simulation with Different Partly Filled Scenarios of Repaired Condition

Numerical simulations of the stress wave responses of crack were carried out using the explicit finite-element program LS-DYNA. All cases are simulated using 2-D axial-symmetric, 4-node elements, and isotropic models. Concrete is modelled as an isotropic material with a Poisson's ratio 0.2 and a density 2300kg/m3. A Young's modulus (E) of 33.1 GPa is chosen to provide a P-wave speed of 4000 m/s [1], the material parameters of epoxy resin and air used in the numerical simulation are also shown in table 1. The time-history of nodal vertical displacements were recorded every 1.334 μs, and 1024 data were recorded in each simulation.

Table 1. Material parameters of the numerical simulation[1].

Material	Wave speed C_P, (m/s)	Density (kg/m^3)	Elastic modulus (GPa)	Acoustic Impedance Z, ($C_P \times \rho$)
Air	340	1.2	-	408
Epoxy Resin	2400	1141	5.602	2.73E+6
Concrete	4000	2300	33.10	9.2E+6

3.1. Numerical simulation of different injection condition of crack repairing

This section examines the stress wave propagation after the epoxy resin material filled in the crack with different partly filled scenarios. In the numerical simulation, the depth of crack is setting to 10 cm and the contact time of force-time function is 20μs in all cases. The variables of different epoxy resin injection scenarios in numerical simulations are shown in table 2 and figure 2 For all the models, the epoxy is filled from the top surface and the air pack of the crack is underneath the epoxy.

Table 2. Variables of different epoxy resin injection scenarios.

Code	Length of Epoxy Resin (cm)	Length of Air (cm)	Remark
E0 cm_A10 cm	0	10	Non- epoxy injection
E2 cm_A8 cm	2	8	Partial epoxy injection
E4cm_A6 cm	4	6	Partial epoxy injection
E6 cm_A4 cm	6	4	Partial epoxy injection
E8 cm_A2 cm	8	2	Partial epoxy injection
E10 cm_A0 cm	10	0	Fully epoxy injection
E0 cm_C10 cm	0	-	Non-crack case

The results of time-history of nodal vertical displacements with the different condition of epoxy resin injection are shown in Figure 2. The amplitude of R_{max} in time-history of nodal vertical displacement waveform is used as the parameter for investigation as shown in Figure 3. The time-histories of nodal vertical displacements are varied with different scenarios of epoxy resin injection. Table 3 summarizes the maximum displacement of the waveform (R_{max}) for all the cases. The maximum value of nodal vertical displacements (R_{max}) is obtained from the non-crack case. The R_{max} of the non-crack case (E0 cm_C10 cm) from table 3 can be compared with the fully epoxy injected case (E10 cm_A0 cm), the value of R_{max} attenuation was 11.4 %. In addition, the maximum displacement of the R-wave decreases with the decrease of the length of epoxy resin injection.

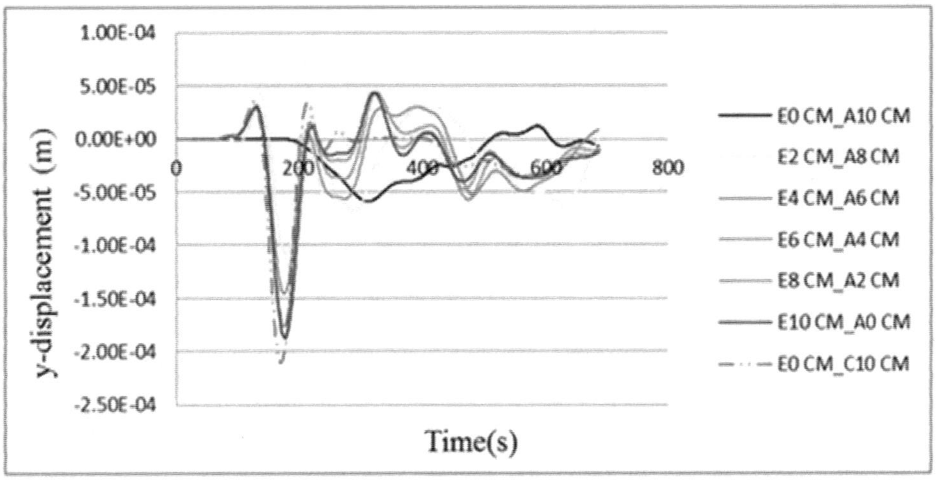

Figure 2. Time-history of nodal vertical displacements with different epoxy resin injection condition

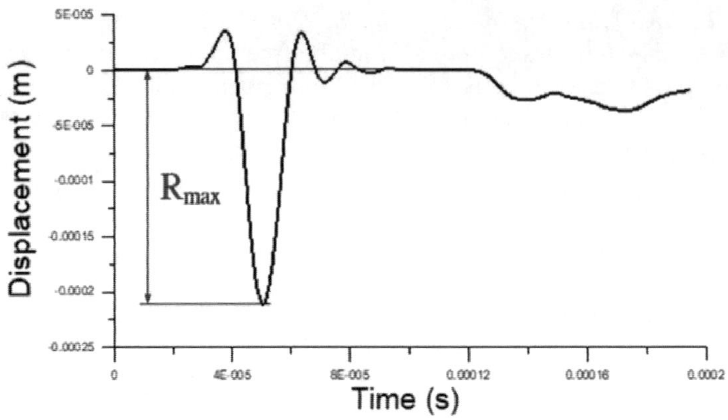

Figure 3. The length of R_{max} in time-history of nodal vertical displacement waveform.

Table 3. Maximum displacement of waveform with different epoxy resin injection condition

..Code	The maximum displacement of a waveform (R_{max}, m)	*Loss percentage of R_{max} (based on E0 cm_C10 cm) (%)	Remark
E0 cm_A10 cm	-0.59E-4	72.0	Non- epoxy injection
E2 cm_A8 cm	-1.09E-4	48.3	Partial epoxy injection
E4cm_A6 cm	-1.46E-4	30.8	Partial epoxy injection
E6 cm_A4 cm	-1.76E-4	16.6	Partial epoxy injection
E8 cm_A2 cm	-1.86E-4	11.8	Partial epoxy injection
E10 cm_A0 cm	-1.87E-4	11.4	Fully epoxy injection
E0 cm_C10 cm	-2.11E-4	0.0	Non-crack case

3.2. Experimental study on specimen

An experimental study is carried out on specimen with different injection condition of the crack in the laboratory. Laboratory concrete specimens containing vertical surface-opening cracks with 11 cm depth were used in the studies. As shown in figure 4, pictures were represented specimens with fully epoxy injection and non-epoxy injection conditions. Transient stress waves are generated by the impact of a spherical steel ball and captured by the transducer. Surface displacement waveform with fully epoxy resin injection of the concrete specimen was shown as figure 5. Surface displacement waveform with non-epoxy resin injection case was shown as figure 6. Compared with the results of numerical simulation, the results of specimens in the laboratory are very similar.

4. Conclusion

Numerical simulation results show that for surface opening crack partially filled by epoxy resin, the maximum displacement of the R-wave can be served as a parameter to evaluate the different injection condition of crack. The maximum displacement of the R-wave decreases with the decrease of the length of epoxy resin in the crack and the same with the results of specimens in the laboratory. The technique has a great potential for evaluation of the repair quality of concrete structural cracks injected by epoxy resin.

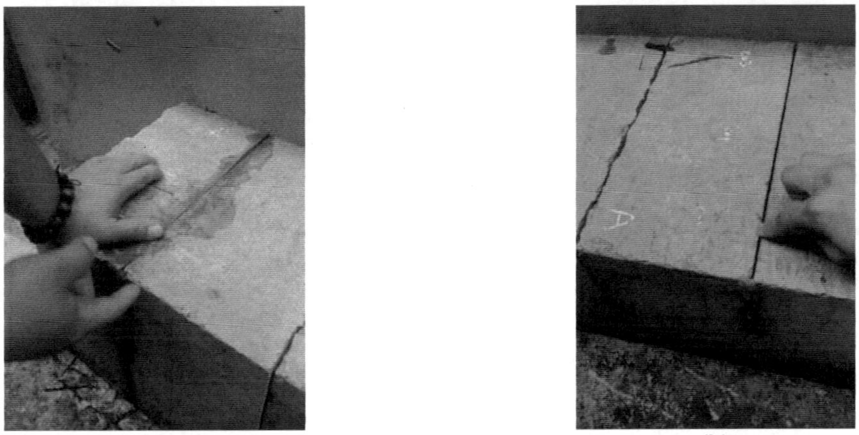

(a) (b)

Figure 4. These two figures have been placed side-by-side to save space. Justify the caption. Pictures of different epoxy resin injection scenarios of specimens (a) Specimen with fully epoxy injection crack ; (b) Specimen with non-epoxy injection crack.

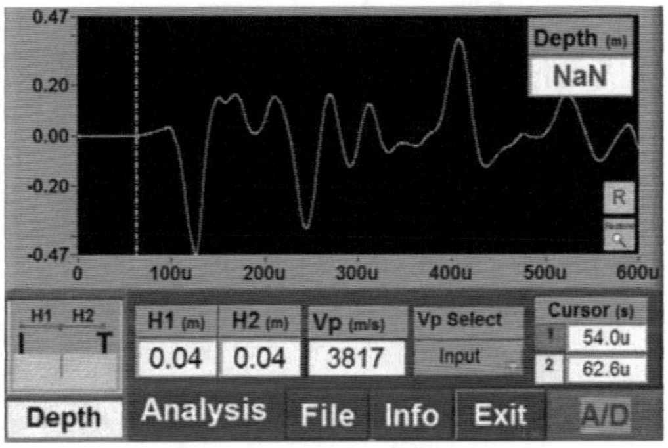

Figure 5. Surface displacement waveform with fully epoxy resin injection of concrete specimen.

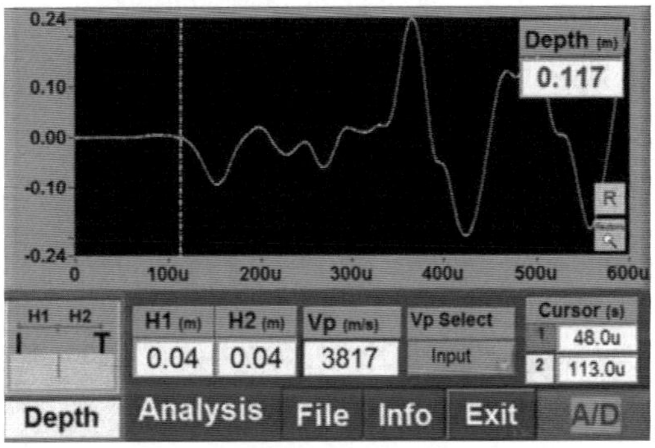

Figure 6. Surface displacement waveform with non-epoxy resin injection of concrete specimen.

References

[1] Lin Y, Liou T, Tsai W H 1999 Determining crack depth and measurement errors using time-of-flight diffraction techniques *Mater. J.* **96(2)** 190-5.

[2] Hsu K T, Cheng C C, Chiang C H, Wang H H 2017 Use of stress wave to evaluate the repair quality of concrete crack *MATEC Web Conf.* **128** 02019 EDP Science.

[3] Zoidis N, Tatsis E, Vlachopoulos C, Gotzamanis A, Clausen J S, Aggelis D G, Matikas T E 2013 Inspection, evaluation and repair monitoring of cracked concrete floor using NDT methods *Constr. Build. Mater.* **48** 1302-8.

[4] Cheng C C, Pei K C, Wu J H 2006 Development of imaging techniques for evaluating the RC plate containing epoxy-repaired delamination *Key Eng. Mater.* **321** 348-51.

[5] Hsu K T, Cheng C C, Lin Y 2008 Use Impact-echo method to evaluate bond of reinforced concrete subjected to early-age vibration *J. Solid Mech. Mater. Eng.* **2(12)** 1528-38.

IOP Conf. Series: Materials Science and Engineering **811** (2020) 012029 doi:10.1088/1757-899X/811/1/012029

Modeling of a Parabolic Trough Using Two Heat Transfer Fluids and an Economic Estimation in the Moroccan Dairy Industry

Meryem EL Badaoui[1], Abdelatif Touzani[1]

[1]Applied Thermodynamics and Solid Combustibles (ATSC), Mohammedia School of Engineers (EMI), Rabat-Morocco

meryemelbadaoui@research.emi.ac.ma, atouzanikia@gmail.com

Abstract. The aim of this article is modeling the parabolic trough (CSP) based on equations of the thermal balances of the different components taking into account all the different modes of heat transfer, as well as a comparison of the temperatures at the outputs in using two heat transfer fluids namely water and synthetic oil. The equations were solved using the finite difference method, and this developed model is validated by the results of the literature. The results obtained show the temperature at the outlet of the fluid, the absorber and the glass as well as the effects of flow, speed and irradiation on the temperature of the fluid, then an economic estimate of the energy consumption at the end.

1. Introduction

During previous work based on energy consumption in Morocco, it has been concluded that the agri-food industry precisely the dairy industry is among the energy-intensive in Morocco. During the production of dairy products, the el-jadida dairy plant requires a huge energy consumption of 247.8 kwh/t of electricity and 209.7 kwh/t of steam [1]. The figure 1 shows the global steam network.

In order to produce this steam, we resort to the use of oil boilers, and following that Morocco has a large solar deposit with more than 3000 h / year of sunshine or irradiation of about 5 kWh/m^2/day as shown in figure 2 below [2], and in the context of Moroccan strategies for the development of renewable energy and the reduction of the energy bill from fossil sources it is within this frame-work that we propose use solar energy including parabolic solar collector technology for direct steam generation.

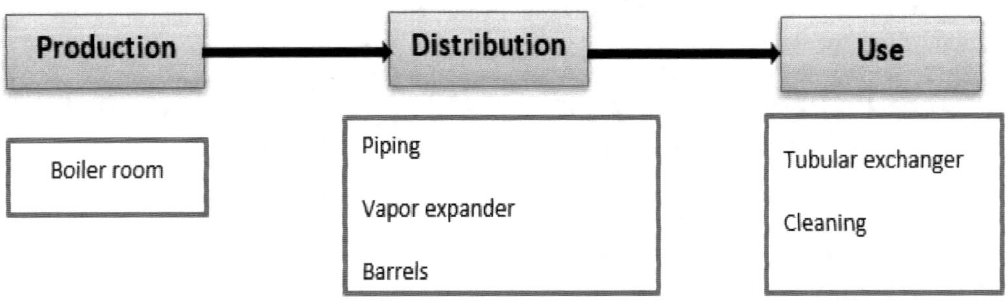

Figure 1. Global network of steam.

IOP Conf. Series: Materials Science and Engineering **811** (2020) 012029 doi:10.1088/1757-899X/811/1/012029

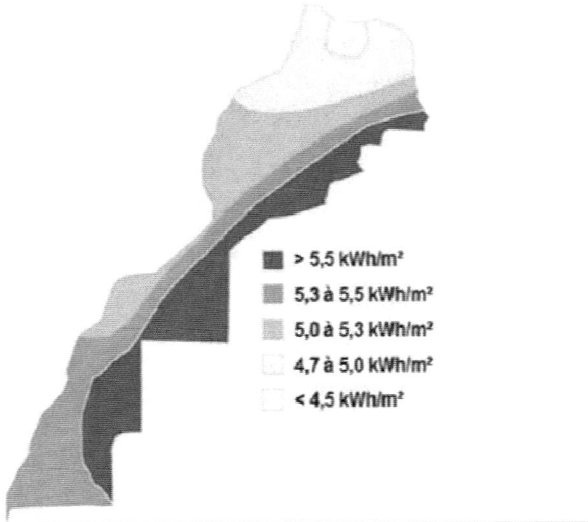

Figure 2. Solar deposit in Morocco.

2. Parabolic Trough

Parabolic trough technology has proven to be the most mature and lowest cost solar thermal technology available today. The parabolic trough consisting of a mirror and a receiver this consists in concentrating the solar rays by orienting the mirror in order to heat a fluid at a high temperature and supply it with an industrial process. The following figure shows the components of a parabolic trough [3].

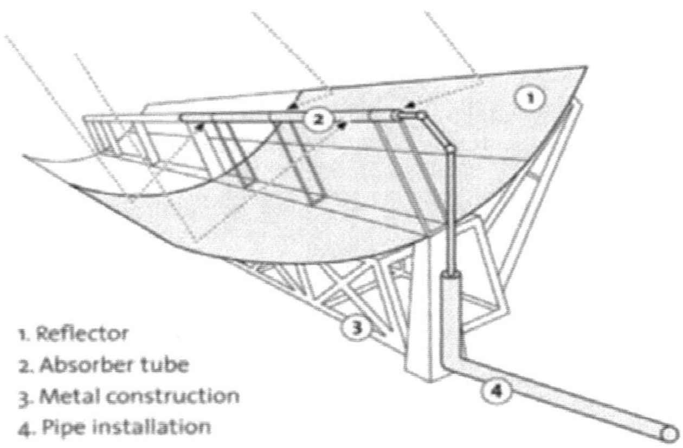

Figure 3. Components of a parabolic trough.

3. Modelisation of a Parabolic Trough

Among the methods common to modeling, we re-run the method of determining the temperature of each element of the trough by a monodimensional discretization. Taking into account all the heat transfer modes, the equations of the energy balance of the trough is as follows [4, 5].

Heat quantity of heat transfer fluid in Watt: $Q_f(x,t) = \rho_f * C_p * D_f * T_f(x,t)$

Quantity of heat absorbed in Watt: $Q_{ab}(t) = \rho * \alpha * \tau * S * G(t)$

Heat exchanges between components.

Between the heat transfer fluid and the absorber: $Q_u(x,t) = h_f * S_{ai} * (T_a(x,t) - T_f(x,t))$

$$h_f = \frac{Nu * \lambda_f}{D_{ai}}$$

Between the absorber and the glass: $Q_{a,v} = Q_{a,v/conv} + Q_{a,v/ray}$

$$Q_{a,v/conv}(x,t) = \frac{2\pi K_{aireff}}{\ln\left(\frac{D_{vi}}{D_{ae}}\right)}(T_a(x,t) - T_v(x,t))$$

$$K_{aireff} = K_{air}.\,0.386\left(\frac{\Pr_{annu}}{\Pr_{annu} + 0.861}\right)^{0.25}\left(F_{cyc}Ra_{Dae}\right)^{0,25}$$

$$Ra_{Dae} = Gr*Pr = \frac{g*\beta*D_{ae}^3*(T_a - T_v)}{\vartheta^2} * Pr$$

$$Q_{a,v/ray}(x,t) = \frac{\sigma\pi D_{ae}(T_a^4(x,t) - T_v^4(x,t))}{\frac{1}{\varepsilon_a} + \left(\frac{(1-\varepsilon_v)}{\varepsilon_v}\right) * \left(\frac{D_{ae}}{D_{vi}}\right)}$$

Between the glass and the environment: $Q_{v,amb} = Q_{v,amb/conv} + Q_{v,amb/ray}$

With : $Q_{v,amb/conv}(x,t) = h_{v,amb}S_{ve}(T_v(x,t) - T_{amb})$

$$Q_{v,amb/ray}(x,t) = \varepsilon_v\sigma S_{ve}(T_v^4(x,t) - T_{amb}^4)$$

Thermal efficiency: $\eta = \frac{\int_0^t Q_u dt}{\int_0^t GSdt}$

3.1. Validation of the model

A modeling of the parabolic trough was developed by Matlab to solve the equations whose resolution is based on the finite difference discretization method. The results of the model developed were compared with the experimental results presented by the Sandia laboratory [6].

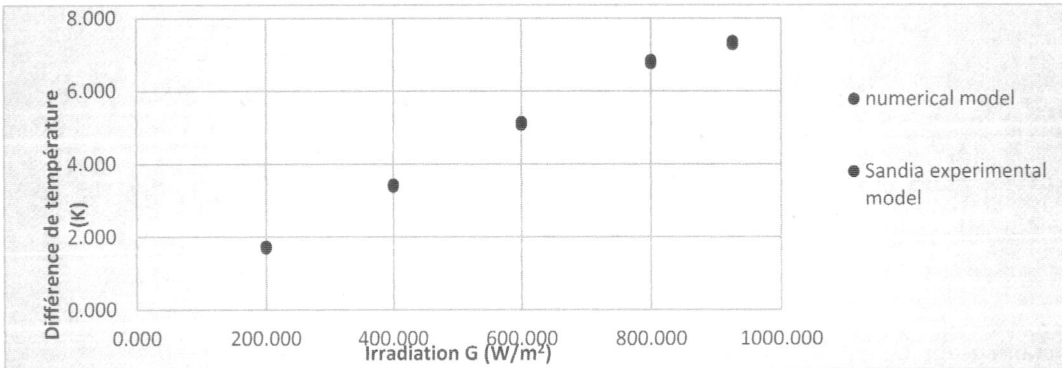

Figure 4. Comparison between the numerical model and the experimental model of the Sandia laboratory.

4. Results and Discussions

The parabolic trough chosen for our study is an S10 parabolic trough from Rackam. This parabolic trough has already been used in the agricultural cooperative of Taroudant (COPAG). The characteristics of this parabolic trough are shown in the table 1 and the figure 5 show us the installation of parabolic through in industry [7].

Table 1. The characteristics of the parabolic trough.

Characteristics	Data	Symbols
Internal / external absorber diameter (m)	0.065/0.07	D_{ai}/D_{ae}
Internal / external glass diameter (m)	0.119/0.125	D_{vi}/D_{ve}
Collector Length (m)	12.19	L
Collector Area (m^2)	13.7	S
Collector nomber	4	
Interception factor	0.92	
Coefficient of reflection	0.92	ρ
Transmissivity of glass	0.97	τ
Absorption coefficient	0.955	α
Wind speed EL-Jadida city (km/h)	5-28	V

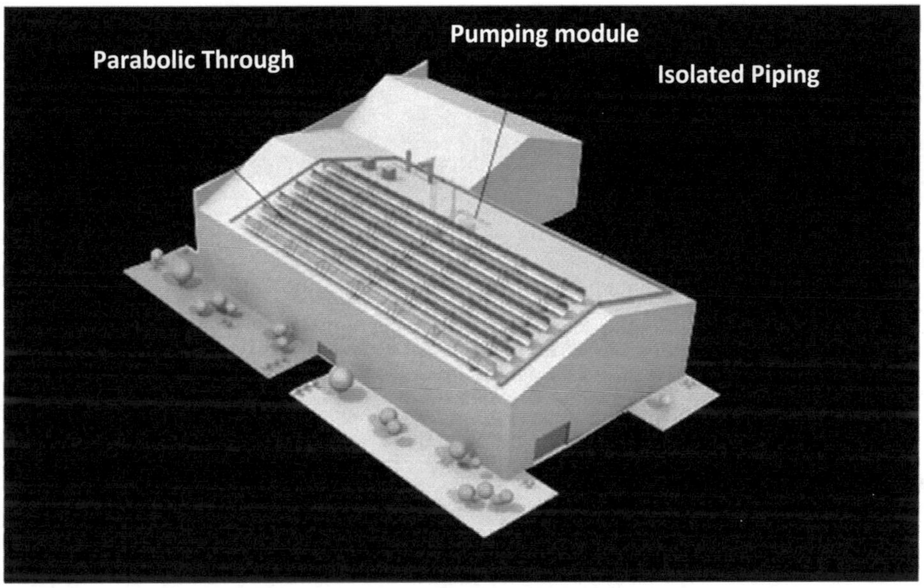

Figure 5. Installation of parabolic through in industry.

4.1. Variation of heat exchange coefficient of heat transfer fluids

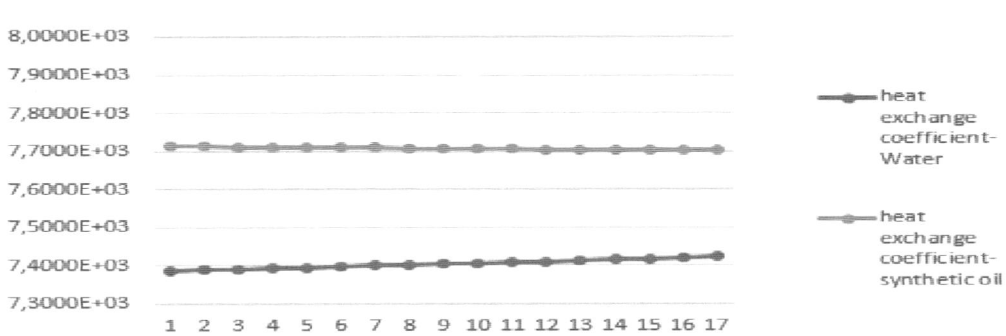

Figure 6. Variation of heat exchange coefficient of heat.

According to the graph, there is, on the one hand, a significant exchange coefficient on the part of the two fluids and, on the other hand, a small increase in this coefficient which translates the heat exchange quality.

4.2. Variation of Grashof number of heat transfer fluids

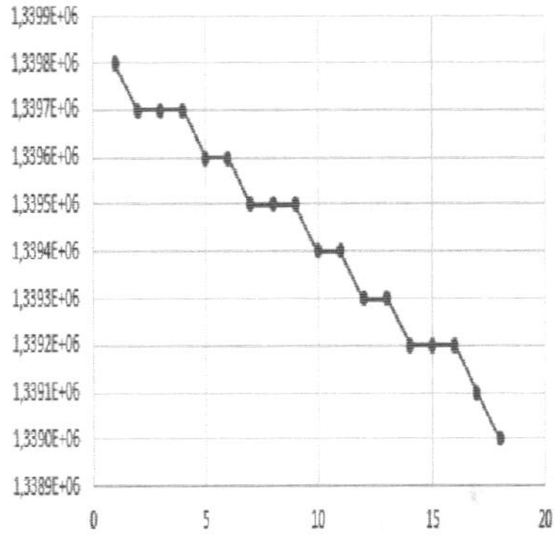

Figure 7. Variation of Grashof Number-Water.

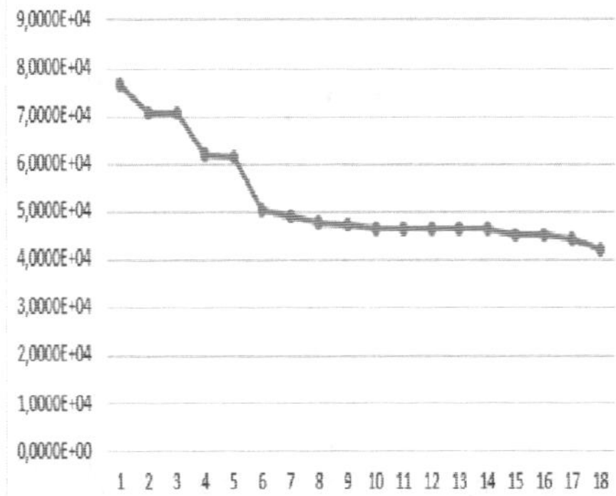

Figure 8. Variation Grashof number-synthetic oil.

There is a decrease in the number of grashof for the two fluids reflecting the presence of reinforced convection.

4.3. Variation of the temperature according to irradiation

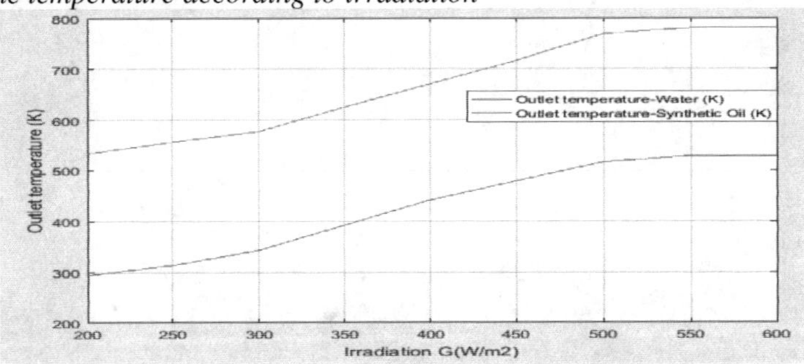

Figure 9. Variation of the temperature of the output according to irradiation.

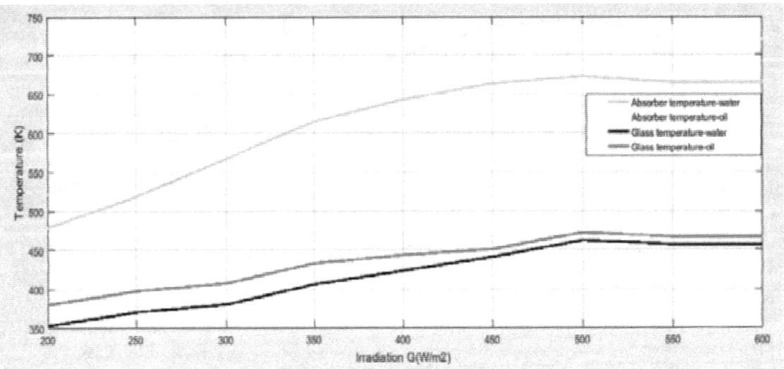

Figure 10. Variation of the temperature according to irradiation.

The figure 9 shows the temperature of the fluids at the exit depending on the irradiation; There is an increase in the two temperatures in addition we also note the output temperature of synthetic oil is greater than that of water.

The figure.10 shows an increase in absorber temperatures and the glass when increasing the intensity of irradiation and we notice that the temperature at the absorber level is the highest temperature while the temperature of the glass is less high compared to the absorber temperature.

4.4. Variation of the temperature according to volume flow rate and wind speed

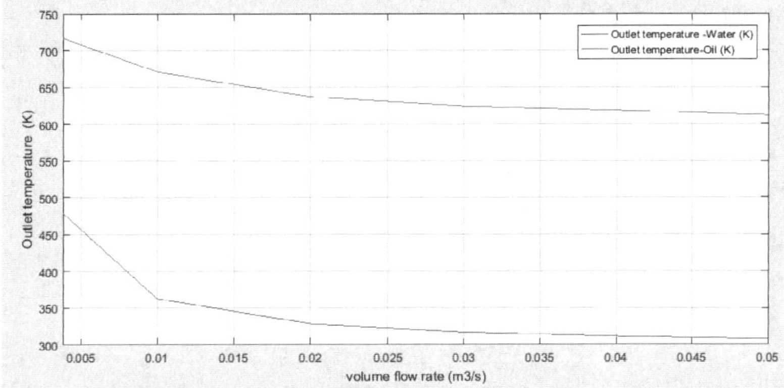

Figure 11. Variation of the outlet temperature according to volume flow rate.

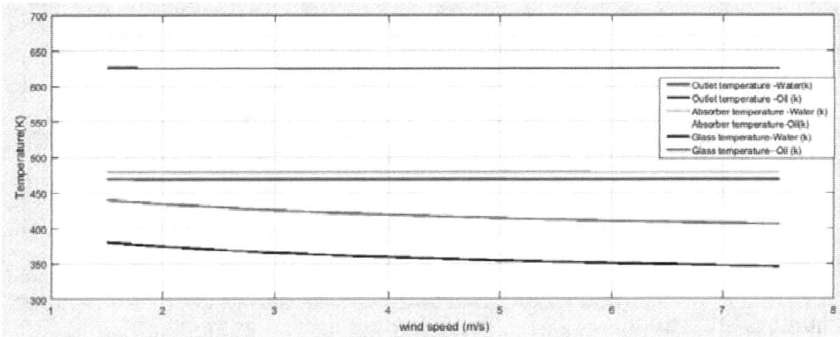

Figure 12. Variation of the temperature according to wind speed.

An inverse proportionality is observed between the fluid temperature and the volume flow rate, this implies an increase in the flow velocity and, for this purpose, impacts the heat exchange process between the fluid and the absorber.

According to the figure12 presented above, we notice an increase of the speed of the wind causing a decrease of the temperature of the glass while the other temperatures are less affected, this concludes to say that the speed of the wind influences the temperature of the glass.

4.5. Variation of the efficiency according to tube length

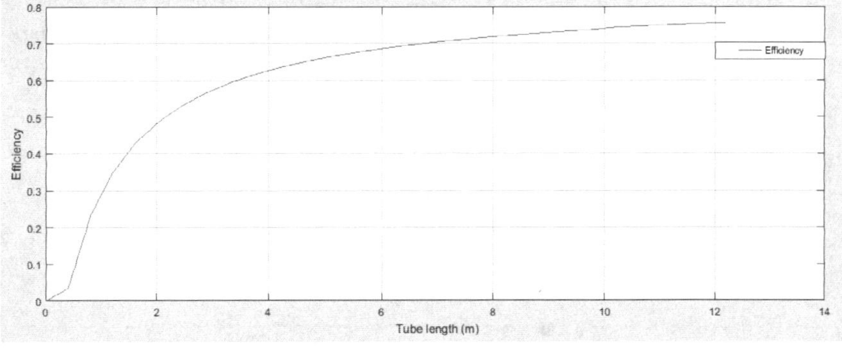

Figure 13. Variation of the efficiency according to tube length.

When the length of the tube increases the efficiency also increases. This increase has limitations related to thermal resistance of materials.

4.6. Economic valuation and environmental impact

The cost of the steam is calculated by: $C_v = \frac{L_v * C_f}{PCI * \eta}$

Table 2. Annual gain steam.

Cost of Parabolic Through (MAD)	7720
Daily gain steam (Kwh/day)	2880
Cost of the steam C_v (MAD/Kwh)	0.106
Annual gain steam (MAD/year)	111427

Table 3. Energetic cost.

Daily fuel consumption (Kwh/day)	60
Cost of fuel (MAD/Kwh)	4.12
Annual energetic cost (MAD/year)	88003

Table 4. Net annual gain.

Annual gain steam (MAD/year)	111427
Annual energetic cost (MAD/year)	88003
Net annual gain (MAD/year)	23424

- **Return on investment**

It calculated using the below formula $ROI = \frac{Investment\ cost}{Net\ annual\ gain}$

The company would be able to cover the costs it has invested over the course of 4 months.

- **Environmental impact**

The fuel consumption emits 0.271kg / Kwh of CO_2 implies an emission of 5790T/year.

5. Conclusion

This work presents a resolution to the issue of hot water production in a dairy industry by proposing to integrate the parabolic cylindro sensors by presenting a numerical simulation of the output temperature. This study favors the choice of synthetic oil as heat transfer fluid in relation to water, but because our aim is to directly generate steam in industrial production and also because of the availability of water and an economy at the level of the purchase price of synthetic oil. It is better to use water as a heat transfer fluid.

Nomenclature

Q_f: Heat quantity of heat transfer fluid (W)	$Q_{a,v/ray}$ Quantity between the absorber and the glass by radiance (W)
Q_{ab} Quantity of heat absorbed (W)	$Q_{v,amb}$ Quantity between the glass and the environment (W)

Q_u : Quantity of useful heat (W)	$Q_{v,amb/conv}$ Quantity between the glass and the environment by convection (W)
$Q_{a,v}$: Quantity between the absorber and the glass (W)	$Q_{v,amb/ray}$ Quantity between the glass and the environment by radiance (W)
$Q_{a,v/conv}$ Quantity between the absorber and the glass by convection (W)	ρ_f Density of the fluid (kg/m^3)
C_p: Massic heat (j/Kg.K)	D_f Flow rate of fluid (m^3/s)
ΔT : Temperature difference (K)	G : Irradiation (W/m^2)
Nu : Nusselt number	h_f: Exchange coefficient (w/m^2·K)
T_a : Absorber temperature (K)	T_f : Fluid temperature (K)
Gr : Grashof Number	S_{ai} : Absorber area (m^2)
T_v : Glass temperature (K)	λ_f : Conductivity coefficient (w/m.K)
Ra_{Dae} Rayleih number	Pr: Prandlt number

Acknowledgment
This work was done in the laboratory "Applied Thermodynamics and Solid Combustibles (ATSC)", Mohammedia School of Engineers (EMI), Rabat- Morocco, under the direction of professor Doctor Abdellatif TOUZANI. We want to thank also Mr. BERRADA the founder and director at Alto Solution and Pr. KADIRI from National School of Applied Sciences -Khouribga for their collaboration.

References

[1] BENHAFID Abdelhakim and EL-GAGHLOULI Abdelhafid, 2016, Reduction of fuel and electricity consumption at the Danone power plant in El-Jadida, page: 30-80
[2] "Morocco renewable energies". [Online]. Available: https://www.fellah-trade.com/fr/developpement-durable/energies-renouvelables-maroc/solaire. [Accessed : 21-Mars-2019]
[3] Caliot and Flamant, 2016, Cylindrical-parabolic concentrator technology, page: 4
[4] Boukelia, Mecibah, and Laouafi, 2016, Performance simulation of parabolic trough solar collector using two fluids (thermic oil and molten salt), Journal of Fundamental and Applied Sciences, 8(2),600-626.
[5] Ghodbane , Boussad BOUMEDDANE and Soulef LARGOT, 2015, Optical and thermal study of a cylindrical-parabolic concentrator on the Algiers site, Algeria, 9th International Congress on Renewable Energy and the Environment,10
[6] "Solar Industrial Process Heat- COPAG" [Online]. Available: https:// https://rackam.com/en/studies/copag/. [Accessed : 03-Mai-2018]
[7] Garcia-Valladares O, Velazquez N 2009 Numerical simulation of parabolic trough solar collector: Improvement using counter flow concentric circular heat exchangers *Int. J. Heat and Mass Transfer* **52(3-4)** 597-609.

Steel Corrosion Behaviour in Acidic Solution for Application in Petrochemical Distillation Systems

M A Fajobi[1], R T Loto[1], O O Oluwole[2]

[1]Department of Mechanical Engineering, Covenant University, Nigeria
[2]Department of Mechanical Engineering, University of Ibadan, Nigeria

fajobi.muyiwa@covenantuniversity.edu.ng

Abstract. The electrochemical reaction response of austenitic 316L stainless steel and carbon steel was examined through weight loss analysis in 1M, 2M and 3M of HCl acid. The results show that austenitic 316L has high corrosion resistance than carbon steel for the test analyzed with the lowest corrosion rate of 0.0018mm/y at 1M of HCl and highest at 0.0053mm/y when compared with carbon steel which has the lowest corrosion rate of 0.0003mm/y for 1M of HCl and highest at 0.0013 mm/y of 3M of HCl solution all at ambient temperature conditions. General corrosion was displayed on the surface of the carbon steel but austenitic 316L was not affected due to the presence of chromium alloy and other alloying elements.

1. Introduction

Steels are generally used as metallic structures for equipment in petrochemical industries such as crude distillation, valves, trays, atmospheric distillation overhead column, condenser, desalter units etc. They contain excellent physical and mechanical properties that can make them survive and withstand high rate of corrosion caused by contaminants from crude oil. Contaminants are the major challenge of petrochemical industries and when they are mixed with water corrosion is formed. Corrosion affects every part of the steel structures of the petrochemical industries, some occur at faster rate while others at slow corrosion rate. Huge losses of revenue have been recorded in petrochemical industry as a result of corrosion maintenance [1-5], which ranges from million dollars to billions of dollars spent yearly. Research showed that stainless-steel [6-8] and carbon steel [9-12] are generally used for petrochemical's structural equipment. Carbon steel are considered for use due to its low cost, good toughness, good formability, weldability, low strength, availability, wear resistance, ductility but low corrosion rate. Most petrochemical industries prefer carbon steel for their equipment because of low cost, which can be replaceable from time to time even when they corrode fast. While stainless-steel equipment in petrochemical industries stay longer for years without been speedy affected by corrosion for high life span and ductility. For instance, austenitic stainless steels, possess high ductility, high passive film protection and high corrosion resistance as a result of the presence of chromium and other alloying elements which shows that they are excellent steel material for petrochemical structural equipment. This research focuses on examining the behaviour of electrochemical reaction of austenitic 316L stainless-steel and carbon steel in HCl acids varied concentration for applications in petrochemical distillation units.

2. Material and Methods

2.1. Material
Austenitic 316L stainless-steel and carbon steel were gotten for this research work and analyzed. The steels have a square dimension of 100 and 100mm respectively.

Table 1. Composition of Carbon Steel [13].

Element	C	Si	Mn	V	Nb	Mo	Cr	Cu	Ni	Al	Fe
% Content	0.125	0.52	1.83	0.078	0.053	0.079	0.121	0.296	0.091	0.043	96.2

Table 2. Composition of 316L Stainless Steel [14].

Element	Cr	Ni	Mo	C	Mg	P	S	Si	N	Fe
% Content	16	10	2	0.03	2	0.05	0.03	0.75	0.1	69.05

2.2. Acid solution test
A specific concentration of the dilute HCl acid solution was prepared with distilled water to dilute the grade of HCl (37%) purity for the test corrosive environment used to represent the petrochemical atmospheric distillation overhead system.

2.3. Preparation of the steel samples
The carbon steel and austenitic 316L stainless steels were machined into test for 3 samples of each steels. Carbon steel with diameter of 160mm and the dimension of the square shaped austenitic 316L steel of 100 mm by 100mm were used for the research. Abrasive of silicon carbide was used to prepare the metallography of the two ends surfaces of the two steels, ranging with the papers size of 80, 120, 220, 800 and1000 grits. The steel samples were polished with diamond liquid, rinsed with acetone, distilled water and dried. They were later stored in the desiccator for weight loss test measurement analysis [15].

2.4. Weight loss analysis

The weight loss experimental setup of the cylindrical shaped carbon steel coupons and square shaped austenitic 316L stainless steel coupons were weighed using a weighing balance. The carbon steel and austenitic 316L stainless steel were both immersed in 200ml of 1M, 2M and 3M of HCl acid. The entire experiment was exposed to the atmosphere and the coupons were fully immersed at ambient temperature for the period of 360 hours. Each sample were analyzed at every 24 hours after washed with distilled water, dried and weighed with the weighing balance [16].

3. Results and Discussion

3.1. Weight loss measurement
The graph obtained from the data of the weight loss measurement for weight loss(*w*) and corrosion rate (mm/y) for austenitic 316L stainless-steel and carbon steel in HCl acids are displayed in figures (i) and (ii) below reveals the corrosion rate against time in the HCl acid media for the two steels. The reaction of HCL acid on the steel led to a dissolution of metallic anodic reaction causing oxidation to occur by release of electrons at the anode. The reaction proceeded at the cathode by reduction process of release of protons. All these processes led to the corrosion reaction of the steels.

IOP Conf. Series: Materials Science and Engineering **811** (2020) 012030 doi:10.1088/1757-899X/811/1/012030

(i)

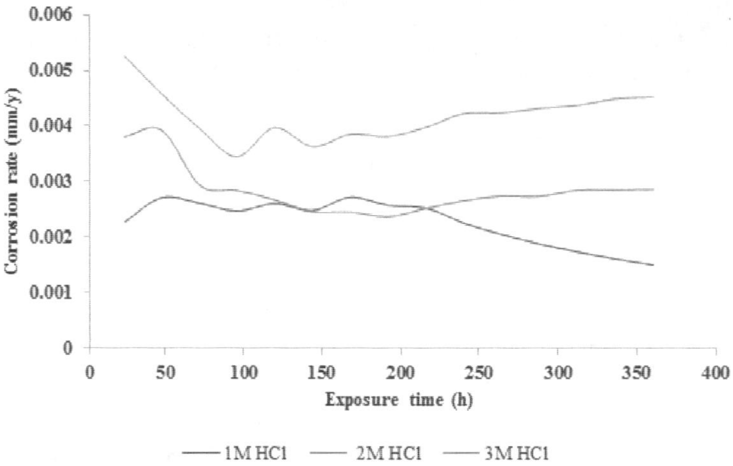

(ii)

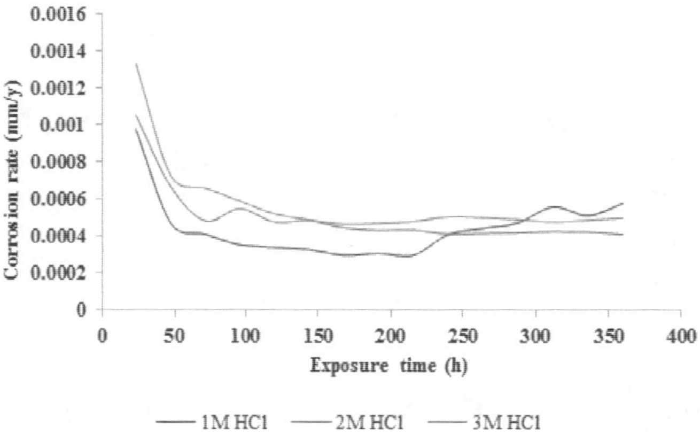

Figure 1. (i) corrosion rate versus exposure time for Austenitic 316L samples in 1M, 2M and 3M HCl acid solution, (ii) is a plot of corrosion rate versus exposure time for carbon steel samples in 1M, 2M and 3M HCl acid solution.

The observation from the graphs showed that the corrosion behaviour of austenitic 316L for the three (3) samples responded differently. The sample with 3M of HCl had the highest corrosion rate all through the exposure time than the other two of 1M and 2M respectively. However, there was reduction in the corrosion rate significantly down the exposure time due to the film protection or passivation on the surface of the austenitic 316L stainless-steel, which slows down the corrosion rate gradually.

The carbon steel performed differently by showing a weak corrosion resistance, which gradually occured throughout the period of exposure time. The 3M HCl sample of the carbon steel had the highest rate of corrosion, which drastically and practically reduced down the reaction of the exposure time. The carbon steel corrosion resistance strength was obviously lower than austenitic 316L stainless-steel with a great difference. This is due to the presence of chromium formation of the alloy properties of the 316L, showing that 316L stainless steel is highly sustainable for structural applications in petrochemical equipment, especially distillation overhead systems.

4. Conclusion

Austenitic 316L stainless-steel showed far stronger and higher corrosion resistance to general corrosion when compared with carbon steel in 1M, 2M and 3M HCl acid solution due to the presence of the alloy composition (chromium, nickel) of 316L. The percentage corrosion damage was absolutely different which showed that austenitic 316L possesses a high formation of passivation or film protection by the alloying element. As a result of these, austenitic 316L showed excellent perform for stable corrosion resistance behaviour than carbon steel and makes it suitable for general purpose of structural applications in petrochemical distillation overhead units. This helps to save cost and its shows high sustainability for eco-friendly petrochemical distillation systems at a long run.

Acknowledgement

The authors appreciate the financial support of Covenant University towards the success of this research.

References

[1] Ramraj S, Anandaraj C 2014 Refining of crude and its corrosion control using distributed control system. *World Sci. Eng. Acad. Soc. Trans Syst. Control* **9** 533-9.
[2] Fajobi M A, Fayomi O S I, Akande I G, Odunlami O A 2019. Inhibitive Performance of Ibuprofen Drug on Mild Steel in 0.5 M of H_2SO_4 Acid. *J. Bio- Tribo-Corrosion*, **5(3)** 79.
[3] Moiseeva L S, Kondrova O V 2005 Biocorrosion of oil and gas field equipment and chemical methods for its suppression. *Prot. Met.* **41(4)** 385-93.
[4] Koch G H, Brongers M P, Thompson N G, Virmani Y P, Payer J H 2005 Cost of corrosion in the United States. *In: Handbook environ. degradation mater.* 3-24.
[5] Study reveals $1.4 billion annual corrosion costs for oil and gas industry. https ://www.ultra filtrex.com/2018/08/study -reveals-1-4- billion-annual-corrosion-costs -oil-gas-industry. Accessed 7 Aug 2008.
[6] El-Yazgi A A, Hardie D 1998 Stress corrosion cracking of duplex and super duplex stainless steels in sour environments. *Corrosion Sci.* **40(6)** 909-30.
[7] Cui Z D, Wu S L, Zhu S L, Yang X J 2006 Study on corrosion properties of pipelines in simulated produced water saturated with supercritical CO2. *Appl. Surf. Sci.* **252(6)** 2368-74.
[8] Smith L, Celant M, Pourbaix A 2000 A guideline to the successful use of duplex stainless steels for flowlines. In: Proc. Duplex America conference, Houston, USA, KCI Publishing DA_2-102, 17-29.
[9] Gutzeit J 2000 Effect of organic chloride contamination of crude oil on refinery corrosion. *Corrosion 2000*. NACE International, Houston.
[10] Breen A J 1974 Primary petroleum distillation plant-selection of materials. Br *Corros. J.* **9(4)** 197-203.
[11] Chambers B, Srinivasan S, Yap K M, Yunovich M 2011 Corrosion in crude distillation unit overhead operations: a comprehensive review. In: NACE corrosion, conference expo, 1-11.
[12] Fajobi M A, Loto R T, Oluwole O O 2019. Corrosion in Crude Distillation Overhead System: A Review. *J. Bio-and Tribo-Corrosion*, **5(3)** 67.
[13] Haruna K, Obot, I B Ankah, N K, Sorour A A, Saleh T A 2018 Gelatin: A green corrosion inhibitor for carbon steel in oil well acidizing environment. *J. Molecular Liquids*, **264** 515-25.
[14] Uns S, Nr, S W 2014. Alloy 316 / 316L. *Specification Sheet,* **1** 9-11.
[15] Nam, N D, Kim, M J, Jang Y W, Kim J G 2010. Effect of tin on the corrosion behavior of low-alloy steel in an acid chloride solution. *Corrosion Sci.* **52(1)** 14-20.
[16] Loto R T, Loto C A 2018 Corrosion behaviour of S43035 ferritic stainless steel in hot sulphate/chloride solution. *J. Mater. Res. Technol.* **7(3)** 231-9.

IOP Conf. Series: Materials Science and Engineering **811** (2020) 012031 doi:10.1088/1757-899X/811/1/012031

Corrosion Inhibition of Rosemary Oil on High Carbon Steel in Sulphuric Acid Medium

O A Odunlami[1], O T Olomukoro[1], R T Loto[2]

[1]Department of Chemical Engineering, Covenant University, Ota, Ogun State, Nigeria
[2]Department of Mechanical Engineering, Covenant University, Ota, Ogun State, Nigeria

Abstract. Rosemary oil was investigated on carbon steel in 0.5M sulphuric acid through weight loss technique. It was observed that weight loss, inhibitor efficiency and corrosion rate differ in concentration of the oil and exposure time. Maximum inhibitor efficiency of 78.45% was obtained at 10ml concentration of the oil. The performance of the oil was associated with the existence of organic compounds such as alkaloids, flavonoids, tannins, among others which form a protective film and act as a barrier against corrosion attack in sulphuric acid environment.

1. Introduction

Corrosion is a process that occurs naturally resulting in the gradual breakdown of a material by reaction with their environment [1]. The corrosion process in metals is inevitable but controllable [2]. The effects of corrosion are seen in the destruction of bridges, marine equipment, automobiles, energy production systems, airplanes, household devices and other structural materials [1]. Corrosion constitute a severe global issue in industries and has a notable influence on a country's development. Billions of dollars are lost annually to corrosion. In the United States, annual metallic corrosion cost rose to over 200 billion dollars [2].

Carbon steels are widely utilised in industrial processes and structural applications since they are cheap and possess the capacity to be easily shaped. Corrosion of carbon steels have gained international attention as a result of their safety and economic effects from their destruction. Carbon steels have low corrosion resistance in acid medium, particularly in sulphuric acid [3]. Sulphuric acids are used for cleaning boilers through the removal of calcareous deposits, pickling, acidizing of oil wells and cleaning of refinery equipment materials [1].

Corrosion in metals can be controlled through cathodic protection, coatings, suitable selection of materials and the use of inhibitors [4]. Inhibitors are chemicals that lessen the effect of corrosion in metals in harsh conditions. Inhibitors work by preventing the destructive effects of corrosive anions in different conditions. Inhibitors are cost effective relative to other corrosion control methods [5]. Organic inhibitors are used in preventing carbon steel from corrosive chemicals [3]. Organic compounds that contain nitrogen, oxygen and sulphur atoms are regarded as efficient corrosion inhibitors. Unfortunately, organic inhibitors are relatively quite expensive and harmful to the environment. Alternatively, inexpensive environmentally friendly inhibitor compounds have been considered and have progressively gained wider recognition [5].

Green inhibitors are eco-friendly inhibitors. These inhibitors are not toxic to the environment [6]. They are biodegradable, biocompatible and readily available materials [2]. Organic green inhibitors such as extracts from plants have been described by different researchers as excellent corrosion inhibitors [2]. Plant extracts include Allium Sativum, Gossipium Hirsutum, Carica Papaya, Telfaria Occidentalis,

Hibiscus Subdariffa, Rosemarinus Officinalis among several others. The performance of rosemary oil in reducing the corrosive effect of sulphuric acid on carbon steel is being investigated.

2. Materials and Methods

High carbon steel samples used for this study were acquired commercially with nominal weight composition of 0.9% C, 0.5% Mn, 0.04% P, 0.05% S and 98.51% Fe [5]. After machining into the desired cylindrical shape, the following dimensions were obtained: 1cm length, 1.2cm diameter and 6.03cm^2 surface area. The carbon steel surface was prepared with abrasive papers.

Rosemary essential oil was used as the inhibiting compound obtained from Jannys beauty salon, Ikoyi, Lagos State. It is a yellowish transparent liquid. Rosemary oil has a density of 0.908 g/mL with a boiling point of 176°C [5].

2.1. Weight loss Experiment

The weight of the cylindrical steel pieces was determined by an electronic weighing balance and dipped completely in 0.5 M sulphuric acid concentration at fixed quantity of rosemary oil with no variation in temperature. The samples were taken out of the solution and analysed every 24 hours for 21 days. During analysis, samples of carbon steel were carefully cleansed with distilled water and acetone, air-dried completely and reweighed. Parameters such as the surface coverage (θ_s), inhibitor efficiency (IE) and rate of corrosion (C) were obtained through the following equations [5].

$$\theta_s = 1 - \frac{WL_n}{WL_i} \tag{1}$$

θ_s denotes the degree of the oil adsorbed for each gram of surface of steel. W_n is the weight loss of the sample in the non-inhibited solution and W_i is the weight loss of the sample in the inhibited solution.

The inhibitor efficiency is calculated as:

$$\%IE = \theta_s \times 100 \tag{2}$$

The rate of corrosion, C expressed in mm/y is calculated as:

$$C = \frac{87.6 * WL}{D * A_S * T_e} \tag{3}$$

Where WL is the weight loss (g), T_e is the exposure time (h), A_s is surface area of steel surface (cm^2), D is the steel density (g/cm^2).

3. Results and Discussion

3.1. Weight-loss results

The results obtained from corrosion rate, inhibitor efficiency and weight loss in sulphuric acid concentration of 0.5M at exposure time of 504 hours are shown in table 1. Figures 3.1(a) portray the weight loss (g) against time (hrs) for carbon steel in 0.5M sulphuric acid solution. The loss in weight of the steel samples shown in figure 3.1(a) increased with the time of exposure and decreased when the concentration of inhibitor (rosemary oil) was increased. The control curve (0ml) contained no inhibitor (rosemary oil) possesses the highest weight loss value. A decline in the weight loss was noticed when the oil was added. Inhibitor quantity of 10ml produced the lowest weight loss. The order of increasing inhibition performance of the oil is as follows: 10ml>8ml>6ml>4ml>2ml. It is clear that the activity of the inhibitor was largely dependent on its concentration. Organic compounds such as flavonoids, tannins among others contained in plant extracts as reported by Mo *et al* [8] and Okafor & Ebenso [9] can be associated with the performance of rosemary oil.

Corrosion rate against time of exposure with different quantity of rosemary oil is shown in figure 1(b). The control experiment showed a higher corrosion rate than other experiments with inhibition addition. The corrosion rate for non-inhibitor and inhibitor solutions initially showed a sharp increase and then decreased progressively with higher time of exposure. Rosemary oil with quantity of 10ml produced the lowest rate of corrosion. The reduction in the corrosion rate was largely determined by the inhibitor concentration. This can be associated with the generation of film on the metallic surface which shields the surface from corrosion attack and therefore reduce the corrosion rate as reported by Nnanna *et al.* [10]

Figure 1(c) shows the inhibitor efficiencies with exposure time at different concentrations of the oil. The increase in the quantity of the inhibitor led to an increase in the efficiency of the oil. The Inhibitor efficiency for non-inhibitor and inhibitor solutions initially showed a sharp increase and decreased and then increased progressively with exposure time. The inhibitor concentration of 10ml produced the highest inhibitor efficiency of 78.45%. Green inhibitors as reported by Amitha and Barathi [11] possesses phytochemical constituents which prevents the diffusion of ions and inhibits any form of corrosion attack on the metallic material. The inhibitor used was considered for use to prevent pollution from occurring in order to have a stainable environment. Research has shown that different mediums such as co [12-14], inorganic inhibitors [15, 16] are sources of pollution to our environment. They are highly harmful to human lives and are not encouraging in order to have an eco-friendly environment.

Table 1. Results of weight loss for carbon steel in 0.5M sulphuric acid solution after 504 hours.

Sample	Inhibitor Conc. (ml)	Weight loss (g)	Surface Coverage (θ_s)	Inhibition Efficiency (%)	Corrosion Rate (mm/y)
A	0	7.4812	0	0	0.0275
B	2.00	2.4383	0.6741	67.41	0.0090
C	4.00	2.0497	0.7260	72.60	0.0075
D	6.00	2.0440	0.7268	72.68	0.0075
E	8.00	2.1840	0.7081	70.81	0.0080
F	10.00	1.6125	0.7845	78.45	0.0059

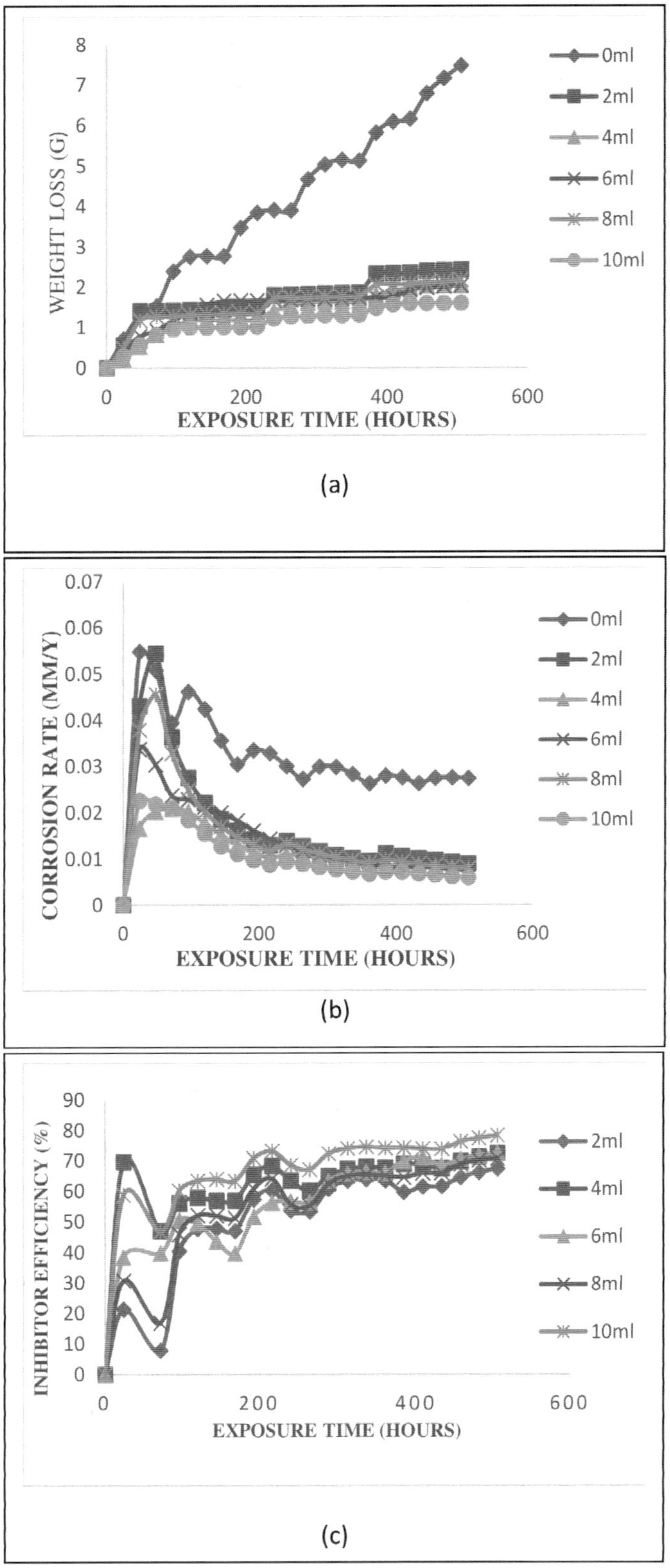

Figure 1. Graph of (a) Loss in weight (g) against time of exposure (b) Corrosion rate against time of exposure time (c) Inhibitor efficiency against time of exposure in 0.5M sulphuric acid medium.

4. Conclusion

Results obtained from weight loss experiment proved the effectiveness of rosemary oil in suppressing the corrosive effects on carbon steel in sulphuric acid environment. At increasing amounts of the oil, the rate of corrosion was reduced.

Acknowledgement

Covenant University's support offered to the success of this research work is appreciated by the authors.

References

[1] Odusote J K, Ajayi O M 2016 Corrosion inhibitive properties of extract of Jatropha curcas leaves on mild steel in hydrochloric acid environment. *Iranian J. Mater. Sci. Eng.* **13(3)** 1-11.

[2] Kesavan D, Gopiraman M, Sulochana N 2012 Green inhibitors for corrosion of metals: A review. *Chem. Sci. Rev. Let.* **1(1)** 1-8.

[3] Al-Nowaiser FM, Abdallah M, El-Mossalamy E H 2011 Rosemary oil as a corrosion inhibitor for carbon steel in 0.5 M sulphuric acid solution. *Chem. and Tech. Fuels Oils* **47(1)** 66-74.

[4] Fajobi M A, Loto R T, Oluwole O O 2019 Corrosion in Crude Distillation Overhead System: A Review. *J. Bio-and Tribo-Corrosion*, **5(3)** 67.

[5] Loto R T 2018 Electrochemical analysis of the corrosion inhibition properties of L-leucine and trypsin complex admixture on high carbon steel in 1M H_2SO_4 solution. *Quimic Aplicada y Analitica*, **47(2)** 12-20.

[6] Fajobi M A, Fayomi O S I, Akande I G, Odunlami O A 2019 Inhibitive Performance of Ibuprofen Drug on Mild Steel in 0.5 M of H_2SO_4 Acid. *J. Bio-and Tribo- Corrosion*, **5(3)** 79.

[7] Loto R T, Loto R O, Joseph O O, Akinwumi I 2015 Electrochemical studies of the corrosion inhibition property of *Rosmarinus officinalis* on mild steel in dilute sulphuric acid. *J. Chem. Pharm. Resear.* **7(7)** 105-16.

[8] Mo S, Luo H, Li N 2016 Plant extracts as green corrosion inhibitors for steel in sulphuric acid. *Chem. Paper*, **40(9)** 1131-43.

[9] Okafor P G, Ebenso E E 2007 Inhibitive action of Carica papaya extracts on the corrosion of mild steel in acidic media and their adsorption characteristics. *Pigment & Resin Tech.* **36(3)** 136-40.

[10] Nnanna L A, Onwuagba B N, Mejeha I M, Okeoma K B 2010 Inhibitive effects of some plants extract on the acid corrosion of aluminium alloy. *Afr. J. Pure Appl. Chem.* **4** 11-6.

[11] Amitha B E, Bharathi B J 2012 Green inhibitors for corrosion protection of metals and alloys: An overview. *Inter. J. Corros.* Article ID 380217, 15 pages doi:10.1155/2012/380217.

[12] Odunlami O A, Elehinafe F B, Oladimeji T E, Fajobi M A, Akeredolu F A, Okedere O B, Fakinle B S 2018 Implications of lack of maintenance of vehicles on ambient air quality. *Int. J. Mech. Eng. Technol. (IJMET)*, **9(9)** 647-54.

[13] Fakinle B S, Odekanle E L, Olalekan A P, Odunlami O A, Sonibare J A 2018 Impacts of polycyclic aromatic hydrocarbons from vehicular activities on the ambient air quality of Lagos mega city. *Environ. Quality Manage.* **27(4)** 73-8.

[14] Odunlami O A, Elehinafe F B, Oladimeji T E, Fajobi M A, Okedere O B, Fakinle B S 2018 Implications of lack of maintenance of motorcycles on ambient air quality. In *IOP Conf. Ser. Mater. Sci. Eng.* **413(1)** 1-8.

[15] El-Meligi A A 2010 Corrosion Preventive Strategies as a Crucial Need for Decreasing Environmental Pollution and Saving Economics. *Recent Patents Corrosion Sci.* **2** 22-33.

[16] A El-Meligi A 2011 Corrosion of materials in polluted environment and effect on world economy. *Recent Patents Corrosion Sci.* **1(2)** 144-55.

IOP Conf. Series: Materials Science and Engineering **811** (2020) 012032 doi:10.1088/1757-899X/811/1/012032

Amine-based Carbon Dioxide Absorption: Evaluation of Species Activity Coefficient for MEA-H_2O System from 303 to 333K

S Rusdi[1, 2], S Ma'mun[1, 2], M Y Qifari[1, 2], R D V Gite[1, 2]

[1]Department of Chemical Engineering, Universitas Islam Indonesia
[2]Jl. Kaliurang Km. 14.5, Yogyakarta 55501, Indonesia

sholeh.mamun@uii.ac.id

Abstract. Acid protonation constant (K_a) is one of the parameters required for thermodynamic modelling. In addition, a rigorous thermodynamic model requires solvent non-ideality properties in order to have a better model representation. This study aims to evaluate species activity coefficients from potentiometric titration data of the MEA-H_2O system at temperatures ranging from 303 to 333K. The results show that the activities of the free MEA and the protonated MEA depend on pH. In addition, the activity coefficients of the protonated MEA determined by the extended Debye-Hückel model were less than unity at different temperatures. This means that the system observed was the non-ideal one.

1. Introduction

The global warming problem has been becoming one of the challenging problems facing the world community. This is caused by the CO_2 emission in the atmosphere which increases annually. Therefore, some parties have been attempting to implement Carbon Capture and Storage (CCS) technology to reduce the CO_2 emission with several scenario options.

Absorption of CO_2 by use of a solvent is a mature technology, however, some researchers have been working on CO_2 capturing solvent development to invent promising solvent candidates, such as alkanolamines, sodium carbonate, amino acid salts (AAS), chilled ammonia, etc. [1-6]. A commercial CO_2 absorbent must fulfil some criteria such as high reaction rate, high net cyclic capacity, low solvent regeneration energy requirement, less corrosive, less volatile, etc. [7-15].

Acid protonation constant (Ka or pK_a) is one of the parameters which is required during kinetic and thermodynamic model developments. The higher the pK_a value of a solvent, the better the solvent performance, because it offers fast kinetic to reduce the dimension of absorption and desorption columns [16]. Aqueous solutions of monoethanolamine (MEA) are the most commonly used as absorbents for the CO_2 absorption. MEA has several advantages compared to other amines, such as high absorption rate, high pK_a, low solvent cost, etc. Determinations of pK_a for the MEA-H_2O system at different temperatures have been reported by [17-22].

A rigorous thermodynamic model requires solvent non-ideality properties, such as species activity coefficients in order to have a better model representation. However, the species activity coefficients data, such as those for the MEA-H_2O system, are still limited in the literature. Therefore, this study aims to evaluate the species activity coefficients from the potentiometric titration data of the MEA-H_2O system at temperatures from 303K to 313K.

IOP Conf. Series: Materials Science and Engineering **811** (2020) 012032 doi:10.1088/1757-899X/811/1/012032

2. Activity Coefficient Model

In aqueous solution, a protonated amine will be decomposed into a free amine and a proton. A protonation reaction and its equilibrium constant are written in equations (1) and (2).

$$AmH^+ \leftrightarrow Am + H^+ \quad (1)$$

$$K_a = \frac{a_{Am}\ a_{H^+}}{a_{AmH^+}} \quad (2)$$

where a and subscripts Am and AmH^+ in equation (2) refer to the activity of the species, amine, and protonated amine, respectively. Equation (2) can then be rearranged to get equation (3) as follows:

$$\boldsymbol{pK_a = pH - log\left[\frac{\gamma_{Am}c_{Am}}{\gamma_{AmH^+}c_{AmH^+}}\right]} \quad (3)$$

where γ and c refer to species activity coefficient and concentration, respectively. The species activity coefficient is then determined by the use of the extended Debye-Hückel equation [23] as written in equation (4). This equation had also been used by [11] to evaluate the species activity coefficient for the 2-((2-Aminoethyl) amino) ethanol-water system.

$$\boldsymbol{\ln\gamma_i = \frac{-2.303A_\gamma z_i^2 I^{1/2}}{1+Ba_i I^{1/2}} + 2\sum_j \beta_{ij} m_j} \quad (4)$$

The first term of equation (4) represents the electrostatic forces while the second one represents the Van der Waals (VdW) forces. Since the effect of the VdW forces is insignificant, their contribution can, therefore, be disregarded. The values of A_γ, a_i, and B can be found elsewhere [24, 25]. As seen in equation (4), the species activity coefficient is also ionic strength I dependence where it is defined as follows:

$$\boldsymbol{I = \frac{1}{2}\sum_j m_j z_j^2} \quad (5)$$

where m_j and z_j are the species molality and the ion charge number, respectively.

3. Materials and Method

The materials used in this work were MEA (min. 98 wt.%, Sigma-Aldrich), sulfuric acid (H2SO4, 96.0 wt.%, Merck), and distilled water.

The experimental setup used in this work was similar to that used by [21, 22] which consists of a 300-mL jacketed reactor, a pH meter, a TFA® digital thermometer, a water bath, a digital balance, a magnetic stirrer, and a 5-mL syringe.

About 0.5 g of MEA was added into a beaker glass containing 50 g of distilled water. The solution was then stirred and a certain amount of 0.1 M H_2SO_4 solution was discreetly added into the beaker glass until an equivalence point was reached. The equivalence point can be evaluated graphically by choosing the lowest pH slope or $(dpH/dV)_{min}$. A detailed procedure can be found elsewhere [21, 22].

4. Results and Discussion

MEA protonation constants were determined at temperatures ranging from 303 to 333K by the use of the potentiometric titration (PT) method. The results can be seen in table 1 in which the pK_a values decrease as the temperatures increase. Based on equation (3), when pH equals pK_a then the activity of free MEA (a_{MEA}) and that of protonated MEA (a_{MEAH+}) are the same. This is confirmed by the experimental results. As seen in figure 1, the activity of protonated MEA decreases as pH increases, while the activity of free MEA increases as pH increases. These values intersect at pH 9.41 which is similar to the experimental pK_a value at 303K. In addition, if the concentrations of MEA and $MEAH^+$ are plotted, they will intersect at pH 9.53 (see figure 1) and this value is higher than the pK_a value from the experiment. This means that the activity coefficient of $MEAH^+$ is not equal to unity, i.e. non ideal system.

Table 1. Experimental data.

	Temperature (K)			
	303	313	323	333
W_{H2O} (g)	50.0093	50.0180	50.0363	50.0213
W_{MEA} (g)	0.5559	0.3257	0.6057	1.0666
$C_{MEA,0}$ (mol.L^{-1})	0.1807	0.1064	0.1966	0.3432
$m_{MEA,0}$ (mol/kg H_2O)	0.1820	0.1066	0.1982	0.3491
Number of data	79	76	89	87
pK_a	9.41	9.18	8.89	8.68

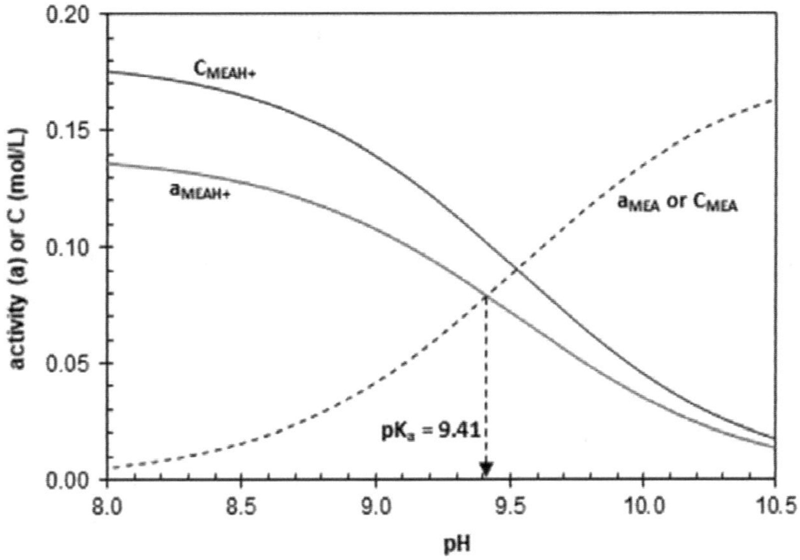

Figure 1. Activities and concentrations of MEA and $MEAH^+$ as a function of pH at 303K.

By neglecting the short-range VdW forces, the activity coefficient of the protonated MEA (γMEAH+) is equal to 0.77 at 303K with the ionic strength of 0.09 (see equation 4). Since the free MEA is a neutral species, its ionic strength will, therefore, be equal to unity (see equation 5). The changes in the activity coefficients and the ionic strengths of the protonated MEA during titration are shown in figure 2. It can be seen from the figure that the activity coefficients of $MEAH^+$ are constant from pH > 4, but they decrease at pH < 4. This occurs because the ionic strengths start to increase as the concentration of ions increases due to dilution of titrant, i.e. H_2SO_4. This has also been confirmed by [16] in which the solution ionic strength could be determined not only by the salt concentration but also due to dilution from the titrant.

Figure 3 shows the species activities at temperatures from 303 to 333K. It can be seen that the activity of free MEA decreases as pH decreases, but the opposite for the activity of $MEAH^+$. In addition, from the figure, it can also be seen that the intersection between the activity of free MEA and that of $MEAH^+$ gives the pK_a values at different temperatures.

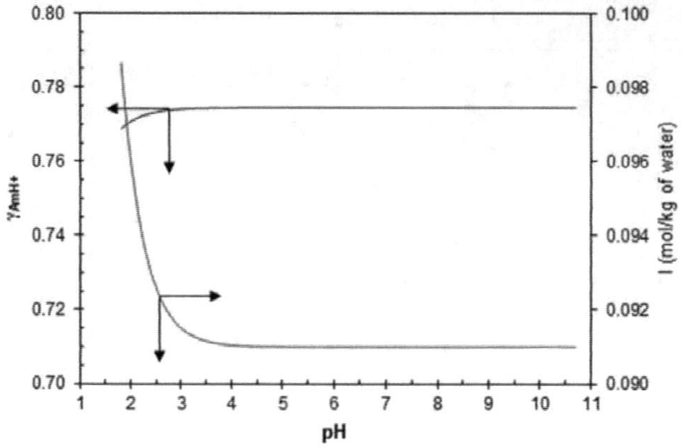

Figure 2. Change in the activity coefficients and the ionic strengths of $MEAH^+$ as a function of pH at 303K.

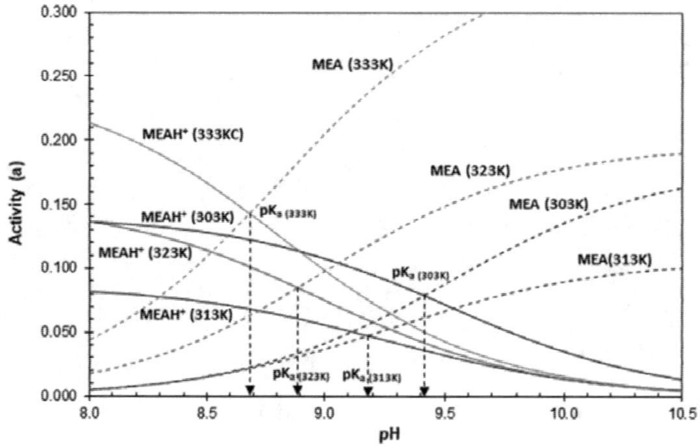

Figure 3. The activity of species as a function of pH from 303 to 333K.

5. Conclusion

The pK_a values of MEA were measured at temperatures from 303 to 333K by the PT method. Evaluation of the spices activity for the MEA-H_2O system has also been conducted at those temperatures. The results show that the pKa values decrease as the temperatures increase. Moreover, the extended Debye-Hückel model was implemented to determine the species activity coefficients. It was observed that the activity of free MEA increases as pH increases while the activity of the protonated MEA decreases as pH increases.

Acknowledgment

The Chemical Engineering Department, Universitas Islam Indonesia, is acknowledged by the authors for the Research Grant 2019.

References

[1] Ma'mun S, Svendsen H F, Hoff K A, Juliussen O 2007 Selection of new absorbents for carbon dioxide capture *Energ. Convers. Manage.* **48(1)** 251-8.

[2] Idem R, Wilson M, Tontiwachwuthikul P, Chakma A, Veawab A, Aroonwilas A, Gelowitz D 2006 Pilot plant studies of the CO_2 capture performance of aqueous MEA and mixed MEA/MDEA solvents at the University of Regina CO_2 capture technology development plant and the Boundary Dam CO_2 Capture Demonstration Plant *Ind. Eng. Chem. Res.* **45(8)** 2414-20.

[3] Lu Y, Ye X, Zhang Z, Khodayari A, Djukadi T 2011 Development of a carbonate absorption-

based process for post-combustion CO_2 capture: the role of biocatalyst to promote CO_2 absorption rate *Energy Proc.* **4** 1286-93.

[4] Alstom Chilled ammonia-based wet scrubbing for post-combustion CO_2 capture (DOE/NETL report No. 401/021507, 2007).

[5] Knuutila H, Aronu U. E, Kvamsdal H. M, Chikukwa A 2011 Post combustion CO_2 capture with an amino acid salt *Energy Proc.* **4** 1550-7.

[6] Ma'mun S, Kim I 2013 Selection and characterization of phase-change solvent for carbon dioxide capture: precipitating system *Energy Proc.* **37** 331-9.

[7] Ma'mun S, Svendsen H. F, Bendiyasa I 2018 Amine-based carbon dioxide absorption: evaluation of kinetic and mass transfer parameters *J. Mechan.l Eng. Sci.* **12** 4088.

[8] Ma'mun S 2014 Solubility of carbon dioxide in aqueous solution of potassium sarcosine from 353 to 393K *Energy Proc.* **51** 191-6.

[9] Hoff K. A, da Silva E. F, Kim I, Grimstvedt A, Ma'mun S 2013 Solvent development in post combustion CO_2 capture-Selection criteria and optimization of solvent performance, cost and environmental impact *Energy Proc.* **37** 292-9.

[10] Ma'mun S, Dindore V. Y, Svendsen H. F 2007 Kinetics of the reaction of carbon dioxide with aqueous solutions of 2-((2-aminoethyl)amino)ethanol *Ind. Eng. Chem. Res.* **46(2)** 385-94.

[11] Ma'mun S, Jakobsen J. P, Svendsen H. F, Juliussen O 2006 Experimental and modeling study of the solubility of carbon dioxide in aqueous 30 mass% 2-((2-aminoethyl)amino)ethanol solution *Ind. Eng. Chem. Res.* **45(8)** 2505-12.

[12] Ma'mun S, Nilsen R, Svendsen H. F, Juliussen O 2005 Solubility of carbon dioxide in 30 mass % monoethanolamine and 50 mass% methyldiethanolamine solutions *J. Chem. Eng. Data* **50(2)** 630-4.

[13] Liu Y, Zhang L, Watanasiri S 1999 Representing vapor–liquid equilibrium for an aqueous MEA-CO_2 system using the Electrolyte Nonrandom-Two-Liquid Model *Ind. Eng. Chem. Res.* **38(5)** 2080-90.

[14] Jou F. Y, Mather A. E, Otto F. D 1995 The solubility of CO_2 in a 30 mass percent monoethanolamine solution *Can. J. Chem. Eng.* **73(1)** 140-7.

[15] Li Y. G, Mather A. E 1994 Correlation and prediction of the solubility of carbon dioxide in a mixed alkanolamine solution *Ind. Eng. Chem. Res.* **33(8)** 2006-15.

[16] Hartono A, Saeed M, Kim I, Svendsen H. F 2014 Protonation constant (pKa) of MDEA in water as function of temperature and ionic strength *Energy Proc.* **63** 1122-8.

[17] Hamborg E. S, Versteeg G. F 2009 Dissociation constants and thermodynamic properties of amines and alkanolamines from (293 to 353)K *J. Chem. Eng. Data* **54(4)** 1318-28.

[18] Kim I, Jens C. M, Grimstvedt A, Svendsen H. F 2011 Thermodynamics of protonation of amines in aqueous solutions at elevated temperatures *J. Chem. Thermodynamics* **43(11)** 1754-62.

[19] Kim I, Grimstvedt A, da Silva E. F 2011 Thermodynamics of protonation of alkanolamines in aqueous solutions *Energy Proc.* **4** 576-82.

[20] Fernandes D, Conway W, Wang X, Burns R, Lawrance G, Maeder M, Puxty G 2012 Protonation constants and thermodynamic properties of amines for post combustion capture of CO_2 *J. Chem. Thermodynamics* **51** 97-102.

[21] Ma'mun S, Kurniawan D, Amelia E, Rahmat V, Alwani D. R 2017 Experimental determination of monoethanolamine protonation constant and its temperature dependency *MATEC Web of Conf.* **101** EDP Science p02001.

[22] Ma'mun S, Chafidz A, Indrayanto E, Setiawan P. K 2019 The effect of ionic strength on protonation constant of monoethanolamine in water at 303K *IOP Conf. Ser: Mater. Sci. Eng.* **1295(1)** 012017.

[23] Debye V. P 1923 The Theory of Electrolyte *Physikalische Zeitschrift* **24** 185.

[24] Manov G. G, Bates R. G, Hamer W. J, Acree S. F 1943 Values of the constants in the Debye–Hückel equation for activity coefficients1 *J. Am. Chem. Soc.* **65(9)** 1765-7.

[25] Kielland J 1937 Individual activity coefficients of ions in aqueous solutions. *J. Am. Chem. Soc.* **59(9)** 1675-8.

IOP Conf. Series: Materials Science and Engineering **811** (2020) 012033 doi:10.1088/1757-899X/811/1/012033

Amino-Acid-Salt-based Carbon Dioxide Capture: Precipitation Behavior of Potassium Sarcosine Solution

S Ma'mun[1, 2, 3]

[1]Department of Chemical Engineering, Universitas Islam Indonesia
[2]Jl. Kaliurang Km. 14.5, Yogyakarta 55501, Indonesia
[3]SINTEF Materials and Chemistry, N-7465 Trondheim, Norway

sholeh.mamun@uii.ac.id

Abstract. This work studied the precipitation behavior of solids in the aqueous solution of potassium sarcosine (KSar) during the CO_2 capture process with the concentration of 5m' (mol/kg solution). The effect of precipitation on the CO_2 capture capabilities was also studied. The results showed that 5m' KSar gives a comparable result to MEA. The relatively high initial CO_2 absorption efficiency combined with the positive effect of precipitation on CO_2 absorption rate and CO_2 capacity have made the 5m' KSar be a promising solvent for the CO_2 capturing process with precipitation.

1. Introduction

Carbon dioxide (CO2) is produced in large quantities from coal-fired power plants, industrial processes, natural gas processing, and transportation. Amine-based CO2 capture processes have been considered as a leading technology to remove CO2 from various flue gases. Aqueous monoethanolamine (MEA) has been used extensively for this purpose because it offers some advantages such as high reactivity, low solvent cost, reasonable thermal stability, and thermal degradation rate [1]. However, due to its drawbacks, e.g. corrosion of equipment, the high energy requirement for absorbent regeneration, and oxidative degradation [2], development on new solvents are therefore required to overcome these problems.

Amino acid salts (AAS) were previously used in acid gas removal as promoters. However, some research on AAS application in CO2 capture processes was conducted. The AAS may, according to [3, 4], be used as alternative solvents to alkanolamines. Due to the presence of an amino function group, aqueous solutions of AAS have similar performance on CO2 reactivity and capacity to amines [5]. Some potassium salts of amino acid, such as potassium sarcosine (KSar), potassium glycine (KGly), and potassium proline (KPro), show faster reaction rate with CO2 than MEA [6-8]. In addition, the AAS also show good degradation-resistance performance in the oxygenated atmosphere [5]. The ionic structure of the AAS makes its volatility to be relatively low even at a high temperature, therefore, it reduces the solvent vaporization losses.

Precipitation is normally avoided during post-combustion processes. In recent years, however, attention on the precipitating systems has increased, such as chilled ammonia processes [9], carbonate systems [10], and AAS [11, 12]. The AAS solutions may have a promising feature of the solid formation when absorbing CO2. When the formed precipitates contain CO2 species such as carbamate and bicarbonate, the selective removal of solids from a reaction mixture will shift the equilibrium towards the product side, thus it may increase the absorption capacity. It was confirmed that, in the

AAS solutions, the formation of solid may increase the total CO_2 loading and reduce the energy consumption during CO_2 desorption [13-15].

In this work, the concentration of KSar used was 5 m' as this system was also tested by [16]. In order to understand the effect of precipitation on the CO_2 capturing efficiency, a LabMax® system equipped with online probes for in-situ Focused Beam Reflectance Measurement (FBRM) and Particle Vision Measurement (PVM) was used in which both the crystallization behavior and the CO_2 absorption profile can be recorded simultaneously.

2. Materials and Method

2.1. Materials

Sarcosine (≥ 98 wt. %) and KOH (≥ 85 wt. %) were used in this work and were obtained from Sigma Aldric while CO2 (purity ≥ 99.99mol %) and N2 (purity ≥ 99.999mol %) were obtained from AGA Gas GmbH. The aqueous solution of 5m' KSar was prepared by neutralizing 5 moles of sarcosine with an equimolar amount of KOH in a total solution weight of 1 kg. The actual solution concentration was then determined by potentiometric titration.

2.2. Apparatus and Procedure

The precipitation behavior and absorption rates of CO2 into the AAS solutions were studied in a commercially available automatic laboratory reactor Mettler Toledo LabMax®. FBRM and PVM probes were used to record precipitation behavior during the experiment. FBRM spectra may monitor the onset of crystallization and the chord length distribution (CLD) of the formed solids and PVM was used to observe the morphology of crystals online. The detailed procedure can be found elsewhere [17].

After terminating the process, a liquid sample was collected and the CO2 concentration in the liquid was measured offline by both a GC and the precipitation-titration method as described by [18, 19], whereas the total alkalinity was determined by an automatic titrator Metrohm 905 Titrando. The absorption rate was then calculated by an equation from [1].

3. Results and Discussion

Absorption of CO2 into 5m' KSar was conducted by the use of LabMax®. Figure 1 shows the changes in absorption rate, pH, Tr, and FBRM spectra in the KSar solution after bubbling 10 vol.% CO2 at the flowrate of 5 L/min. The solution temperature was kept at 40.0 °C for at least 20 min to ensure the pH and FBRM spectra stable before loading CO2. The onset of bubbling gases was recorded as time zero. The temperature was maintained at 40.0 °C for the entire experiment with slight fluctuation at the start of loading gas. The pH, with the initial value of 12.6, kept decreasing after loading CO2, accompanying with the decline of absorption rate with the initial absorption rate of 7.96x10–4mol/(kg·s) which is much higher in comparison to the initial absorption rate in 5m' MEA, i.e. 4.25x10–4mol/(kg·s) [1, 20]. The onset of precipitation was detected by a sudden increase of FBRM spectra (blue line in figure 1) at 148 min, resulting in an upward shift of CO2 absorption rate (pink line in figure 1). As mentioned by [16], the shift could be due to the blockage of the gas line or by the formation of solids during CO2 capture. Since the tube was used to feed the gas in this work instead of a sparger, the problem of the blockage was avoided; the increase of absorption rate can, therefore, be confirmed as a result of the occurrence of precipitation. This was also confirmed by [17] in which the Vapor-Liquid-Solid Equilibrium (VLSE) data of 5m' KSar at 40°C showed that the CO2 partial pressures decreased in the precipitation region.

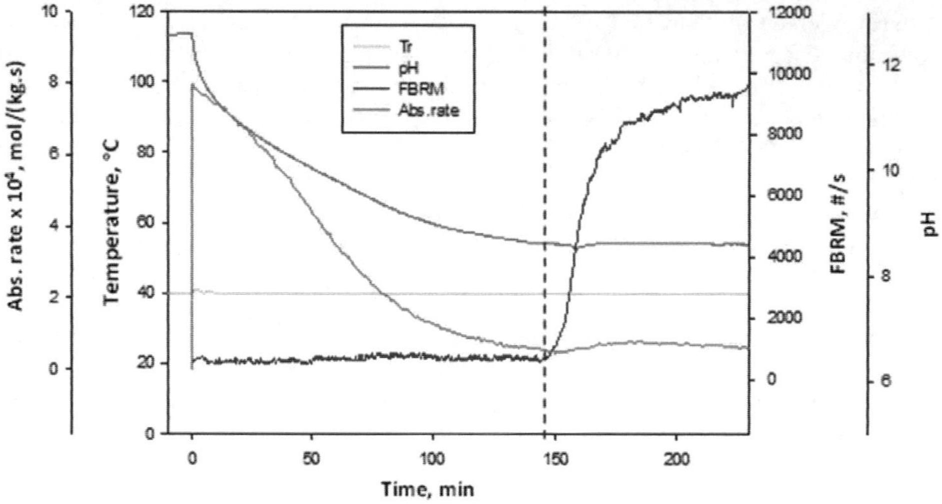

Figure 1. The changes of absorption rate, pH, T_r, and FBRM spectra by absorbing 10 vol.% CO_2 at the gas mixture flowrate of 5 L/min in a 5m' KSar solution at 40.0°C.

When the absorption rate is plotted against the CO_2 loading as shown in figure 2, the change of absorption rate caused by precipitation is clear. Precipitation occurred at the CO2 loading of about 0.56, where the absorption rate showed the lowest point at 0.38×10-4mol/(kg.s), following an evident increase up to 0.68×10-4mol/(kg·s). The crystal growth, taken by PVM, can be seen in figure 3. It should be noted that even though the precipitation led to a rise in the CO2 absorption rate, it took a relatively long time (over 2 hours) to form solids, in which case a very large absorption column would be required. In addition, the absorption rate had reduced to a quite low value before precipitation. These limitations could affect the use of 5m' KSar as a candidate for CO2 capture with precipitation.

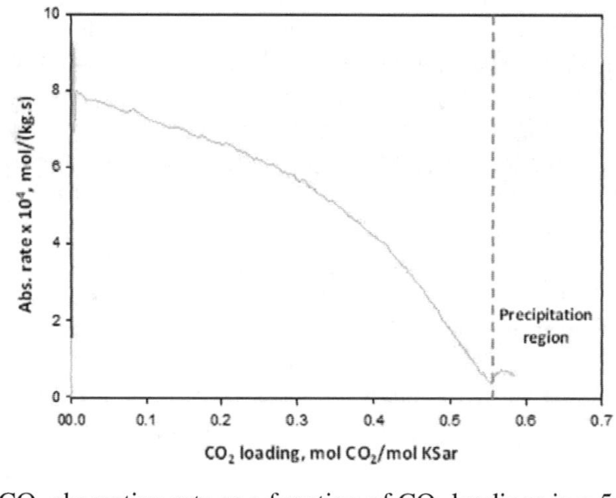

Figure 2. The CO_2 absorption rate as a function of CO_2 loadings in a 5m' KSar solution by loading a 10 vol.% CO_2 at the gas mixture flowrate of 5 L/min at 40.0°C.

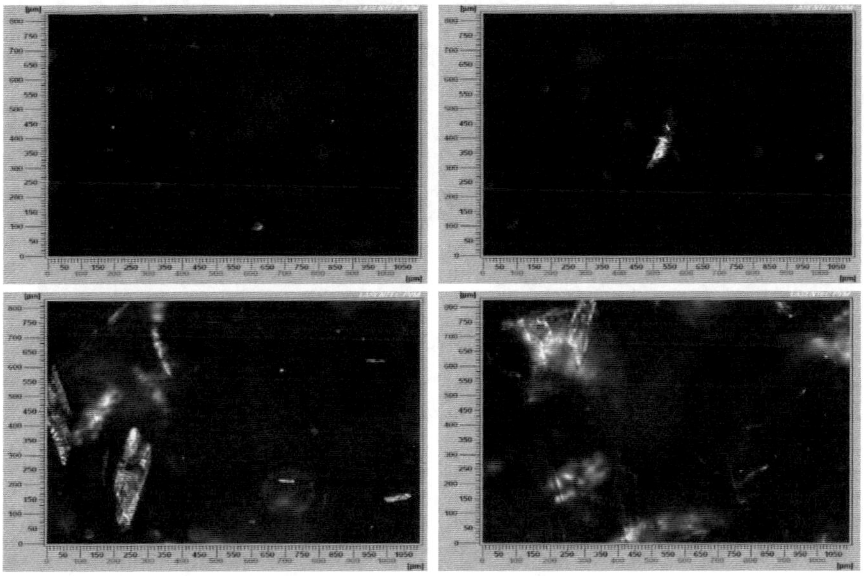

Figure 3. The crystal growth pictures taken by PVM at 40.0°C.

4. Conclusion

This work studied the effect of precipitation in the aqueous solution of 5m' KSar on the CO2 absorption efficiency. It was found that the 5m' KSar solution exhibited a high initial CO2 capture rate in comparison to 5m' MEA solution. Although the occurrence of precipitation led to an upward shift on the absorption rate and CO2 capacity, it would take too long to obtain crystals and the absorption rate at that condition was very low. Thus, the KSar solution was not considered as a good candidate for the CO_2 capture process with precipitation.

References

[1] Ma'mun S, Svendsen H F, Hoff K A, Juliussen O 2007 Selection of new absorbents for carbon dioxide capture *Energy Conversion Manag.* **48(1)** 251-8.

[2] Goff G S, Rochelle G T 2004 Monoethanolamine degradation: O_2 mass transfer effects under CO_2 capture conditions *Industr. Eng. Chem. Res.* **43(20)** 6400-8.

[3] Kumar P S, Hogendoorn J A, Versteeg G F, Feron P H M 2003 Kinetics of the reaction of CO_2 with aqueous potassium salt of taurine and glycine *AIChE J.* **49(1)** 203-13.

[4] Hamborg E S, Niederer J P, Versteeg G F 2007 Dissociation constants and thermodynamic properties of amino acids used in CO_2 absorption from (293 to 353)K *J. Chem. Eng. Data.* **52(6)** 2491-502.

[5] Hook R J 1997 An investigation of some sterically hindered amines as potential carbon dioxide scrubbing compounds *Industr. Eng. Chem. Res.* **36(5)** 1779-90.

[6] Aronu U E, Svendsen H F, Hoff K A 2010 Investigation of amine amino acid salts for carbon dioxide absorption *Int. J. Greenhouse Gas Control.* **4(5)** 771-5.

[7] van Holst J, Versteeg G F, Brilman D W F, Hogendoorn J A 2009 Kinetic study of CO_2 with various amino acid salts in aqueous solution *Chem. Eng. Sci.* **64(1)** 59-68.

[8] Song H J, Park S, Kim H, Gaur A, Park J W, Lee S J 2012 Carbon dioxide absorption characteristics of aqueous amino acid salt solutions *Int. J. Greenhouse Gas Control* **11** 64-72.

[9] Darde V, Thomsen K, van Well W J M, Stenby E H 2009 Chilled ammonia process for CO_2 capture *Energy Procedia.* **1(1)** 1035-42.

[10] Endo K, Stevens G, Hooper B, Kentish S E Anderson C 2011 A process and plant for removing acid gases *WO Patent.* 2011/130796 A1.

[11] Ma'mun S 2014 Solubility of carbon dioxide in aqueous solution of potassium sarcosine from 353 to 393K *Energy Procedia* **51** 191-6.

[12] Wang X, Akhmedov N G, Hopkinson D, Hoffman J, Duan Y, Egbebi A, Resnik K, Li B 2016 Phase change amino acid salt separates into CO_2-rich and CO_2-lean phases upon interacting with CO_2 *Appl. Energy.* **161** 41-7.

[13] Feron P H M, ten Asbroek N 2004 New solvents based on amino-acid salts for CO_2 capture from flue gases *Greenhouse Gas Control Technol. 7* 1153-8.

[14] Lim J, Kim D H, Yoon Y, Jeong S K, Park K T, Nam S C 2012 Absorption of CO_2 into aqueous potassium salt solutions of l-Alanine and l-Proline *Energy Fuel* **26(6)** 3910-8.

[15] Sanchez-Fernandez E, Mercader F M, Misiak K, van der Ham L, Linders M, Goetheer E 2013 New process concepts for CO_2 capture based on precipitating amino acids *Energy Procedia* **37** 1160-71.

[16] Aronu U E, Ciftja A F, Kim I, Hartono A 2013 Understanding precipitation in amino acid salt systems at process conditions *Energy Procedia* **37** 233-40.

[17] Ma'mun S, Kim I 2013 Selection and characterization of phase-change solvent for carbon dioxide capture: precipitating system *Energy Procedia* **37** 331-9.

[18]Ma'mun S, Nilsen R, Svendsen H F, Juliussen O 2005 Solubility of carbon dioxide in 30 mass % monoethanolamine and 50 mass % methyldiethanolamine solutions *J. Chem. Eng. Data.* **50(2)** 630-4.

[19] Ma'mun S, Jakobsen J P, Svendsen H F, Juliussen O 2005 Experimental and modeling study of the solubility of carbon dioxide in aqueous 30 mass % 2-((2-aminoethyl)amino)ethanol solution *Industr. Eng. Chem. Res.* **45(8)** 2505-12.

[20] Ma'mun S, Svendsen H F, Bendiyasa I 2018 Amine-based carbon dioxide absorption: evaluation of kinetic and mass transfer parameters *J. Mech. Eng. Sci.* **12** 4088.

IOP Conf. Series: Materials Science and Engineering **811** (2020) 012034 doi:10.1088/1757-899X/811/1/012034

A Systematic Mapping Study of Innovative Cloud Applications

I Odun-Ayo[1], T Williams[1], Olamma Iheanetu[1], Modupe Odusami[2], Sherenne Bogle[3]

[1]Department of Computer and Information Sciences, Covenant University, Ogun State, Nigeria
[2]Department of Electrical and Information Engineering, Covenant University, Ogun State, Nigeria
[3]Humboldt State University, Arcata CA, USA

isaac.odun-ayo@covenantuniversity.edu.ng

Abstract. Cloud computing is a business paradigm wherein computers and computing related services are provided by Cloud Service Providers to consumers either as software, development platform, or infrastructure. Innovative applications are growing in a productive manner on the cloud landscape. Innovative applications are being developed for use in the area of e-learning, automotive processes, cloud containers and machine learning. The objective of this paper is to conduct a systematic mapping study of innovative cloud applications and experiences. The systematic map provided a structured overview of research work carried out and the frequency of publications, presenting them pictorially in form of a map. The obtained results showed that 7.34% of the publications were on development of innovative cloud applications in terms of model. Architecture and modelling, and simulation in relation to model were both at 13.76%, 11.93% of the papers respectively, while 8.26% of the articles were on deployment in terms of process. Architecture had most publication in the area of solution research, with 15.2%. For articles published on deployment and development, most were on solution research with 8.80% and 14.40% respectively. The outcome of this study will be beneficial to practitioners in the industry and academic researchers alike.

1. Introduction

The Cloud is a parallel and distributed computing system consisting of a collection of interconnected and often virtualized computers, which are dynamically provisioned and presented to customers as a single computing resource based on pre-agreed service levels between the Cloud Service Providers (CSPs) and the Cloud customers [1]. Although the CSPs are striving to provide very efficient and reliable services on the cloud, there are also issue of service level agreements [2]. Traditionally, there are three Cloud service models, namely: Software-as-a-Service (SaaS), Platform-as-a-Service (PaaS) and Infrastructure-as-a-Service (IaaS). These services are offered to clients in one of four deployment models - private, public, community and hybrid [3]. Cloud computing now offers some unique experiences and affect lives in completely new ways. In addition, cloud computing is very broad and there is a plethora of subjects such as virtualization and Cloud identity management requiring significant discussion. Such issues are directly or indirectly related to the focus of this study [4, 5].

There are numerous research works on going in the area of innovative Cloud applications and experiences targeted at both private individuals and corporate organizations alike. The Cloud-based

Accounting solution is a mobile SaaS that allows the use of mobile phones to access accounting data. Containerization presented in [6] for instance allows for application abstraction which results in application portability and rapid deployment. Cloud as a heating source in [7] uses heat generated from super computers to warm homes. Cloud-based car parking in [8], uses the Cloud to manage parking lots. In [9], the Cloud is used for remote cars diagnosis and to proffer solutions in event of a breakdown. With the development of private educational cloud, web based lecturing and collaboration are now a possibility [10]. In [11], a location-based augmented reality application is presented which uses mobile phones and geographic location information to provide transform the users' environs and provide a life-like digital experience.

With the increasing use of cloud computing, there has also been an increase in the need for applications which would provide cloud-based services. Some examples of innovations currently in the cloud includes intelligent databases, internet of things, advanced identity-based security, container operations, etc. There are various benefits of using cloud applications, this has also been acknowledged by cloud computing users. With a lot of innovations evolving in cloud computing in terms of cloud applications, there is the need to properly study and conduct research on the technology driving these applications in other to avoid challenges like data insecurity and enhance service quality and performance. In addition, it is important to study cloud innovations because individuals, corporations and government agencies benefit from these innovations. Researchers will therefore be motivated to improve on existing innovations or develop new innovations. The systematic mapping process is usually done in facets to allow for consideration of all aspects. Three facets were employed in this study namely; the topic, contribution and research facets. The topic facet was used to extract core aspects of innovative cloud applications. The research facet focused on the types of research carried out, while the contribution facets focused on the method and model applied.

The motivation for this work is that more researches are still required in the areas of innovative cloud applications. Hence, the problem this study seeks to address is identifying areas where there are dearth of papers and pointing it out to upcoming researchers. The purpose of this paper therefore, is to conduct a systematic mapping study of innovative cloud applications and experiences. The major contribution of this review is that a visual map was created with percentages indicating the extent of work that has been done in various research types and research contributions such as framework and model in this field of study. The remaining part of the paper is organized as follows: section 2 examines related work, section 3 discussed the materials and methods, while the results and discussion are presented in section 4. Finally, the paper is concluded in section 5 with recommendations for future studies.

2. Related Work

[12] examines how researchers should conduct the process of systematic mapping and identify how the guidelines should be updated based on lessons learned from existing systematic maps and systematic literature review guidelines. In the affirmative, the authors conducted a systematic mapping study of systematic maps, considering some practices of systematic review guidelines. They discovered that in the large number of the studies conducted, multiple guidelines were used and combined which lead to different ways in conducting systematic mapping studies. [13] assesses the impact of systematic literature review which are the recommended evidence-based software engineering methods for aggregating evidence. The authors used a manual search of Journals and conference proceedings. In total, of the 20 relevant studies, eight addressed research trends rather than technique evaluation. Seven systematic literature reviews addressed cost estimation. The quality of systematic literature reviews was fair with only three scoring less than 2 out of 4.

[14] is a systematic mapping study of cloud policy languages and programming models which based their work on the concepts of [12]. This study provided six classes of study in the areas of survey, framework, privacy, paradigms, accountability and reliability, and security in relation to the focus of study. The selected studies were applied on the contribution facet in relation to tool, method, and model. In addition, the selected studies were used on the research facet which dealt with evaluation, validation, solution, experience and opinion research. [15] is Cloud Design and Deployment Models: A Systematic Mapping Study. The classification scheme in relation to design and deployment models discussed six topics which are design, service deployment, implementation, configuration, privacy and security. The

selected studies were applied on the contribution facet based on metric, tool, method, and model. In addition, the selected studies were applied to the types of research in terms of evaluation, experience, opinion validation and solution research.

[16] is A Systematic Mapping Study On Cloud-Based Mobile Application Testing. The classification scheme in relation to cloud-based testing discussed four topics which are security testing, compatibility testing, functional testing and GUI testing. The selected studies were applied on the contribution facet based on framework, metric, tool, method, and model. In addition, the selected studies were applied to the types of research in terms of evaluation, validation and solution research. The study also examined Testing-as-a-Service using some contribution category consideration. From literature examined, there has been no work focused specifically on systematic mapping study of Cloud native applications.

3. Materials and Methods

3.1. The systematic mapping process

A systematic mapping study employs a visual representation process to grant insight into materials published in a field of study. This systematic mapping study was carried out using the formal guidelines for a systematic mapping study as described in [12]. It is defined in [17] as a repeatable process for extracting and interpreting available materials related to a research objective. The first step is the definition of research questions in which the scope of the study is outlined. A search is conducted for primary studies in the proposed area of study. The publications found are then screened to determine their relevance to the study. The next step is the key wording, which involves using the abstracts of the paper for designing a classification scheme. The last step is the process of data extraction, which results in the creation of the systematic map. These steps were rigorously followed in this paper. With respect to the title of this paper and in order to meet the objectives as well as answer the research questions, a total of 125 papers were deemed relevant out of an initial search comprising of 1,449 papers. This study covered the period 2013 – 2018.

3.2. Conduct Search for Primary Studies

An appropriate research question is essential for a comprehensive study, because the essence of a systematic map is to have an insight into the quality and type of research activities that have been done in a given field of study. These issues determine how the research questions are crafted in order to accurately capture the essence of the study. In this particular study, the research questions are as follows:

--RQ1: What areas in Innovative Cloud Applications and Experiences are addressed and how many articles cover the different areas?

--RQ2: What types of papers are published in the area and what particular evaluation and novelty did they introduce?

--RQ3: What research approaches do these studies apply and what contributions were enunciated?

Typically, searching for papers usually involves exploring major digital libraries. However, this can also be accomplished by manually searching through books, printed materials, conference proceedings and journals. For this study, only papers available on online databases were considered, thus excluding book and printed resources. The primary search focused on searching relevant literature from online database resources using the search string below. Thereafter, a backward snowballing was carried out to enhance the search process [18]. The search utilized five digital libraries which are ACM, IEEE, Science Direct, Springer and Scopus.

The search string for this study was designed in terms of outcome, population, comparison and intervention. The keywords used in the search string were taken from every aspect of the structure of the title of the study. For this study on innovative cloud applications and experiences, the search string used on the selected databases is:

(TITLE-ABS-KEY (Cloud) AND (TITLE (Innovative) AND (TITLE-ABS-KEY (Applications) OR TITLE-ABS-KEY (Experience)))).

The searches were performed using the customized search string above on document metadata to ensure that relevant studies were not omitted. For this study on innovative cloud applications and experiences all the result from the relevant databases relating to cloud and computer science were examined. The goal of the selection criteria was to find and include all papers relevant to the field of study. It is usual to use the inclusion and exclusion criteria to eliminate papers that are not relevant to a study. The inclusion and exclusion criteria were also used to remove papers that were not relevant to the research questions. Abstracts can usually be considered here as they provide concise yet sufficiently detailed information about the main focus of a paper. However, abstracts are not usually included in papers on editorials, summaries, tutorials, presentation slides, panel discussions and prefaces; hence these types of publications were excluded from the study. In building the classification scheme, a thematic analysis approach, which aimed at identifying, analyzing and reporting themes within the studies under consideration was adopted [19].

4. Results and Discussion

4.1. Topic Category and Contribution Facet

The contribution facet focus on discussion in the areas of framework, model, tool, evaluation, metric and method [20]. The topics category was central to this study. The topics that were expected during the classification scheme of innovative cloud applications and experiences are: deployment, advantages and disadvantages, development, architectures, modelling and simulation and orthogonal. The list of primary studies used for checking the topics against the types of contributions is at table 1. The systematic map of innovative cloud applications and experience is depicted in figure 1. On the X-axis of the left half are the results of the contribution facet. This category depicted the contributions to a study. The result showed that 41.28% of the publications discussed model in relation to innovative cloud applications and experience. While those on metric were 1.83%; tool was 19.27%; method was 23.85% and process were 13.76%.

From the figure 1, model discussion contributed 41.28% of the paper reviewed in this facet. The breakdown showed that 2.75% of model discussion were on deployment, 3.67% on advantages and disadvantages, 7.34% is on development, while architectures and modelling and simulation facets each has 13.76%. Other results on the contribution facets as it relates to topics are shown on figure 1.

Table 1. Primary Studies for Topic and Contribution facet.

Contribution Facet / Topic	Metric	Tool	Model	Method	Process
Deployment		48, 56, 76, 77, 104, 106,122,	19, 21, 60,	12, 13, 17, 34, 54, 107, 110,	1, 4, 8, 20, 22, 27, 36, 44, 123
Design		78, 81, 121, 93, 96, 113,	38, 73, 79, 91,	3, 40, 51, 52, 61, 71, 74, 80, 82, 94, 120, 124, 125	75
Development		14, 16, 18, 30, 33, 39, 59	28, 29, 35, 37, 41, 45, 46, 47, 50, 57, 86, 87, 89, 90	7	95, 70,
Architecture		58, 62, 66, 67, 83, 84, 85, 102, 105	2, 5, 6, 9, 10, 11, 15, 23	97, 111	25, 31, 114, 115,
Modelling and Simulation			42, 43, 49, 65, 68, 69, 72, 100, 117, 118, 119, 116, 99, 108, 109		
Percentage	**0.00%**	**22.22%**	**41.67%**	**22.22%**	**13.89%**

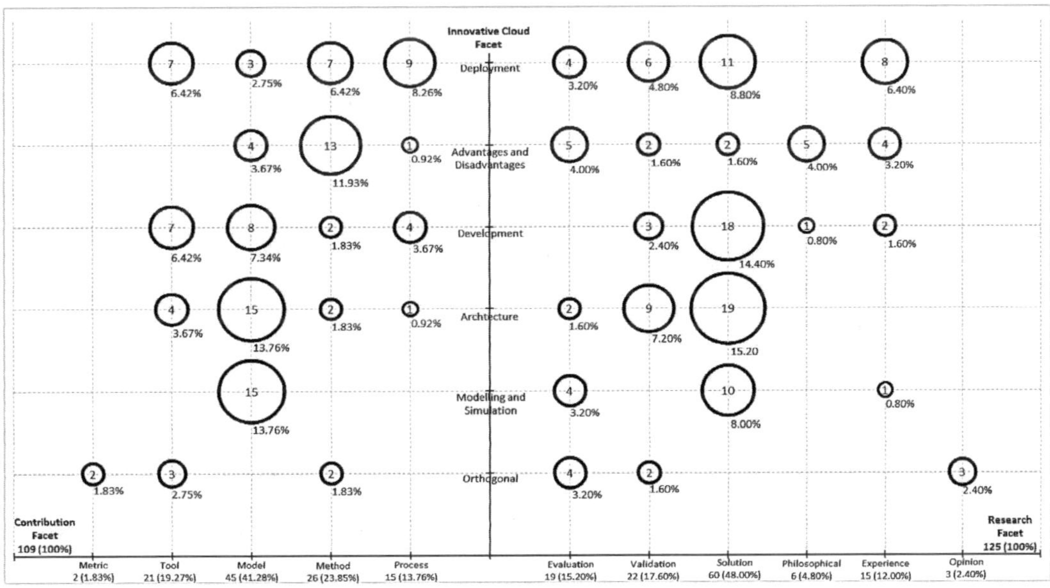

Figure 1. A Systematic Map on Innovative Cloud Applications and Experiences.

4.2. Topic Facet and Research Category

The categories and description of research works as listed in [21] was utilized for the third facet. They are validation research, evaluation research, solution proposals, philosophical papers, opinion paper and experience paper. The list of primary studies used for examining the topics against the types of research is at table 2. Depicted on the X-axis of the right half of figure 1 are the results of the type of researches conducted within the focus area. The results showed that 48% of the surveyed papers proposed solution, while 15.20%, were evaluation-based publications. 17.60% were validation-based, 4.80% were philosophical, 12.0% were experience and 2.40% were opinion papers. On the figure, Solution proposal contributed 48% of the paper reviewed in terms of type of research. Out of this 48%, 8.8% discussed deployment, 1.6% were on advantages and disadvantages, 14.4% were on development, 15.2% were on architectures and 8.0% were on modelling and simulation.

Table 2. Primary studies for topic and research facets.

Research Facet / **Topic**	Evaluation	Validation	Solution	Philosophical	Experience	Opinion
Deployment	76, 77, 104, 106,	54, 55, 53, 56, 122, 123	1, 4, 8, 12, 13, 17, 19, 21, 26, 36, 44		20, 22, 27, 34, 48, 60, 107, 110,	
Design	73, 91, 93, 96, 113, 78, 81, 120, 121	71, 79, 80, 82	3, 38	52, 74, 94, 124, 125	40, 51, 61, 75	101, 103, 98
Development		30, 33, 39	14, 16, 18, 28, 29, 35, 37, 41, 45, 46, 47, 50, 57, 59, 86, 87, 89, 90	7	95, 70,	
Architecture	63, 64,	58, 62, 66, 67, 83, 84, 85, 102, 105	2, 5, 6, 9, 10, 11, 15, 23, 24, 25, 31, 32, 88, 92, 97, 111, 112, 114, 115,			
Modelling and Simulation	116, 99, 108, 109		42, 43, 49, 65, 68, 69, 72, 117, 118, 119,		100	
Percentage	**15.20%**	**17.60%**	**48.00%**	**4.80%**	**12.00%**	**2.46%**

4.3. Major Findings

The result of the analysis carried out and presented on figure 1, made it easy to identify which areas had more emphasis based on the frequencies of publications. From figure 1:

a. It can be seen that 7.34% of the publications were on development of innovative cloud applications in terms of mode. Architecture and simulation in relation to model were both at 13.76%, 11.93% of the papers respectively. Finally, 8.26% of the articles were on deployment in terms of process.

b. Similarly on the right half there were more publications on advantages and disadvantages in terms of evaluation research which was 4.0%, more articles on advantage and disadvantage in terms of evaluation and philosophical at 4.0% and 3.2% respectively. Architecture had the most publications in the area of solution research, with 15.2%. For articles published on deployment and development, most were solution research with 8.80% and 14.40% respectively.

c. To the best of the researchers' knowledge, there were no articles identified by this study on modelling and simulation except under model which had 13.76%. There were also no articles on modelling and simulation in terms of validation and philosophical researches. Few articles were seen on orthogonal with the most being evaluation research at 3.20%. Similarly, no facets made contributions to Metric except Orthogonal. With respect to architecture, research work was only focused on evaluation, validation and solutions.

d. Generally, it was identified from the map that there were more publications on evaluation, validation and solution areas of research; and very few on philosophical and opinion.

The unique thing about a systematic map is that there are several interesting aspects of the results generated. The visual appeal of the map helps to summarize and provide information to potential researchers on what areas to focus on. There is no doubt that the bubble plot results will be quite useful. Suffice to mention, a systematic map without a systematic literate review has a unique value in itself, as it clearly helps to identify research gaps for further studies.

The relevance of this is that researchers at all levels and industries practitioners can use this as a starting point to conduct further studies. This study provided six classes of topics namely: deployment, advantages and disadvantages, development, architectures, modelling and simulation, and orthogonal in relation to innovative cloud applications and experience. In addition, the six classes of study can be discussed either in terms of tool, model, method, metric and process or in terms of evaluation, validation, solution, philosophical and opinion research. These areas amongst others are therefore recommended for future research. The list of primary studies would also assist intending researchers. The important lessons learnt in this study is that research work is a continuum and it is inexhaustible.

5. Conclusion

This systematic mapping study has been able to identify some areas where there was less emphasis in terms of innovative cloud applications and experience based on the categories used in the scheme. This paper has therefore contributed to knowledge by indicating different aspects of the study where there were gaps. The gaps that have been identified are recommended for further studies. It is expected that it would serve as a broad guide into topics that can be researched on in the area of innovative cloud applications and experience. Further research could also be carried out to validate this study or resolve contradictory issues. In summary, this study created a systematic map of innovative cloud applications and experience that can be beneficial to the cloud community. This study will help the researchers to uncover the critical gaps of innovative cloud applications and experience that many researchers were not able to explore. Thus, expanding the frontiers of knowledge in cloud computing.

References

[1] Buyya J, Bromberg K, Goscinski A 2011 Cloud computing: principles and paradigm. *John Wiley & Sons* 4-11.

[2] Odun-Ayo I, Udemezue B, Kilanko A 2019 Cloud Service Level Agreements and Resource Management, Advances in Science, Technology and Engineering Systems Journal **4(2)** 228-36.

[3] Mell P, Grance T 2011 The NIST Definition of Cloud Computing, NIST Special Publication 800-145.

[4] Odun-Ayo I, Idoko B, Abayomi-Zannu T 2019 Virtualization in Cloud Environment: Bandwidth Management, Advances in Science, Technology and Engineering Systems Journal **4(2)** 221-7.

[5] Abayomi-Zannu T, Odun-Ayo I 2019 Cloud Identity Management – A Critical Analysis, *Int. MultiConf. Engineers Comput. Scientists,* 170-5.

[6] Pahl C, Brogi A, Soldani J, Jamshidi P 2017 Cloud Container Technologies: A State-of-the-Art Review, *in IEEE Transactions on Cloud Computing* pp 14.

[7] Liu J, Goraczko M, James S, Belady C, Lu J, Whitehouse L 2011 The Data Furnace: Heating Up with Cloud Computing, HotCloud pp 5.

[8] Pham T, Tsai M, Nguyen D, Dow C, Deng D 2015 A Cloud-Based Smart-Parking System Based on Internet-Of-Things Technologies, *in IEEE Access* pp 1581-91.

[9] Amarasinghe M, Kottegoda S, Arachchi A, Muramudalige S, Bandara H, Azeez A 2015 Cloud-based driver monitoring and vehicle diagnostic with OBD2 telematics, *in Advances in ICT for Emerging Regions (ICTer) IEEE* pp. 243-9.

[10] Matthew S 2012 Implementation of Cloud Computing in Education – A Revolution, *Int. J. Comput. Theory Eng.* **4(3)** 473.

[11] Meenalishi S, Shriram K, Ritesh A, Santhosh C 2015 An Innovative App with for Location Finding with Augmented Reality Using CLOUD, *Procedia Comput. Sci.* **50** 585-9.

[12] Petersen K, Vakkalanka S, Kuzniarz L 2015 Guidelines for conducting systematic mapping studies in software engineering: An update, Information and Software Technology, **64** 1-18.

[13] Ahmad A, Brereton P, Andras P 2017 A systematic mapping study of empirical studies on software cloud testing methods. *In IEEE Int. Conf. Software Quality, Reliability Security Companion,* pp 555-62.

[14] Kitchenham B, Brereton O P, Budgen D, Turner M, Bailey J, Linkman S 2009 Systematic literature reviews in software engineering–a systematic literature review, Information and Software Technology, **51(1)** 7-15.

[15] Odun-Ayo I, Goddy-Worlu R, Yahaya J, Geteloma V 2019 A systematic mapping study of cloud policy languages and programming models, *J. King Saud Univ. – Comput. Inform. Sci.*

[16] Odun-Ayo I, Goddy-Worlu R, Samuel V, Geteloma V 2019 Cloud Design and Deployment Models: A Systematic Mapping Study. *BMC Reseaerch Notes* **12(1)** 436.

[17] Ya'u B I, Salleh N, Nordin A, Alwan A A, Idris N B, Abas H A 2019 A systematic mapping study on cloud-based mobile application testing. *J. Inform. Communication Technol.* **18(4)** 485-527.

[18] Muhammad A C, Muhammad A B 2014 A Systematic Mapping Study of Software Architectures for Cloud Based Systems, Software System section, IT University of Copenhagen.

[19] Wohlin C 2014 Guidelines for snowballing in systematic literature studies and a replication in software engineering. Paper presented at the Proceedings of the 18th international conference on evaluation and assessment in software engineering.

[20] Braun V, Clarke V 2006 Using thematic analysis in psychology. *Qualitative res. psychol.* **3(2)** 77-101.

[21] Wieringa R, Maiden N A, Mead N R, Rolland C 2006 Requirement Engineering Paper Classification and Evaluation Criteria, A Proposal and a Discussion, *In Requirement Eng.* **11(1)** 102-7.

IOP Conf. Series: Materials Science and Engineering **811** (2020) 012035 doi:10.1088/1757-899X/811/1/012035

Development of a Human Powered Pedal Washing MACHINE

M A Fajobi[1] *, E Y Salawu[1], J Azeta[1], J O Dirisu[1], O O Ajayi[1], P Onwordi[1]

[1]Mechanical Engineering Department, Covenant University, Ota, Ogun State, Nigeria

fajobi.muyiwa@covenantuniversity.edu.ng

Abstract. Human powered pedal washing machine has been designed using materials suitable for its application. An existing bicycle was used as the pedal and other parts of the machine such as the drum, the frame was fabricated using galvanized steel. The material selection was put into consideration, such as corrosion because of the machine's involvement with water. The human powered pedal washing machine was tested with a used dirty Laboratory coat. The pedaling washing was first done for the first 15 minutes, using water and sunlight detergent soap. There was a partial cleaning of the Laboratory coat, though the coat was very dirty. Another round of washing was done after which the first washing water was drained out through the outlet part of the machine. The second washing was excellent because the Laboratory coat was clean. The human powered pedal washing machine performed well with all the designed parts functioning well. This is an eco-friendly machine, maintenance cost free, energy conservation machine and highly sustainable for underdeveloped nations of the world.

1. Introduction

Cleaning of cloths by washing is a very important activity in human life which requires much energy to do. Washing of clothes with our hands has been in existence, which is as old as man and for centuries ago, before washing machines were invented. Humans make use of streams, nearby rivers or get water from wells to get their clothes washed by hand. The process of hand washing proved to be both consuming time and energy lose, but in order to address those two reasons, the innovation of a machine for washing came up, known as washing machine. Washing machine helps to make life easier and safe human energy. Research by Bhatawadekar et al. [1] revealed that washing machine is purposely constructed for cleaning cloths by using water as fluid and using of energies such as mechanical, chemical and thermal. The first washing machine by Alva Fisher was created in 1901, which uses electricity engine and propelled by the cylinder of a used engine [2]. Washing machines are designed to be electricity powered, whereby the clothes to be washed receives transfer of mechanical energy [3].

Machining is an innovative idea for technology advancement which involves designing of parts and assembling them together to form a whole [4]. The use of materials is not left out during machining of product design because they are the actually the raw ingredients used for making up any product into real life existence. Materials in engineering are of different types (iron, stainless steel, carbon steel etc) and they are chosen for use based on the application of the product to be designed for. They are considered for use due to the properties they possess. For example, good mechanical properties (hardness, strength, ductility, toughness, wear resistance) are contained in ductile iron [5], mild steel, stainless steel etc. There are many factors that are put into consideration during machining and design of products in mechanical engineering. Apart from the materials to be selected, protection of the material is also important, for instant corrosion is another part that was put into consideration for this research design project.

However, this research work focuses on the development a human powered pedal washing machine that does not require electricity power to work but only requires human activity to keep fit and helps to save electricity energy. This work is based on the principles of pedal motion, where the barrels are rotated causing motion washing in the inner barrel [4-6]. This principle was also used for a shredder equipment that was designed in a bicycle form and powered humanly [7], and for generating electricity produced by Suhalka et al [8].

2. Materials and Method

2.1. Material used

Material selection on the specific materials to be utilized for any application is often founded on the basis of such a material's properties, suitability of use in the environment of operation, performance, design consideration, cost and compatibility with other materials [9]. Recently, there has been an increase of consideration in the fact that materials are used for designed based on the particular properties utilized for a particular operation, enhanced due to the wide range of materials in existence today.

This aspect of manufacturing products is crucial to obtain an excellent performance from such products, due any design which does not have the proper consideration for its materials based on study and analysis of the specific functionality of each material in the operational procedure of the machine will fail before time. For example, corrosion was put into consideration for this design because of its involvement with water. Water is an excellent agent that aids corrosion, most especially in metals. Corrosion is an unlike natural event that occurs between materials (mild steel, aluminium, stainless steel) and its environment, leading to their gradual destruction [10-14]. This is caused by the flow of current of the metal involved due to the presence of air and any other water agent means.

Table 1. Materials used and reasons for selection.

Part	Material selected	Reason
Small aluminum pulley	Aluminum	Aluminum is non-corrosive and resistant to rust, so any splash of water won't affect it
Completely assembled bicycle	-	-
Shaft	Galvanized steel	Resistant to corrosion and durable
Outer cylinder	Galvanized steel	Resistant to corrosion and durable
Inner cylinder	Galvanized steel	Resistant to corrosion and durable
Mesh material	Galvanized Steel	Resistant to corrosion and durable
Bearing (2)	Galvanized steel	Resistant to corrosion and durable
One-way valve	Plastic and Galvanized steel	Resistant to corrosion and durable

2.2. Methods

Design considerations for the cylinder

The force carried, F_c = force of the water carried + force of the clothes carried

=F(water) + F(clothes)

For the design, we are only required to fill one-fifth of the cylinder

Therefore, the volume of water becomes; $= \frac{volume\ of\ cylinder}{5}$

$$volume\ of\ water = \frac{\pi r^2 h}{5}$$

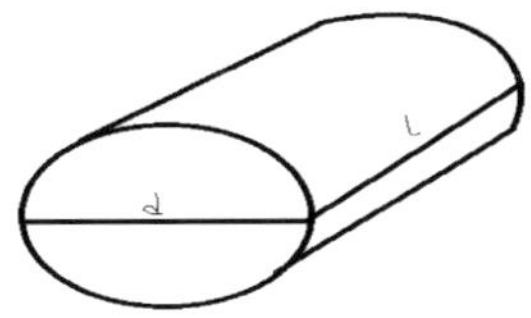

$$r = \frac{d}{2}$$
$$h = l$$

where d= 45.72 cm or 0.4572m

$d = 45.72cm = 0.4572m$ and

$l = 50cm = 0.5m$ (Length of the cylinder).

Therefore;

$$mass\ of\ water\ carried = volume * density\ (water)$$
$$weight = 9.81 * mass$$
$$weight = 9.81 * \frac{1}{5} * \pi * \left(\frac{0.4572}{2}\right)^2 * 0.5 * 1000 = 161.05N$$

$Average\ mass\ of\ one\ shirt = 0.125kg$ [15]

$Assuming\ 7\ shirts, weight = 0.125 * 9.81 * 5 = 6.131N$

$$thus, F_c = 161.05 + 6.131 = 167.18N$$

3. CAD Modelling Parts and Assembling

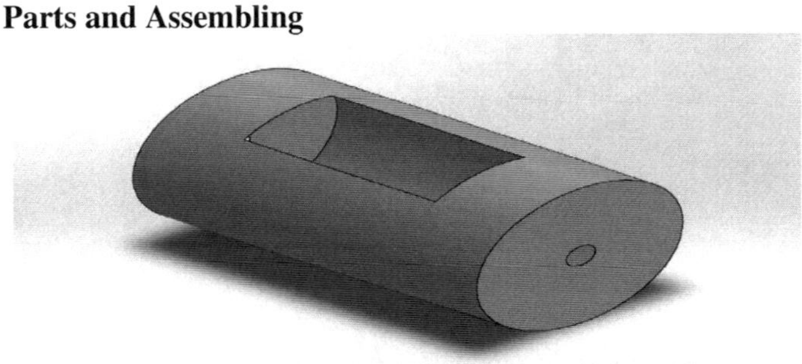

Figure 1. Cad Modelling for the cylinder shell.

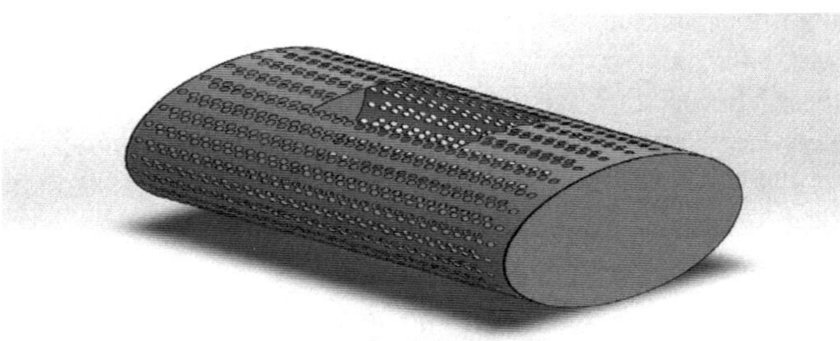

Figure 2. Cad Modelling for the perforated cylinder.

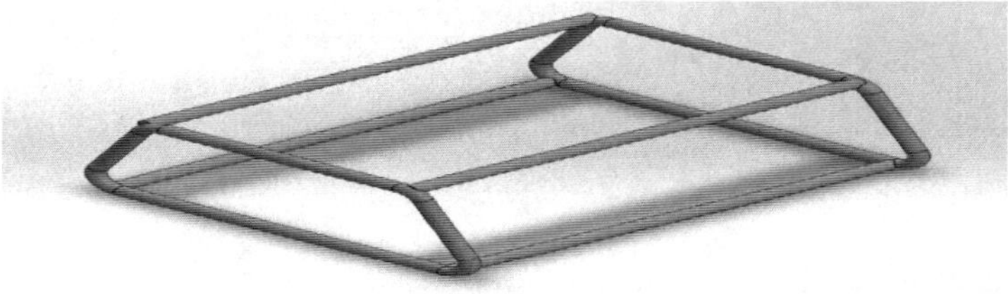

Figure 3. Cad Modelling for the cylinder stand.

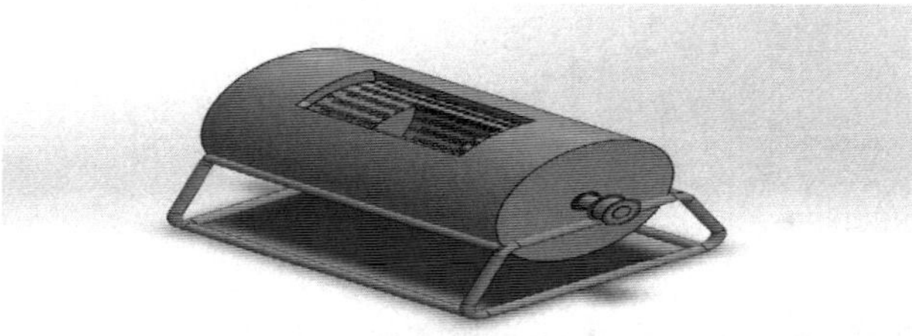

Figure 4. Cad Modelling for the assembling of the cylinder and the stand.

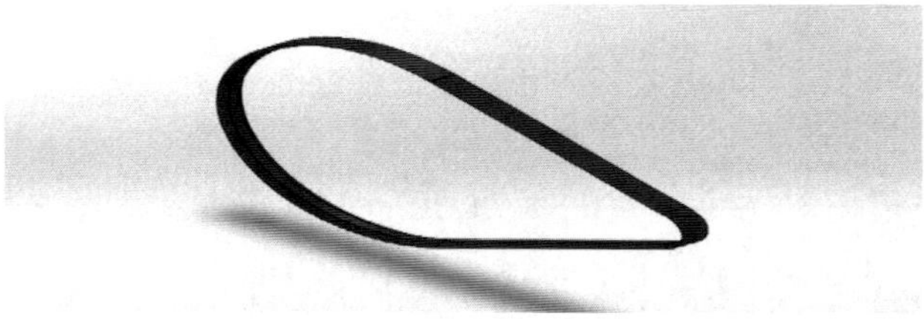

Figure 5. Cad Modelling for the belt.

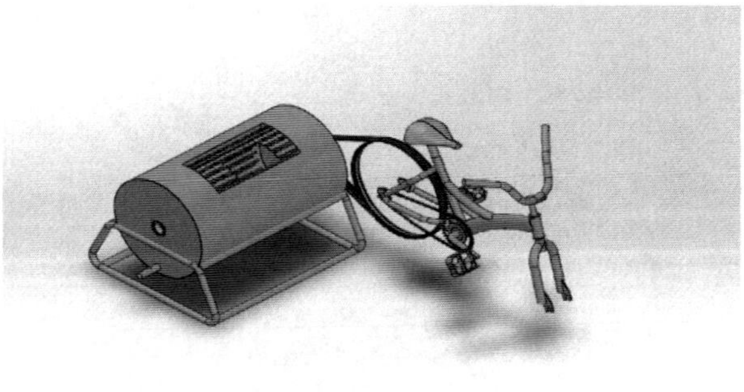

Figure 6. Cad Modelling of Assembling.

4. Results and Discussion

The human pedal washing machine was tested by using a Laboratory coat used from the mechanic workshop, with sunlight detergent soap and water. The pedal was powered and cycled for 15 minutes continuously. The result showed that part of the Laboratory coat was cleaned with the detergent water changing to black colour. The soap water was drained out from the out let tap and replaced with another water and detergent. The washing process was repeated for another 10-15 minutes. There was an improvement in the washing of the coat, which gave a clean wash after another round of cycling, which showed that the washing machine worked successful.

Figure 7. Complete Assembly of the Human powered pedal washing machine.

5. Conclusion
Human powered pedal washing Machine has been designed with various parts assembled together, using mostly galvanized steel to achieve the machine fabrications. The various CAD modelling of each parts were shown. The human powered pedal washing machine was tested and showed excellent performance in the washing after some minutes of continuous pedaling, which demonstrated the good assembling of the machine mechanisms. The human powered pedal washing machine serve as a washing machine for domestic use, mainly for house use, especially where there is no electricity and also for general exercise of individuals during use. It is easy to use and low cost of maintenance at the long run, energy sustainable (no electricity) and also its eco-friendly.

Acknowledgement
The authors appreciate the financial support of Covenant University towards the success of this research.

References
[1] Bhat Bhatawadekar G, Salman B, Chiplunkar N, Devrukhakar S, Akashdeep S 2015 Design and Fabrication of Pedal Powered Washing Machine. *Int. J. Eng. Res. General Sci.* **3(1)** 1307-11.
[2] Tawanda M, Tererai J M, Charles M 2017 Design and Fabrication of a Pedal Powered Washing Machine. *Int. Conf. Indust. Eng. Operat. Manage.* 355-66.
[3] Pal Pandian P, Salinpaul V 2015 Design and Fabrication of Pedal Power Top Loaded Washing Machine. *Int. J. Innovat. Res. Sci. Eng. Technol.* **4(8)** 6977-83.
[4] Okokpujie I P, Salawu E Y, Nwoke O N, Okonkwo U C, Ohijeagbon I O, Okokpujie K O 2018 Effects of Process Parameters on Vibration Frequency in Turning Operations of Perspex Material. *World Congress Eng.* **2** 1-8.
[5] Abioye A A, Abioye O P, Ajayi O O, Afolalu S A, Fajobi M A, Atanda P O 2018 Mechanical and microstructural characterization of ductile iron produced from fuel-fired rotary furnace. *Int. J. Mech. Eng. Technol. (IJMET).* **9(1)** 694-704.
[6] Akshay A, Shubham K, Ashwin D, Akshay M, Kolhe B P 2016 Pedal Operated Washing Machine. *Global J. Eng. Sci. Res.* 70-3.
[7] Azeta J, Salawu E Y, Fajobi M A, Ajayi O O, Dirisu J O, Onwardi P N, Aworinde A K 2019 Technical Innovations for Developing Countries: the case of a human-powered fruit blender. *Int. J. Mech. Eng. Technol. (IJMET),* **10(3)** 673-9.
[8] Suhalka R, Khandelwal M C, Sharma K K and Sanghi A 2014 Generation of Electrical Power using Bicycle Pedal. *Int. J. Recent Res. Rev.* **7(2)** 63-7.
[9] Ajayi O O 2019 MCE 412: Machine Design I, [PDF slides and Power Point slides]. Retrieved from https://www.coursehero.com/file/12917106/Machine-Design-MCE-412-Lecture-Note/
[10] Fayomi O S I, Akande I G, Oluwole O O, Daramola D 2018 Effect of water-soluble chitosan on the electrochemical corrosion behaviour of mild steel. *Chem. Data Collect.* **17** 321-6.

[11] Fayomi O S I, Atayero A A, Mubiayi M P, Akande I G, Adewuyi P A, Fajobi M A, Ayara W A 2019 Mechanical and opto-electrical response of embedded smart composite coating produced via electrodeposition technique for embedded system in defence application. *J. Alloys Compound.* **773** 305-13.

[12] Omotosho O A, Oluseyi O A, Ojo S F, Vincent O I 2012 Assessing the deterioration behaviour of mild steel in 2 M sulphuric acid using Bambusa glauscescens *Int. J. Appl. Eng. Res. Dindigul,* **2(2)** 406-18.

[13] Fajobi M A, Loto R T, Oluwole O O 2019 Corrosion in Crude Distillation Overhead System: A Review. *J. Bio-and Tribo-Corrosion*, **5(3)** 67.

[14] Loto R T, Tobilola O 2018 Corrosion inhibition properties of the synergistic effect of 4-hydroxy-3-methoxybenzaldehyde and hexadecyltrimethylammoniumbromide on mild steel in dilute acid solutions. *J. King Saud Univ.-Eng. Sci.* **30(4)** 384-90.

[15] Varadaraju R, Srinivasan J 2018 Design of sports clothing for hot environments. *Appl. ergonomics,* 1-8.

IOP Conf. Series: Materials Science and Engineering **811** (2020) 012036 doi:10.1088/1757-899X/811/1/012036

Contribution of Nano Crumb Rubber on Recycled Asphalt Concrete As Pedestrian Paving Block

S L Annisa[1], S P Hadiwardoyo[1], R J Sumabrata[1], D Iskandar[1]

[1]Civil Engineering Department, Faculty of Engineering, Universitas Indonesia, Depok, 16424, Indonesia

sigit@eng.ui.ac.id

Abstract. Stripping the road surface layers of asphalt concrete caused by various considerations that have been damaged by the aging process of asphalt. This stripping material will continue to grow in sufficient quantities if the recycling process is not utilized. The recycling process can be done by restoring the performance of asphalt or aggregate and can even occur both to meet the specifications required originally. In this research, asphalt concrete mixture has been used where the asphalt element has undergone an aging process so that it needs to be rejuvenated or softened. The addition of used oil has been used in the process of softening the hot Asphalt Concrete Wearing Course (ACWC) type. The process of increasing the performance of recycled blends has used Nano Crumb Rubber (NCR) material. The test results show that the ratio of the old asphalt content and used oil as a softener is obtained with a ratio of 11: 4 to obtain the required level of softening. This composition has produced a bitumen penetration value of 60, and the addition of NCR 2.5% and 5% to the asphalt content. Testing for recycled asphalt mixes is carried out with Marshall Test, Indirect Tensile Strength (ITS), and water absorption testing for recycled mixtures for paving blocks. The test results show the addition of NCR by the dry method has increased the strength of the mixture at both ambient and higher temperatures of 35 °C and 45 °C.

1. Introduction

Transportation infrastructure, especially road construction, has a significant influence on the environment. Road construction is a subsector that requires significant energy consumption because the total volume of material has been used on the road every year [1]. In Indonesia, flexible pavement is a type of road pavement that is mostly used on the highway. The many uses of asphalt have caused a lot of pavement waste called the Reclaimed Asphalt Pavement (RAP). This material is the residual waste of pavement that has been damaged or has expired. Not only from road construction, the high volume of vehicles that makes the use of tires increases can also cause waste of used tires. Negative impacts that require alternative steps to reduce these impacts. One of them is the utilization of road construction wastes such as RAP and used tires.

There are two main factors that influence the use of asphalt recycling (RAP) are economic savings and environmental benefits [2]. The use of RAP for road construction is also increasing because it is considered to be more cost-effective and included in environmental aspects that are sustainable [3]. Estimates of the average percentage of RAP use in asphalt mixtures have increased from 15.6 percent in 2009 to 20.4 percent in 2014 [4]. Based on research that has been done, asphalt mixes that use RAP with high levels can reduce energy use and greenhouse gases by 12.2% [5]. RAP is most commonly used as a substitute for virgin asphalt and aggregate material in recycled asphalt pavement, but is also used as a stabilization of basic aggregates and road embankments [6]. One of the main problems in the use of RAP is that the material contained has experienced aging which can affect pavement performance [7]. Although the use of RAP helps in reducing the utilization of virgin asphalt binder and increases rut resistance, the old binder material in RAP is considered as a potential contributing factor responsible for pavement thermal failure and failure of fatigue cracking [8]. The use of high levels of RAP in asphalt mixtures tends to decrease fatigue resistance [2]. There is a need to improve properties on asphalt mixture with RAP to obtain a good quality asphalt mixture, one of which is the addition of crumb rubber. From the research that has been done, the addition of crumb rubber to the recycled asphalt mixture increases fatigue resistance [8].

Based on the problems and research results that have been described above, this study has conducted research on the effect of the use of reclaimed asphalt pavement and crumb rubber on the strength of the paving block. From the results of previous studies, the crumb rubber shape modification has changed into nano-sized material to determine its effect on the strength of the paving block. The use of reclaimed asphalt pavement and crumb rubber in the manufacture of paving blocks, based on previous studies where the mixture test showed results that tend to be lower when compared to asphalt mixtures with virgin material. So that the recycled asphalt mixture is more suitable for light weight roads such as pedestrian roads or bicycle lanes. Paving blocks can simplify the technique of utilizing recycled asphalt mixture because the mixture has been made in the form of paving blocks. Such a process will simplify the road construction process because it does not require the use of heavy equipment such as rollers and long processes such as the laying and compaction of asphalt mixtures.

2. Materials and Methodes

1.1. Reclaimed Asphalt Pavement (RAP)

The recycled asphalt material that has been used is a material from the surface layer which is derived from the dredging of the Toll Road Jakarta Outer Ring Road (JORR). Asphalt mixture on the surface layer has Asphalt Concrete Wearing Course (AC-WC) specifications. RAP material tends to be of poor quality because the aggregate and asphalt contained has been damaged by the aging process. Therefore, it is necessary to add new aggregates and asphalt to improve the characteristics of recycled materials. Table 1 shows the results of characteristic tests for virgin asphalt and recycled asphalt material.

Table 1. Asphalt Characteristic.

Asphalt Characteristics					
No.	Type of Testing	Method	Indonesian Asphalt 60/70	Virgin Asphlat	Aged Asphalt
1	Penetration (25 C, 5 s, 100 g)/0.1 mm	SNI 2432:2011	67.75	62.5	8
2	Loss on heating (TFOT)	SNI 06-2440-1991	0.034	0.108	0.35
3	Penetration after TFOT	SNI 06-2456-1991	58.5	61	5.8
4	Softening Point (° C)	SNI 2432:2012	51.5	45	65.3
5	Flash Point (° C)	SNI 2433:2011	263	306	322.6
6	Solubility in Trichloroethylene (%)	AASHTO T44-03	99	99	89.8
7	Ductility (cm)	SNI 2432:2012	>100	107	12.3
8	Specific Gravity	SNI 03-6893-2002	-	1.048	1.07

1.2. Nano Crumb Rubber (NCR)

Crumb rubber is a material obtained from used tires in the form of small granules which are made in a certain size for material modification in asphalt mixtures. At present, the use of crumb rubber in asphalt mixes is increasing because it has the advantage of increasing the value of indirect tensile strength and skid resistance [9]. In addition, the addition of crumb rubber also increases fatigue resistance and deformation in road pavement [8].

There are two methods that can be used to mix crumb rubber into asphalt mixture, namely the wet and dry method. The wet method is done by first mixing crumb rubber with asphalt. The dry method is done by first mixing crumb rubber with aggregate. In this study, crumb rubber was modified to nano size to determine its effect on the mixture. The process of mixing nano crumb rubber (NCR) has been carried out by a dry method where NCR is added during the aggregate heating process.

1.3. Asphalt Mixtures
In this study, there are three types of specimens to be made namely two types of specimens for Marshall testing and one other type for Indirect Tensile Strength (ITS). There are 2 types of Marshall tests namely Marshall standard and Marshall immersion. To find the optimum asphalt content, variations of asphalt without nano crumb rubber with 0%, 5%, 5.5%, 6%, 6.5% and 7% are used. Then to find the optimum levels of nano crumb rubber with nano crumb rubber with variations in levels of 0%, 5%, 5.5%, 6%, and 6.5%. The third type is specimens with the best mix of asphalt and nano crumb rubber for paving block and ITS testing with temperature variations of 25 ° C, 35 ° C and 45 ° C.

1.4. Marshall Test
Marshall testing is an asphalt heat mixture test to get the optimum asphalt content and to determine the characteristics of the mixture. This test is carried out to determine the stability and plasticity (flow) of the asphalt mixture. Stability is the ability of asphalt mixtures to accept loads up to plastic discharge. Plastic discharge is a change in shape caused by a load until it reaches the collapse limit.

1.5. Indirect Tensile Strength Test
Indirect Tensile Strength testing for asphalt mixtures is carried out by placing a load on the cylindrical specimen perpendicular to the diameter of the specimen with a certain rate of deformation and temperature. The maximum load when failure occurs is used for the calculation of Indirect Tensile Strength on the asphalt mixture. This test is also carried out to evaluate the quality of the asphalt mixture and is used to determine the potential damage due to water when the test results are obtained from different specimens, namely when immersed and not soaking the test specimen [10]

2. Result and Discussion

2.1. Modified Asphalt
The addition of NCR in the asphalt mixture was carried out by the dry mix method in which NCR was added during the heating process of RAP material. The variation of NCR content used in this test is 0%, 5%, 5.5%, 6%, and 6.5% as in table 2. From the results of the Marshall test it was known that the largest value of Void in Mineral Aggregate (VMA) with a NCR content of 6%. The value of Void Filled with Asphalt (VFA) for modified asphalt mixture is smaller than asphalt mixture without NCR. This is due to the VFA being replaced by the NCR. The VIM value of the modified asphalt mixture has a higher than virgin asphalt mixtures, so that this mixture has properties more flexible and porous. The Marshall stability of the modified asphalt mixtures with NCR 2.5% has a higher strength. The effect of NCR content on the flow value of the modified asphalt mixture causes the flow value to decrease, while the Marshall quotient value is lower than the virgin asphalt mixture so that the modified asphalt mixture is more flexible. From the optimum NCR content that has been obtained, the variation of NCR content in asphalt mixtures in making ITS testing samples and paving blocks is 0%, 2.5%, and 5%. Table 3 shows the results of the Marshall test.

From table 3 and table 4 can be seen the change in the characteristics of asphalt mixture modification conditions before and after immersion. Table 5 has calculated the Residual Index value as a comparison of Marshall stability value from the effect of soaking for 24 hours at 60 °C. From this value, it can be seen the level of asphalt mixture susceptibility to temperature and water.

Table 2. Asphalt Modified Test Result.

NCR content	0.00%	5.00%	5.50%	6.00%	6.50%	AC-WC Specification
Void in Mineral Agreggate (VMA)	16.07	17.64	18.14	18.27	17.79	Min. 15
Void Filled With Asphalt (VFA)	80	71.58	69.05	68.5	70.69	Min.65
Void in Mixtures (VIM)	3.23	4.88	5.61	5.77	5.22	Min.3,5 Max.5
Marshall Stability (kg)	1416.4	1315.6	1236.6	1166.2	1190	Min.800
Flow	3.13	3.45	2.85	2.82	2.8	Min.3
Marshall Quotient (MQ)	453.44	381.38	439.15	416.01	426.93	Min.250

Table 3. Modified Asphalt Mixture Test Results.

NCR content	0%	2.50%	5%
VMA (%)	16.07	17.14	17.5
VFA (%)	80	74.07	72.22
VIM (%)	3.23	4.46	4.88
Stabilitas (Kg)	1416.4	1425.8	1315.6
Flow (mm)	3.13	3.3	3.45
Marshall Quotient (Kg/mm)	453.44	432.24	381.38

Table 4. Marshall Immersion Test Results.

NCR content	0%	2.50%	5%
VMA (%)	16.83	17.05	18.07
VFA (%)	75.6	74.46	69.48
VIM (%)	4.11	4.36	5.54
Stabilitas (Kg)	1391.6	1396.7	1275.2
Flow (mm)	2.95	2.88	2.98
Marshall Quotient (Kg/mm)	504.4	373.56	475.85

Table 5. Residual Index of Asphalt Mixtures.

NCR (%)	Marshall *Standard*	Marshall *Immersion*	Residual Index (%)
0	1416.42	1391.64	0.983
2.5	1425.76	1396.74	0.980
5	1315.64	1275.22	0.969

2.2. *Indirect Tensile Strength (ITS) Test Results*

Indirect Tensile Strength (ITS) testing has been carried out at the Bandung Road and Bridge Testing Center using the Universal Material Testing Apparatus (UMATTA), and the sample preparation process at the Structure and Material Laboratory of the Department of Civil Engineering, University of Indonesia. Sample measurements were carried out three times to obtain height and diameter data from the sample. Tests on test specimens are carried out by giving vertical direction shock loads to the sample.

The shock load given depends on the temperature of the test. For testing at 25 ° C a shock load of 2000 N - 2500 N has been given, while for temperatures of 35 ° C and 45 ° C, the shock load given is 900 N - 1500 N. This loading is the result of a computer simulation based on Poisson theory, and the sample will deform in the horizontal direction, then the instrument will read the ability of the sample to return to its original condition or called deformation recovery. Figure 1 is the result of testing from Indirect Tensile Strength (ITS).

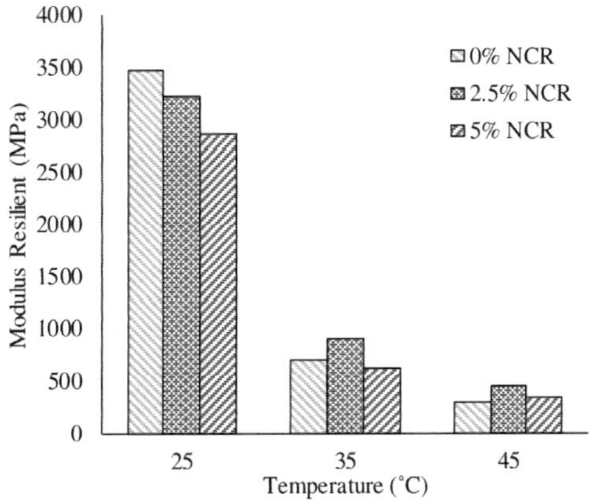

Figure 1. Average Resilient Modulus Value with different temperatures.

From the figure 2, the greatest resilient modulus for 25 ° C is in the virgin asphalt mixture without NCR. However, resilient modulus values with 2.5% NCR content have higher values at the test temperature at 35 ° C and 45 ° C. This has shown that the modified asphalt mixture with NCR has a stronger ability to higher temperatures. However, on the contrary the addition of 5% NCR has reduced the value of resilient modulus. This shows the presence of optimum NCR content in the asphalt mixture at 2.5%.

Figure 1, the lowest horizontal deformation value in the virgin asphalt mixture at a test temperature of 25 ° C. However, for testing at 35 ° C and 45 ° C, the lowest deformation value was at 2.5% NCR content. Whereas for specimens with 5% NCR content the value of horizontal deformation is higher. This shows that there is an optimum limit of NCR content in the asphalt mixture. The chart above shows that the NCR content is 2.5% stronger resisting horizontal deformation at high temperatures.

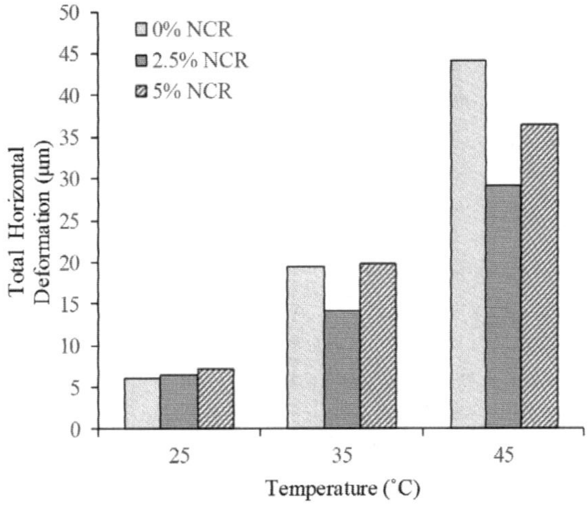

Figure 2. Total Horizontal Deformation with different temperatures.

IOP Conf. Series: Materials Science and Engineering **811** (2020) 012036 doi:10.1088/1757-899X/811/1/012036

2.3. *Absorption Paving Block Test Results*

Paving block testing is carried out to determine the effect of RAP material and NCR added material on the characteristics of paving blocks. The process of making RAP paving block samples is done differently from cement concrete paving blocks. The process of making this RAP paving block refers to the making of a Marshall test sample. RAP material is heated to 120 ° C, the next step is mixing with NCR at 150 ° C. The process of compacting the modified RAP asphalt mixture is incorporated into the paving block mold. Compaction is carried out when the temperature is in the range of 110 ° C-120 ° C. Compaction is done manually by referring to Marshall compaction which does 75 collisions for each side. Compaction is carried out by means of an iron plate and after compacting the paving blocks are removed from the mold for testing the absorption rate.

Table 6. Test Results of Paving Block Water Absorption.

Water Absorption				
	Sample 1	Sample 2	Sample 3	Average
NCR content 0%				
Wet weight (kg)	2411.2	2381.2	2401.2	2.26
Dry weight (kg)	2355.6	2322	2357.4	
Water absorption (%)	2.36	2.55	1.86	
NCR content 2,5%				
Wet weight (kg)	2423.4	2409	2409	2.02
Dry weight (kg)	2373	2353	2372.2	
Water absorption (%)	2.12	2.38	1.55	
NCR content5%				
Wet weight (kg)	2402.4	2422.4	2402.6	1.34
Dry weight (kg)	2368.8	2388.8	2374	
Water absorption (%)	1.42	1.41	1.2	

From table 6, it can be seen that the relationship between the value of water absorption and NCR content is inversely proportional. This is indicated by the increasing NCR content in the modified asphalt mixture, the water absorption value will be smaller. This is due to the more NCR content, the more cavities that are filled by NCR. So that the absorption of water in the mixture becomes lower. Based on the Indonesian National Standard (SNI), the three specimens have met the standard, namely the water absorption value is below 8%.

In this study, testing for paving blocks is different from paving blocks made of concrete material. Paving block forming material used in this study is a mixture of asphalt so that the material will have different characteristics with concrete paving blocks where to determine the strength of the compressive strength test. As for paving blocks with asphalt binding material, testing is carried out to determine the characteristics and strengths carried out by Marshall test and Indirect Tensile Strength (ITS) tests.

3. Conclusion

Based on research and analysis conducted, the conclusions of this study are as follows.

- The asphalt content contained in the RAP material is 5.79% obtained from the extraction process. The average gradation of reclaimed asphalt pavement (RAP) aggregate material is still within the ACWC specifications which indicates that the RAP material has a good aggregate gradation and there is no addition of new aggregate to RAP material.
- The value of optimum asphalt content (KAO) with new material obtained from Marshall test results is 6%. So that the new asphalt content that needs to be added to the RAP material is 0.205%.

- Optimum nano crumb rubber (NCR) content obtained from the results of the Marshall test and Indirect Tensile Strength (ITS) and paving block testing is 2.5%.
- From the test results of water absorption. the results show that the recycled asphalt mixture with nano crumb rubber (NCR) as a paving block material has a low absorption rate of less than 8%.

Acknowledments

This research was funded by the Direktorat Riset dan Pengabdian Masyarakat Universitas Indonesia through the PIT-9 Grant 2019, funded by contract no. NKB-0076/UN2.R3.1/HKP.05.00/2019. The authors thank the Pusjatan Bandung Laboratory of the Ministry of Public Works and Public Housing Republic of Indonesia, and the Structure and Material Laboratory of the Universitas Indonesia Civil Engineering Department.

References

[1]. Yang R, Kang S, Ozer H, & Al-Qadi I L, 2015, Environmental and Economic Analysis of Recycled Asphalt Concrete Mixtures Based On Material Production and Potential Performance. *Resource, Conservation, and Recycling.*

[2] Izaks R, Haritonous V, Klasa I, & Zaumanis M, 2015, Hot Mix Asphalt With High RAP Content, *Procedia Engineering*, Volume 114, Pages 676-684.

[3] Arshad M, & Ahmed M F, 2017, Potential use of reclaimed asphalt pavement and recycled concrete aggregate in base or subbase layer of flexible pavement. *Construction and Building Materials*, 83 - 97.

[4] Zhao S, & Liu J, 2018, Using Recycled Asphalt Pavement In Construction of Transportation Infrastructure : Alaska Experience, *Journal of Cleaner Production*, Volume 177, Pages 155-168.

[5] Aurangzeb Q, Al-Qadi I, Ozer H, & Yang R, 2014, Hybrid life-cycle assessment for asphalt mixtures with high RAP content, *Resources, Conservation and Recycling*, Volume 83, Pages 77-86.

[6] Sultan S A, & Guo Z, 2016, Evaluating the performance of sustainable perpetual pavements using recycled asphalt pavement in China. *International Journal of Transportasion Science and Technology.*

[7] Chen X, & Wang H, 2018, Life cycle assessment of asphalt pavement recycling for greenhouse gas emission with temporal aspect. *Cleaner Production*, 148 - 157.

[8] Xiao F, Amirkhanian S N, Shen J & Putman B, 2009, Influence of crumb rubber size and type on reclaimed asphalt pavement (RAP) mixtures, *Construction and Building Materials* 23, 1028–1034.

[9] Eskandarsefat S, Sangiorgi C, Dondi G & Lamperti R, 2018, Recycling Asphalt Pavement and Tire Rubber : A Full Laboratory and Field Scale Study, *Construction and Building Materials*, Volume 176, Pages 283-294.

[10] Vasconcelos K, Bernucci L B & Chaves J M, 2012, Effect of Temperature on the Indirect Tensile Strength Test of Asphalt Mixtures, *5th Eurasphalt & Eurobitume Congress*, Istanbul.

Temperature Effect on the Deformation of the Recycled Hot-Mix Asphalt Concrete with Nano Crumb Rubber as an Added Material with Wheel Tracking Machine

S P Hadiwardoyo[1], R H Aryapijati[1], R J Sumabrata[1], D Iskandar[1]

[1]Civil Engineering Department, Faculty of Engineering, Universitas Indonesia, Depok, 16424, Indonesia

sigit@eng.ui.ac.id

Abstract. The use of materials for the construction of roads can be done by recycling asphalt concrete mixtures that have reached fatigue or aging conditions. The results from stripping the asphalt concrete layer on the surface layer can be reused after processing by recycling the material. The recycling process is often intended to preserve the environment from natural damage due to the processing of material for asphalt concrete mix. To improve the performance of the recycled mixture, additives have been widely used and the use of building materials as additives can also be obtained from building material waste. One of the waste that continues to increase is the waste of vehicle tires. In this research used tire waste that has been mashed up to Nano size. The use of finer used tire sizes is intended to facilitate the process of mixing with asphalt. Recycling performance of hot mix asphalt concrete on deformation performance has been used rutting test using Wheel Tracking Machine at a temperature of 27°C, 40°C and 60°C. The test results showed a decrease in the rate of deformation in the asphalt concrete recycle mixture using Nano crumb rubber added material.

1. Introduction

The use of materials for buildings including road construction has been oriented towards environmental sustainability efforts. Rock material mining permits are increasingly restricted so that material recycling technology is being developed. Pavement material recycling technology that can produce new pavement material as a result of the saving of material, money, and energy. Materials of old pavement can still be utilized even though this pavements layer has reached at the end of its service life. The utilization of asphalt mixture that has been used, over the years, mixed with new aggregates and asphalt has been able to produce new asphalt mixture. This way has proved economical and effective in protecting the environment [1]. Increased number of road construction material needs as a consequence of the increase of the number of vehicles and the load axis of truck vehicles. Road construction should be able to serve traffic during the service period.

Long-term sustainability programs with technological advancements have allowed longer material life so that they can reduce costs over the service life. Also, the use of local materials, low-energy technology, recycling processes and waste material reduction programs as part of this program [2]. Road surface layers are a part of road construction that receives direct repetitive traffic loads so that structural damage occurs faster than other layers of the road structure. Restoration of the performance of the surface layer can be done by overlaying the old layer or by stripping the surface layer and replacing it with a new asphalt mixture. Recycle asphalt pavement (RAP) is a technique for mixing old asphalt binding and aggregate by adding new material produced by hot mix-asphalt (HMA). This method is known as the most common recycled material in flexible pavement.

The material additive is needed in high RAP content, due to crack performance, increased stiffness of asphalt mixture and difficulty in the compaction process [3]. Damage to the asphalt mixture can occur due to the aging process of asphalt in the mixture during the service period or changes in grain size due to repeated loading. The recycling process is the recovery of asphalt content in the mixture or the restoration of the aggregate composition. Asphalt pavement recycling has the advantage of saving raw materials. Asphalt pavement recycling has the advantage of saving raw materials. This is part of one of the sustainable strategies that have been developed and used in the road pavement industry on several years. The use of RAP can reduce the need for transportation of virgin asphalt binder, especially when it must be transported from foreign countries over long distances [4].

One technique to reduce environmental pollution, waste tire materials and waste engine oil has been widely used in asphalt mixtures for road pavement construction. The use of crumb rubber from tire waste has been widely used in asphalt mixes as an alternative to making modified asphalt, with the aim to improve the performance of the asphalt mixture and prevent plastic deformation. The addition of crumb rubber modifiers (CRM) to the asphalt mixture has increased the viscosity and elasticity of asphalt and at the same time, increased their resistance to the aging process [5]. Utilization of CRM in aggregate asphalt mixtures using certain techniques so that Crumb Rubber (CR) can be homogeneously mixed in asphalt mixture as a modified asphalt mixture. The use of CR on road pavement has been tried in many ways, especially when CR is mixed directly into asphalt (wet process) or, by mixing with aggregate during aggregate heating. This process is carried out before the addition of asphalt in the hot-mix asphalt (dry process) [6]. The properties of CR in asphalt mixtures at various temperatures are uncertain, this is due to the various effects of CR interactions with asphalt binders, largely influenced by the chemical properties of asphalt binders and the CR content in asphalt mixtures, particle size, texture, temperature and interaction time [7].

In this study, the RAP processing uses waste engine oil (WEO) to soften aged asphalt and add CR that has been processed to Nano size as a modified asphalt mixture. Virgin asphalt has been used in the Asphalt Concrete Wearing Coarse (ACWC) specification for hot-mix asphalt as a reference for evaluating the performance of modified asphalt mixtures.

2. Materials and Methods

2.1. Aggregate and Binders

In this study, recycled asphalt pavement was produced by 60/70 pen asphalt binder and a nominal maximum aggregate size of 19 mm. The RAP obtained from the Jakarta Outer Ring Road toll road has been extracted in the laboratory to determine the asphalt and aggregate characteristics. RAP asphalt binder is extracted through trichloroethylene solvent and obtained using a centrifuge extractor. The extraction results produced data from the studied RAP showed that the asphalt content was 5.8%, and had a very low penetration value of less than 10. These data were obtained from all three samples that had been tested. From the results of this test can be calculated the need for new asphalt and waste oil engine content into the RAP. Tests to control the performance of this recycled asphalt mixture have used the Asphalt Concrete Wearing Coarse (ACWC) specifications with the grain structure as shown in figure 1. The arrangement of the RAP aggregate is formed from the ACWC specification by adding new aggregates.

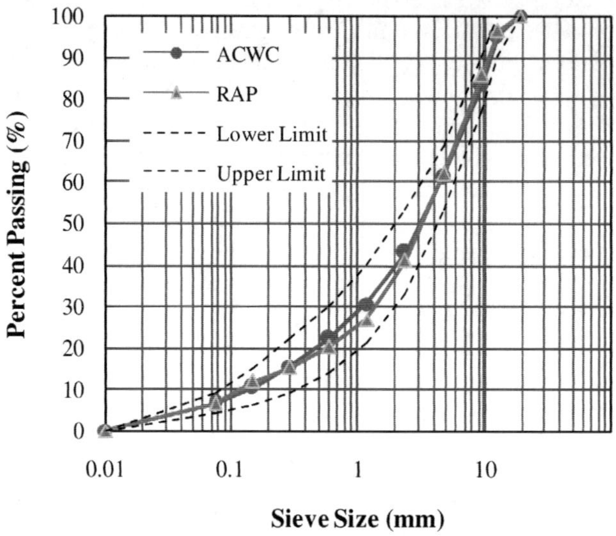

Figure 1. Gradation curve of Aggregate (AC-WC and RAP).

Pen 60/70 virgin asphalt was chosen as an additional binder for all specimens containing RAP. The binder specifications that have been used in this study are presented in table 1. Furthermore, to improve the resistance of the asphalt mixture to cracking, the binding material is modified with CR.

Table 1. Specification of binders.

Asphalt Characteristics					
No.	Type of Testing	Method	Indonesian Asphalt 60/70	Virgin Asphlat	Aged Asphalt
1	Penetration (25 C, 5 s, 100 g)/0.1 mm	SNI 2432:2011	67.75	62.5	8
2	Loss on heating (TFOT)	SNI 06-2440-1991	0.034	0.108	0.35
3	Penetration after TFOT	SNI 06-2456-1991	58.5	61	5.8
4	Softening Point (° C)	SNI 2432:2012	51.5	45	65.3
5	Flash Point (° C)	SNI 2433:2011	263	306	322.6
6	Solubility in Trichloroethylene (%)	AASHTO T44-03	99	99	89.8
7	Ductility (cm)	SNI 2432:2012	>100	107	12.3
8	Specific Gravity	SNI 03-6893-2002	-	1.048	1.07

2.2. Waste Engine Oil

Based on physical testing of RAP, the results show that RAP tends to have a rigid nature. This is evident from the very low penetration and ductility values. To reduce these rigid properties, waste engine oil (WEO) is needed as a softener material for RAP. The mixture of RAP and WEO in the ratio of 11: 4 gives the penetration results according to specifications, the value is in the range of 60 to 70. The percentage of volatile components in WEO tolerate under high temperatures during the production process. In other words, waste oil is highly qualified for applications in the asphalt mixing process because of its stability at temperatures above 220°C [3, 8].

2.3. Crumb Rubber

In this research, used tire material as waste has been crushed into fibre particles through 3 stages. The first stage used tires are destroyed by the ambient process at room temperature into small particles, in this initial process the particles still have metal and fabric so they must be separated. This separation process is carried out in the second process, and in this process also carried out particle size separation, where the size of large particles will be reprocessed. The third process is the separation of metals with magnets and the separation of fabric particles with air. The three stages of the Crumb Rubber milling process are carried out at PT Daurindo Indonesia, Bogor Indonesia. The process of mixing CR with asphalt with a method is very difficult to get a modified asphalt mixture where CR can spread homogeneously in asphalt. The change in shape from CR in the form of fibres to finer-sized granules is expected to complete the mixing process more evenly. This advanced grinding process has been carried out in laboratories that are still on a small scale.

2.4. Recycle Asphalt Pavement

The use of recycled materials for road pavement has been widely carried out in several countries. This is an effort to save the use of new material by making maximum use of the strength of the remaining old material by adding new material. RAP is common recycled materials that are used in asphalt mixtures due to the ability of these asphaltic materials to partially substitute the need for virgin aggregates and asphalt binder. For an asphalt concrete (AC) mix, the percentage of virgin asphalt binder that is substituted by recycled asphalt binder from RAP is known as the asphalt binder replacement (ABR) ratio [9].

In this study the asphalt content of RAP material was 5.8% of the total mixture using the ACWC specification, as a comparison a new asphalt mixture was made with an optimum asphalt content of 6% of the total weight of the asphalt mixture. The RAP mixture from the cutting of the road surface has been mixed with WEO of 36.67% of the aging asphalt content or 2.18% of the total weight of the RAP. The mixture of RAP and WEO has been stored for 24 hours in a closed container. New aggregate material has been added to meet the ACWC middle aggregate specifications as shown in figure 1. RAP changes colour darker and wetter than the previous RAP (figure 2). The next step is the RAP mixing process by heating the RAP + WEO mixture at 160°C, and then adding NCR before adding virgin asphalt.

(a) (b)

Figure 2. Before (a) and after (b) the WEO addition.

2.5. Wheel Tracking Machine

As a result of repetitive vehicle wheel paths on the road surface can cause a decrease in road surface, asphalt mixture is not able to accept the traffic load. On the road lane often dominated by heavy vehicles, the shape of the wheel path deformation will be faster.

The wheel tracking machine (WTM) test produces data to determine the resistance of asphalt mixes due to wheel path deformation, and this test also simulates the traffic loading or vehicle wheel load on the asphalt layer repeatedly until the wheel path deformation. Deformation on the wheel path is often found on flexible pavement, especially in areas with high temperatures. The wheel path deformation is formed when tensile stresses are generated due to high temperatures [10].

In this study, sample tests using WTM were carried out at varying temperatures of 27°C, 40°C and 60°C. The tool has a wheel load of 6.4 ± 0.15 kg/cm^2 which moves across the surface of the sample to reach 1,260 passes in a time of 60 minutes for a set of short cycles and can be extended by continuing to test the standard WTM cycle as needed [11]. The sample for the WTM test was formed with a size of 300 mm x 300 mm x 50 mm with the density value referring to the density of the Marshall test sample. A mixture of virgin asphalt and RAP mixture for the WTM test has been established with an optimal asphalt content of 6%. The addition of NCR to the RAP of 1.25%, 2.5%, and 3.75% of the asphalt content.

3. Result and Discussion

3.1. Optimum Asphalt Content

All testing of virgin asphalt mixture and modified asphalt mixture have been carried out with an optimum asphalt content of 6% on aggregate asphalt mixture. The asphalt mixture has been designed according to standard Marshall mix design procedures. The specimen was compacted using 75 blows on each side of a cylindrical sample diameter of 101.6 mm and 63.5 cm thick. Physical properties of the mixture such as air void (VA), voids filled with asphalt (VFA), voids in Marshall mineral aggregate (VMA), stability and flow.

3.2. Deformation rate (DR) and Dynamic Stability (DS)

Figure 3 shows the changes in DR and DS values at temperature variations of 27°C, 40°C, and 60°C. From figure 3 it can be observed that the change due to the influence of temperature is quite significant, at a temperature of 27°C the slowest deformation speed and the highest Dynamic Stability is achieved at NCR addition of 1.25%, even the performance of virgin asphalt mixture is at the lowest. However, at an increase in temperature up to 40 °C there appears to be a change wherein the modified asphalt mixture with NCR is 2.5% as the best performing asphalt mixture. Even when the temperature is added up to 60°C the best modified asphalt mixture performance is achieved from the addition of 2.5% NCR. From these results it can be seen that the addition of NCR to the modified asphalt mixture for low temperature and high temperature requires different amounts of NCR content.

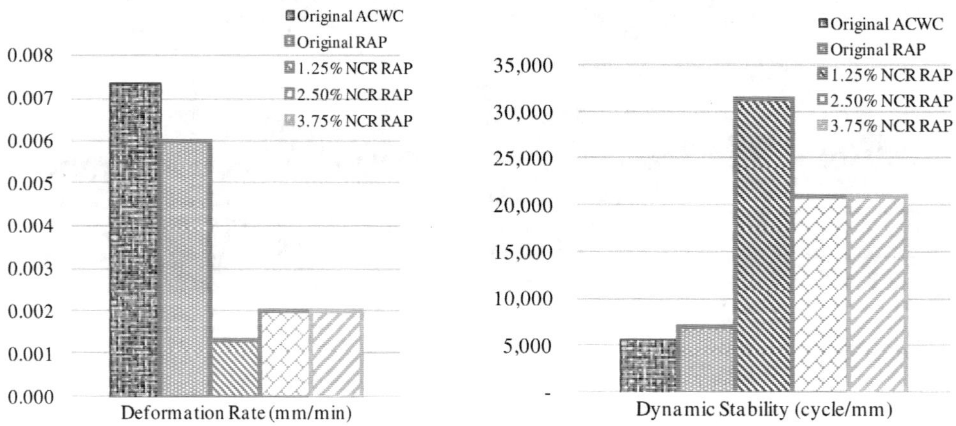

a. Temperature of 27°C.

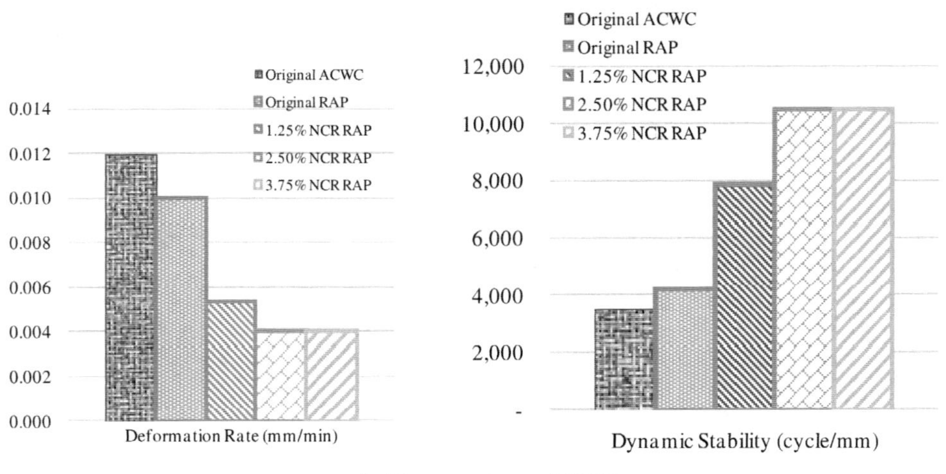

b. Temperature of 40°C.

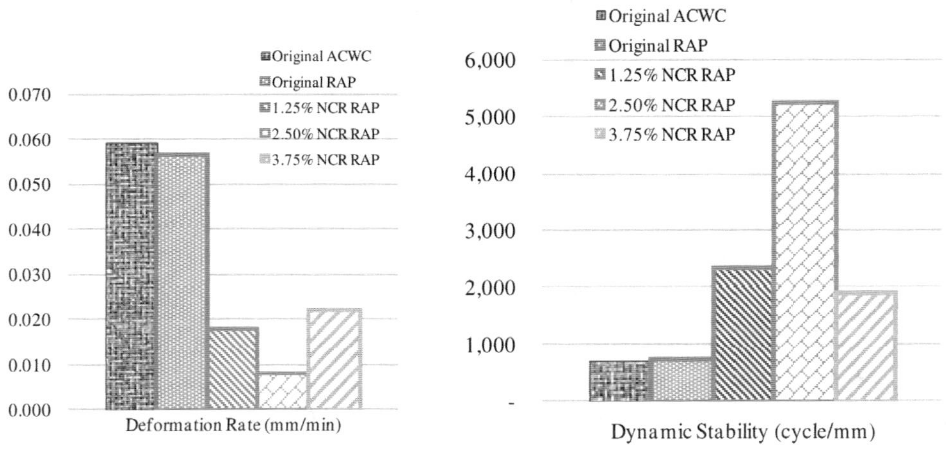

c. Temperature of 60°C.

Figure 3. Deformation Rate and Dynamic Stability on temperature of 60°C.

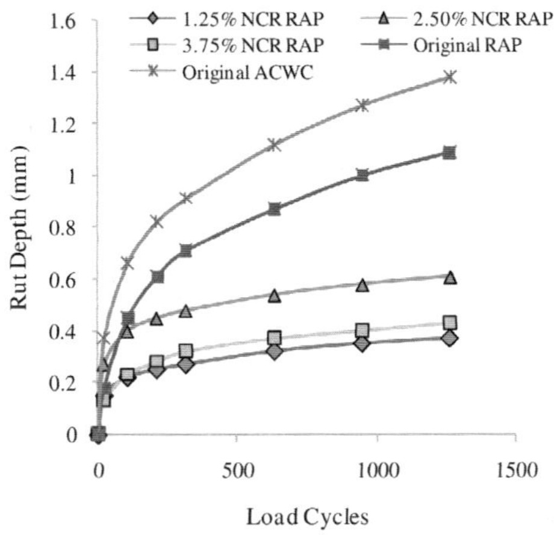

a. Rut depth on temperature of 27 °C.

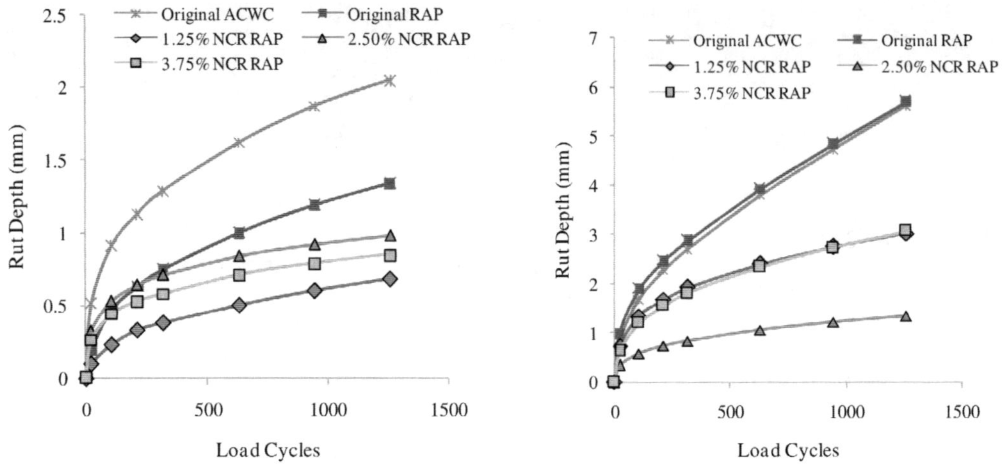

b. Rut depth on temperature of 40°C. **c.** Rut depth on temperature of 60°C.

Figure 4. Rut depth on different temperature.

3.3. Impact of temperature changes on rut depth of the asphalt mixture

From figure 4, it can be observed that there is a faster change in the virgin asphalt mixture, the initial cycle changes very quickly. Deformation continues to increase with increasing loading cycle until the end of the 1,260th cycle. Changes in temperature of 27°C and 60°C there was a significant change in the rut depth value in the mixture of virgin asphalt and each modified asphalt mixture. At 60°C the same performance was seen between virgin asphalt mixture and RAP asphalt mixture without NCR material. The addition of 2.5% NCR to asphalt content showed the best performance with deformation depth <1.5 mm in the 1,260th cycle. The addition of 2.5% NCR appears to be more stable to the effects of temperature changes of 40°C to 60°C.

4. Conclusion

From this research, we can conclude several conclusions about the use of RAP with NCR added as follows:

- The addition of WEO to the RAP mixture has increased the penetration value so that it can facilitate the RAP processing.
- The use of NCR in RAP mixture for low temperature and high temperature needs is different, at a temperature of 27°C the addition of NCR 1.25% shows the best performance results.
- At a temperature of 60°C has shown a very significant change is the addition of NCR 2.5% seems the most stable due to the impact of rising temperatures.
- The addition of NCR to the RAP mixture has shown better performance compared to the virgin asphalt mixture or RAP asphalt mixture without NCR.

Acknowledgments

This research was funded by the Direktorat Riset dan Pengabdian Masyarakat Universitas Indonesia through the PIT-9 Grant 2019, funded by contract no. NKB-0076/UN2.R3.1/HKP.05.00/2019. The authors thank the Pusjatan Bandung Laboratory of the Ministry of Public Works and Public Housing Republic of Indonesia, and the Structure and Material Laboratory of the Universitas Indonesia Civil Engineering Department.

References

[1] Xiao F, Amirkhanian S N, Shen J, Putman B 2009 Influences of crumb rubber size and type on reclaimed asphalt pavement (RAP) mixtures *Construc. Build. Mater.* **23(2)** 1028-34.

[2] Franesqui M A, Yepes J, Garcia-Gonzalez C, Gallego J 2019 Sustainable low-temperature asphalt mixtures with marginal porous volcanic aggregates and crumb rubber modified bitumen *J. Cleaner Produc.* **207** 44-56.

[3] Jahanbakhsh H, Karimi M M, Naseri H, Nejad F M 2019 Sustainable asphalt concrete containing high reclaimed asphalt pavements and recycling agents: Performance assessment, cost analysis, and environmental impact *J. Cleaner Produc.* 118837.

[4] Chen X, Wang H 2018 Life cycle assessment of asphalt pavement recycling for greenhouse gas emission with temporal aspect *J. Cleaner Produc.* **187** 148-57.

[5] Moreno F, Sol M, Martín J, Perez M, Rubio M C 2013 The effect of crumb rubber modifier on the resistance of asphalt mixes to plastic deformation *Mater. Des.* **47** 274-80.

[6] Pettinari M, Dondi G, Sangiorgi C, Hededal O 2014 The effect of Cryogenic Crumb Rubber in cold recycled mixes for road pavements *Construc. Build. Mater.* **63** 249-56.

[7] Lee S J, Akisetty C K, Amirkhanian S N 2008 The effect of crumb rubber modifier (CRM) on the performance properties of rubberized binders in HMA pavements *Construc. Build. Mater.* **22(7)** 1368-76.

[8] Majidifard H, Tabatabaee N, Buttlar W 2019 Investigating short-term and long-term binder performance of high-RAP mixtures containing waste cooking oil *J. Traff. Transp. Eng. (English Edition).*

[9] Yang R, Kang S, Ozer H, Al-Qadi I L 2015 Environmental and economic analyses of recycled asphalt concrete mixtures based on material production and potential performance *Reso. Conserv. Recyc.* **104** 141-51.

[10] Moreno F, Rubio M C, Martinez-Echevarria M J 2011 Analysis of digestion time and the crumb rubber percentage in dry-process crumb rubber modified hot bituminous mixes *Constuc. Build. Mater.* **25(5)** 2323-34.

[11] Kartika L, Hadiwardoyo S P, Sumabrata R J 2019 Rutting deformation of gap-graded hot-mix asphalt with added of waste tire rubber *AIP Conf. Proc.* **2114(1)** 040013.

A Using of Solar Roof Top for Electrical Energy Conservation in Educational Organization

J Sarasook[1]

[1]Lecturer, Nakhon Phanom University, Nakhon Phanom, Thailand

jassadat77@gmail.com

Abstract. This research presents solar roof top for electrical energy conservation. The case study is implemented in educational building in a campus of university in Thailand. To get the goal of electricity cost reducing under limited budget, the solar roof top is used in some building. This research uses surveys the electrical load data and considers the selecting of building following on the budget. The chosen building is the building of accounting department which the majority of electrical loads are air conditioner and lighting system. From the data, it is observed that electrical load consisted of 27000 W of air condition or about 74%, 4536 W of lighting or 12%, 3850 W of computer or 11% and other about 1160 W or 3% which including refrigerator, kettle and so on. The total load is 36546 W. Therefore, the 12 kW solar roof top is installed. Solar cell in type of crystalline silicon is used. The system consists of the 300 W solar panel amount 17 panels amount 2 sets the total of 34 panels, 2 commercials on grid inverter 6 kW. From the experimental results, it is found that the electrical consumption decreased 20.46% approximately 49103 kWh a year or about 5,729 dollars a year. The payback period is 4.01 years.

1. Introduction

Presently it is clearly that the electrical energy is an important thing for all people on the world. The electricity is the energy source which used for running the many activity; factory, business, transportation, healthy and so on. That means it is needed for country development [1]. According to Statistics of Electricity domestic consumption, it is found that electricity is consumed much more than the past [2]. However, the main source for electricity production still be the fossil energy; coal gas oil and etc. From the energy situation, it is investigated that the fossil energy is used rapidly so in a short period becoming energy crisis [3]. A choice way to solve this problem is using the energy effectively. Not only that but also the alternating energy should be used more. This way is correspondent with [4] which presents both energy conservation and developing the renewable energy continuously.

Presently many countries around the world focus on energy conservation. Thailand is an awareness country for energy situation. It can be seen that Thailand declared Energy Conservation Act 1992 which the designed building and factory have needed to do energy conservation [5]. For the ways of electrical energy conservation, they are several solutions, for example building energy management system (BMS), automatic light control, system control only peak demand period and so on. The interesting another solution is using of solar cell for energy saving.

According to Power Development Plan [6], it is found that the goal of solar cell energy is approximately 10,000 MW while the data in [7] shows that it is used just 1419 MW in 2015. Therefore, Thai government promotes home-use solar panels. Nowadays the cost of solar cell is more reduced than 40% in 5 years ago [8]. Hence the reasons why several factories and organizational building with greater energy consumption interest in installing solar panels for energy saving and solar self-generating energy [9, 10]. However, some organization cannot install completely all system because it needs the high budget but it need not instead all building in the area. Another choice it system can replace just some building. By this way, the cost of investment can limit and the cost for electrical energy still reduces. Moreover, several building can do it following budget. Although it is that, the electrical load need survey and carefully design system.

Therefore, this research presents the applying solar roof top for electrical energy conservation. The case study focuses on the building of college. The location is in Nakhon Phanom Province, Thailand.

2. Experimental Setup

The research is done in a college. The policy for energy conservation is using solar roof top for one building from several building because of budget limitation. So the data of electrical loads is needed. After that the design is considered. The research results would head to find benefit from this policy.

2.1. Location

The location is set in an educational organization; a college in Nakhon Phanom province, North East of Thailand. The location refers to Latitude 17°37'33.2"N and longitude 104°15'10.0"E. In this area has average temperature of 16 – 35 °C. The location is near the equator which provides uniform solar intensity throughout a year. The peak density of direct radiation in area is approximately 1,350 – 1,400 kWh/m^2–year. The area receives an average of 8 sunshine-hours per day. The area of plant covers 40,000 m^2 (9.88 acre). The transformer 250 kVA is installed. The building in this area is about 9 building. The 2 floor building is considered. In considered building consists of 8 class rooms, 6 office rooms, an electrical control room, 2 toilets and 1 broadcasting room. The working time is 6 days a week; Monday to Saturday. It is close on Sunday. But the lighting outside building is turn on every in the night.

2.2. The electrical loads in observed building

To design the solar roof top system, the electrical loads were surveyed. The details all of load is shown in table 1 and figure 2. The most load is air conditioner about 27000 W or 74%. The next are lighting, computer and other at approximately 4536 W (12%), 3850 W (11%) and 1160 W (3%) respectively. Therefore, the total load is 36546 W.

Table 1. The Electrical Loads in Observed Building.

The List of Load	Power
Air conditioner	27000
Lighting	4536
Computer	3850
Other	1160
Total	36546

Figure 1. The location of experiment.

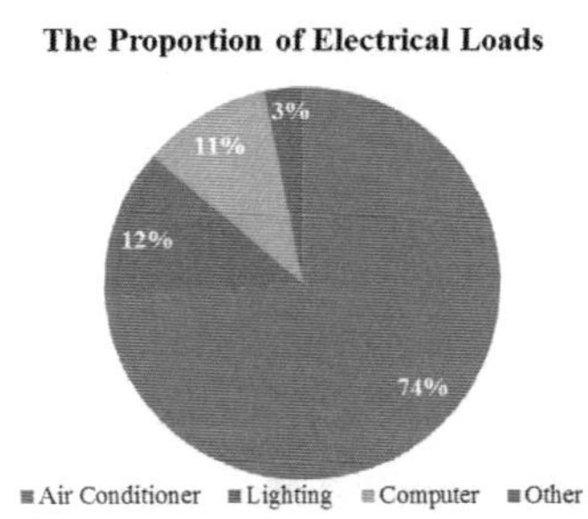

Figure 2. The proportion of electrical loads.

2.3. Specification of device

Nowadays, the silicon solar cells have 2 types; thin film and crystalline silicon solar cell. In this research the crystalline type was chosen because its characteristic is suitable for Thai climate. Moreover, the efficiency of module (watt per area) of crystalline is higher than thin film. So the space needs for installation is less a half of thin film. The crystalline is use by considering from Tier 1 manufacturing certified by PV Tech. It is warrantee the degradation percentage is decreased not over 20% for 25 years.

The inverters presently have two types; central solar inverter and string solar inverter. String solar inverters of 5 kW amount 2 sets were chosen in this research. String solar inverter is small inverter which is suitable of office and home. Because of 2 sets installation, if some inverters fail, the rest can still generate sufficient power required.

The 17 solar panels amount 2 sets (total of 34 solar panels) were installed on the building. They are placed heading to South direction. They were used as pole supporting for solar panel by angle 15O to south as shown in figure 3. The structure made of steel with hot dip galvanizing. The devices to stick made of stainless.

2.4. System diagram

The 10 kW solar roof top was installed on the building. The diagram of system is shown in figure 4. The system consists of 5 kW solar module amount 2 sets. Each set of solar module includes 300W 17 solar panels in type of crystalline. They are connected with grid tie inverter thought out protection sets (fuse and circuit breaker). After that they is connected to main distribution board. The monitoring and data logger is used to get the power generated by VSD (inverter) which they have software.

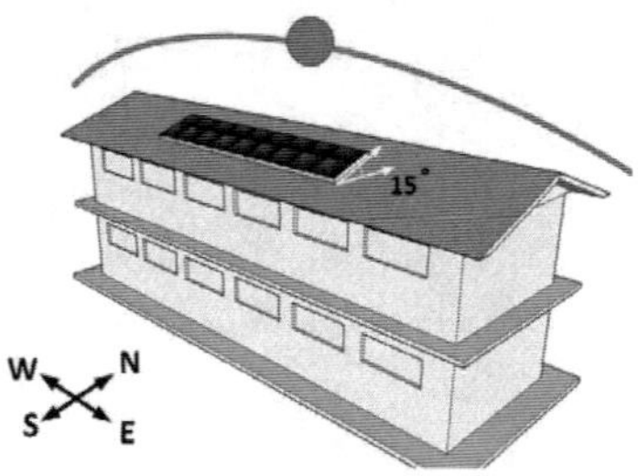

Figure 3. The installation of solar roof top.

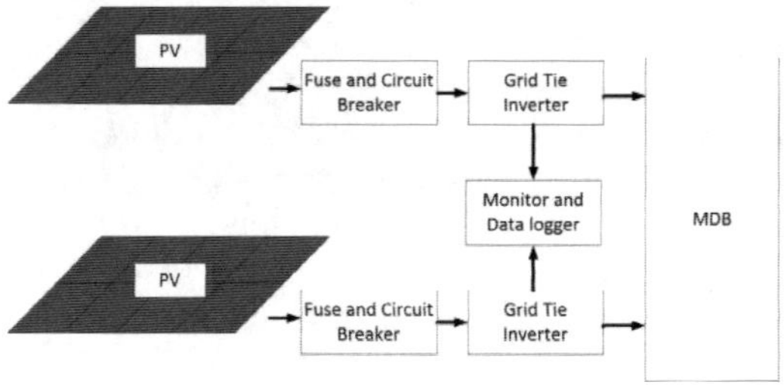

Figure 4. The diagram of solar pump systems.

2.5. Test procedure

The research aims to apply solar roof top for electrical energy conservation. The period of research is succeeded for 2 years. The experiment started since October 2015 to September 2017. The monthly electricity costs were gotten for 2 years. The cost of 12 first months is before performance of solar roof top installation. After installation, the cost of electricity is gotten again to comparing for 12 next months. The solar roof top is stand alone for only one building and no has storage battery. Hench, it distributes the electricity just only on daylight. While in night, the building used electricity form PEA (Provincial Electricity Authority) which is regional seller organization of Thailand. Therefore, analysis in this research is done from real information and economic analysis is performed from the real data at that research time run.

3. Experimental Results

3.1. The Monthly Expense for Utility

Considering monthly expense in 2016 as shown in figure 5, it is found that the college expense for electricity is the highest approximately 86%. The next is water, internet and phone expense at about 8%, 5% and 1% respectively. It is shown that the main expenses for utility in this case is electricity. Therefore, when this cost of electricity is reduced, the money spend is low.

To understand the idea for energy conservation, it can be observed by figure 6. It is found that the total cost varies on the cost of electricity clearly. While another expenses; phone, water and internet is very lower than once. Moreover, they are quite constant. Therefore, to reduce cost, the electrical energy should be focused.

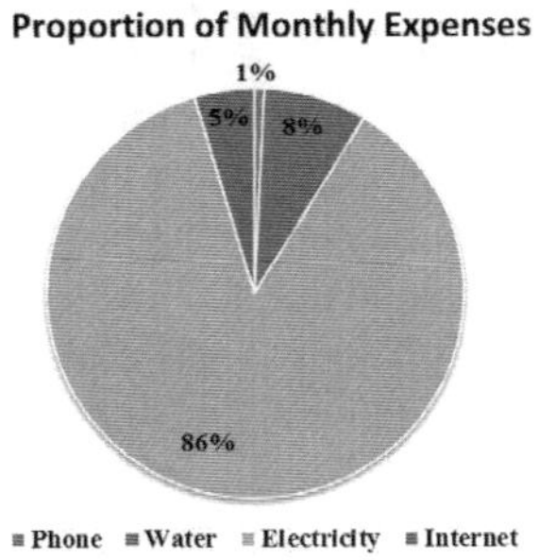

Figure 5. The diagram of solar pump systems.

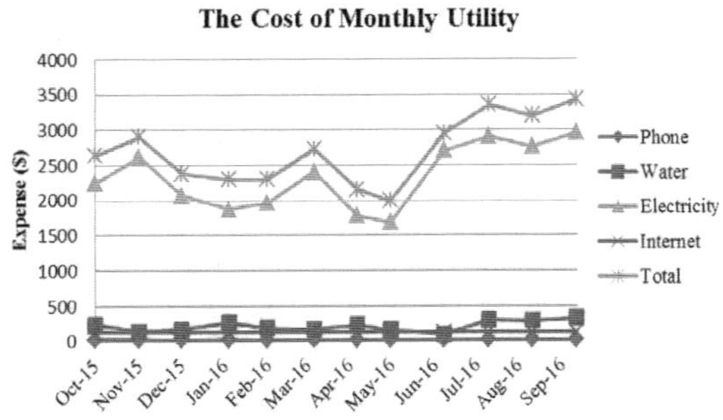

Figure 6. The diagram of solar pump systems.

3.2. The cost of electricity saved

For the cost of electricity in a period which in this research did 2 years. From the figure 7, it can be seen that the electrical consumption in the month for before and after perform is difference outstanding. The electrical consumption of after run project is lower clearly than before.

Considering the data, it is the results of performance. It is found that the average the used electricity after installed the solar roof top is less than before. The used electricity is reduced from 239959 kWh to 190856 kWh which is conserved result is about 49,103 kWh. The cost of electricity is decreased from 27995$ to 22266$ which saved cost is about 5,729$.

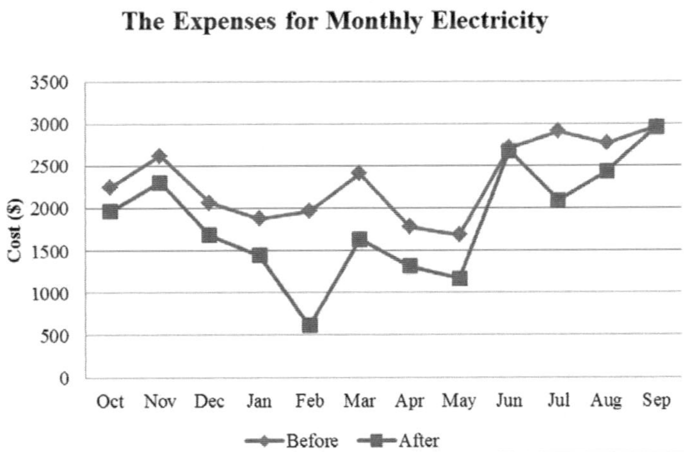

The List	Before	After	%Saving
Cost per year ($)	27995	22266	20.46

Figure 7. The cost of utility for each month.

3.3. The Economic Analysis

To understand investment, the economic analysis is needed. The details are shown in table 2. The investment cost in this project is limited 700,000 Bath or. Therefore, by the cost, the 2 sets of 5 kW solar roof top were chosen. From in formation form electricity bills all 2 years, it shows the consumed units (kWh) is decreased about 49103 kWh a year. It is found that the saved electricity cost is

171585.61 Bath a year. Therefore, this organization would get payback form this investment in 4.08 years. In conclusion, it can be said that this project is interested.

4. Discussion

From the research results, it is found that policy of electrical conservation by using solar roof top is get target well. The way to use this policy, personnel responsible for energy needs understand the behaviour of electricity consumption. Considering this case study, the policy is done in building of educational organization. Generally main expense is electricity cost which it is approximately 86% of all expense. Therefore, it can be seen that if the electricity cost is reduced, the monthly expense is more decreased. Moreover, the next important point should be focused is the type of load. For the educational organization, the majority of electrical loads are air condition and lighting and at 74% and 12% of all loads. For general educational building, these loads are run in daylight. So it is corresponding to solar cell run. However, some load of lighting is turn on in night hence the conventional source of electricity still needed. Although some load run in night, majority of loads run on the day light. The next interesting point, this research aims to present the solution of conservation by replacing solar cell instead some part. Really it should be replaced completely, but it is difficult because the budget is limited. However, this research shows it can save the cost of electricity 20.46% years while investment is just some expensed only. The payback period is just 4.01 years.

Table 2. The Economic Analysis.

The List	The Value
Investment (I)	23000 $
The Saved Energy	49103 kWh
The Saved Cost	5,729 $
Payback Period	4.01 Years

5. Conclusion

From the results presented in this paper, it is concluded as following.

1. The main expense in educational organization is electricity cost which it is approximately 86% of all expense.
2. The majority of electrical loads in educational organization are air condition, lighting and computer at 74%, 12% and 11% of all loads respectively.
3. The policy of solar roof top can save the electrical cost and energy is about 20.46 %.
4. The investment cost is 23000 $. The solar roof top is 10 kW. The project has payback period in 4.01 years.

In conclusion, the solar roof top project is saved energy and the cost therefore this way is interested.

References

[1] Thongchaisuratkrul C 2011 Energy management in buildings *Bangkok: King Mongkut's University of Technology North Bangkok.*

[2] Yearbook, G E S 2016 Electricity Domestic Consumption *Available on: https://yearbook. enerdata. net/world-electricity-production-map-graph-and-data. html# electricitydomestic-consumption-data-by-region. html* [Date accessed 20.10. 16].

[3] Gluck S 2018 Making energy cultures visible with situational analysis *Energy Res. Social Sci.* **45** 43-55.

[4] Plasto J W 2001 Energy services for an electricity industry based on renewable energy *Power Eng. J.* **15(5)** 239-47.

[5] Sarasook J 2018 The factors effect on electrical energy conservation for designed building and factory *GMSARN Int. J.* **12(2018)** 19 – 23.

[6] Energy report, M E T 2018 Power development plan *Available on: http://www.egat.co.th/index.php?option=com_content&view=article&id=325&Itemid=207* [Date accessed 23.08. 18].

[7] Tongsopit S 2015 Scaling Up Solar PV: A roadMap for Thailand *Bangkok: Energy Research Institute Chulalongkorn University.*

[8] Dolf G 2012 Renewable energy Technologies: Cost analysis series *Bonn*: *The International Renewable Energy Agency* **1**1-50.

[9] S.K. Yadav and U. Bajpai 2018 Performance evaluation of a rooftop solar photovoltaic power plant in Northern India *Energy for Sustainable Development* **43 (2018)** 130–138.

[10] Jyotish K P and Dharma B 2017 Rooftop solar plant installation of 25KW *Imperial J. Inter R* **3(2)** 349-362.

IOP Conf. Series: Materials Science and Engineering **811** (2020) 012039 doi:10.1088/1757-899X/811/1/012039

A Study of Inspection for Electrical Safety Systems Case Study Four Locations in Thailand

S Charoenchit[1], C Thongchaisuratkrul[2]

[1]Ph.D. Candidate, King Mongkut's University of Technology North Bangkok, Bangkok, Thailand
[2]Assistant Professor, King Mongkut's University of Technology North Bangkok, Bangkok, Thailand

surinalongkorn@hotmail.com

Abstract. This research presents study of inspection for electrical safety in electrical system. The four locations are observed. The objective of this research are to study the problems which may be occurred in electrical system and introduce the prevent ways. The four locations consist of 2 factories of paper and 2 animal food factories. For factory of paper, the total electrical load is 18,450 kVA which consists of 10 transformers and another 13,450 kVA which consists of 10 transformers. For the animal food 800 kVA, another is 1000 kVA which each location is a transformer. The inspected systems include transformer, main distribution board, ground and electrical wire systems. The check instruments are test oil maker, thermo scan, earth clamp tester, earth ground tester, infrared tester and outlet tester. The all data from test is analysed to compare among of four locations. For experimental results, it is of found that for transformer, the oil insulation transformer is lower than standard and silica gel is deteriorated and broken box, rubber shield and rubber gasket is broken and damage. For main distribution board, the ground value is lower than standard and the temperature of junction for breaker is high and more dust. For ground system, the ground value is lower than standard and broken. For wire system, the electrical is damaged because of heat and animal.

1. Introduction

Nowadays electrical system is very important especially in factory. According to statistic electricity consumption of Thailand, it is found that the quantity increases in the rate of 22% in ten years [1]. Moreover, according to 20 years' power electricity development plane, it shows that the need is more and [2]. It means that the using of electricity is increased. Therefore, the safety system is become more important especially in the big system for example factory, big building and so on [3]. The electrical system should be inspected by professional which corresponding with standard [4-6]. The safety system is the good system which effect on stable of system also. Not only that but also safe system prevent accident; fire or damage, breakdown production. Hence it is seen that the safety is more needed.

The factory is more expand continuously [7]. Therefore, the electricity system is more system which generally it consists of medium distribution about 11kV–36kV, transformer, distribution line, main distribution board, wire system, ground system [8]. According [9, 10], the electrical system be need inspected. In 2015, it is found that there is about 50% of factory and building not passes inspection because the system is not standard [11]. These factories and building is not proved by electricity seller following standard MEA (Metropolitans Electricity Authority) and PEA (Provincial

Electricity Authority) [5]. That mean the factories are not run the process which it affects production process. Not only that but also they effect on export product to abroad and income to the country.

Although, in factory or building, they have technicians for checking the electrical system, sometime may be not enough because they lack of experience. Moreover, some information or point which should be focus may be neglected. The important thing is the technician should be known that problem and the solving.

Therefore, this research presents inspection for electrical safety in electrical system. The four locations are observed. The objective of this research are to study the problems which may be occurred in electrical system and introduce the prevent ways.

2. The Experimental Procedures

The research is test on four factories. The inspected sectors consist of transformer, main distribution board, ground and electrical wire. The testing is done by three engineers who test together. The standard instruments are calibrated before using. The check instruments are test oil maker, thermo scan, earth clamp tester, earth ground tester, infrared tester and outlet tester.

2.1. The observed four locations

To testing, four factories were considered; the locations are shown in figure 1-figure 4. The first location as shown in figure 1 is paper factory. Its electrical load is about 18,450 kVA. It is 10 transformers. This factory locates on Prachin Buri. The second location as shown in figure 2 is paper factory. Its electrical load is about 13,450 kVA. It is 9 transformers. This factory locates on Samut Prakan. The third location as shown in figure 3 is animal food factory. Its electrical load is about 800 kVA. It is 1 transformer. This factory locates on Chachoengsao. The forth location as shown in figure 4 is animal food factory. Its electrical load is about 1,000 kVA. It is 1 transformer. This factory locates on Bangkok.

2.2. Inspected system

The inspected systems in this research consist of four types; transformer, main distribution board, ground and electrical wire. For transformer, the checking lists include oil, high voltage bushing, low voltage bushing, silica gel, rubber shield, rubber gasket, transformer tank. For main distribution broad, the checking lists include physical box, circuit breaker, bus bar, junction, wire, ground, installation safety distance, thermo scan and lighting. For ground system, the checking lists include physical ground wire, ground value, ground junction, wire size and ground rod. For electrical wire, the checking lists include physical wire, type and size, thermo scan, junction and installation.

Figure 1. The four factories.

Figure 2. Inspected systems.

IOP Conf. Series: Materials Science and Engineering **811** (2020) 012039 doi:10.1088/1757-899X/811/1/012039

2.3. Instrument Tester

The instrument testers in this research consist of 6 types; test oil maker, thermo scan, earth clamp tester, earth ground tester, infrared tester and outlet tester. The model of equipment is shown in table 1 which shows details of testing and tested system. Figure 9 shows all 6 instruments.

Table 1. The lists of instrument tester.

Instruments	Testing	System
test oil maker	Dielectric strength	Oil transformer
thermo scan	Heat	High voltage transformer
earth clamp tester	Ground	Ground of Transformer
earth ground tester	Ground	Ground of MDB
infrared tester	Heat	CB, bus bar, wire
outlet tester	Phase	MDB, wire

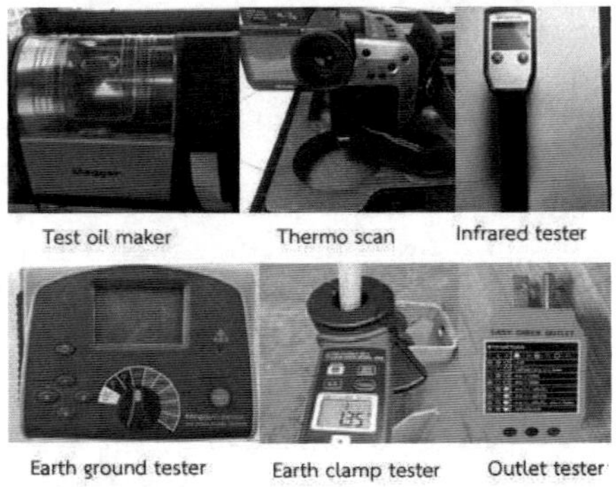

Figure 3. The Instrument tester.

3. Experimental Results

3.1. Transformer

From the inspection of 4 factories, it is found that all abnormal occurred in transformer it shows in table 2. The example of damage is shown in figure 4. From data in figure 8, it is found that the problems which always more found in the transformer include Dielectric strength of oil is lower than standard, Silica gel is deteriorated, Rubber shield is broken and Rubber gasket is broken about each of 17%. The next are Transformer tank occur rust, Oil leak from transformer tank and Bushing broken approximately each of 9%. The finally it is Silica gel is broken at 4%. The first four problems are shown in figure 4. Theses problem is found in all factories.

Figure 4. The abnormal in transformer.

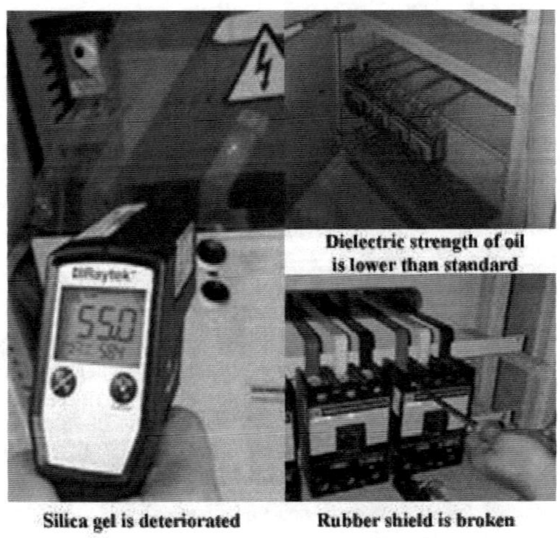

Figure 5. The abnormal in MDB.

IOP Conf. Series: Materials Science and Engineering **811** (2020) 012039 doi:10.1088/1757-899X/811/1/012039

Figure 6. The abnormal in ground.

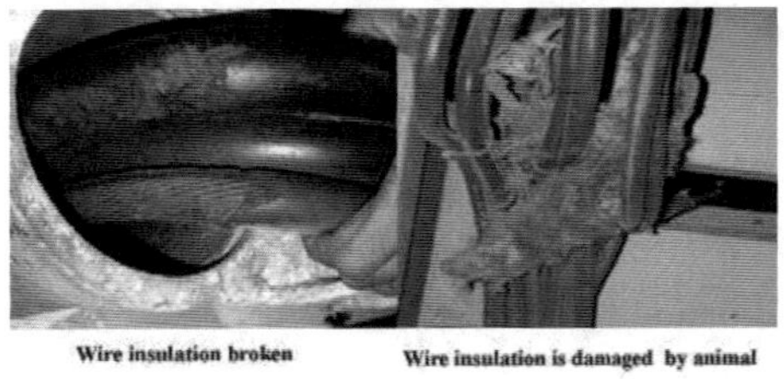

Figure 7. The abnormal in wire system.

3.2. Main Distribution Board

From the inspection of 4 factories, it is found that all abnormal occurred in main distribution board, it shows in table 3. The example of damage is shown in figure 5. From data in figure 9, it is found that the problems which always more found in the main distribution board include more dust and terminal loose about each of 34%. The next is nut of air circuit breaker loose approximately 25%. The finally it is Case of air circuit breaker heat at 8%. The first three problems are shown in figure 5.

3.3. Ground system

From the inspection of 4 factories, it is found that all abnormal occurred in ground system, it shows in table 4. The example of damage is shown in figure 6. From data in figure 10, it is found that the problems which always more found in the ground system are ground injection broken about 50%. The next are ground test value is less than standard and ground wire heat approximately each of 25%. The first problems are shown in figure 6.

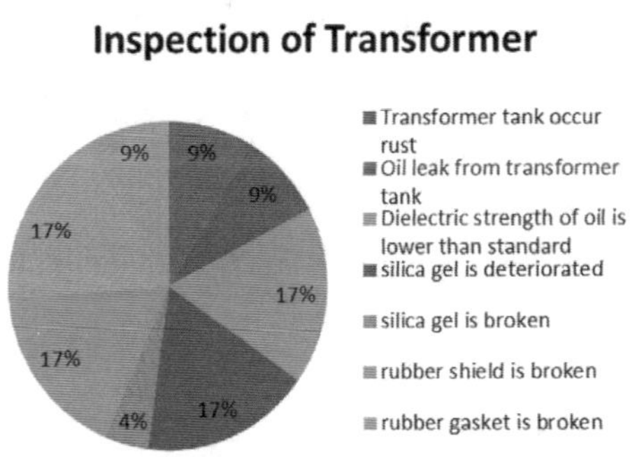

Figure 8. Inspection of transformers.

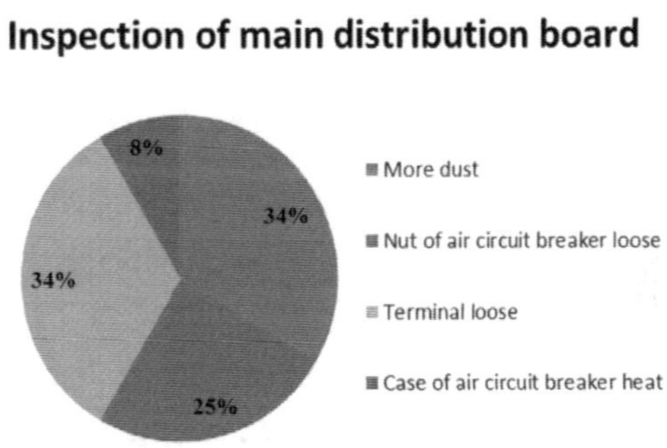

Figure 9. Inspection of Main Distribution Board.

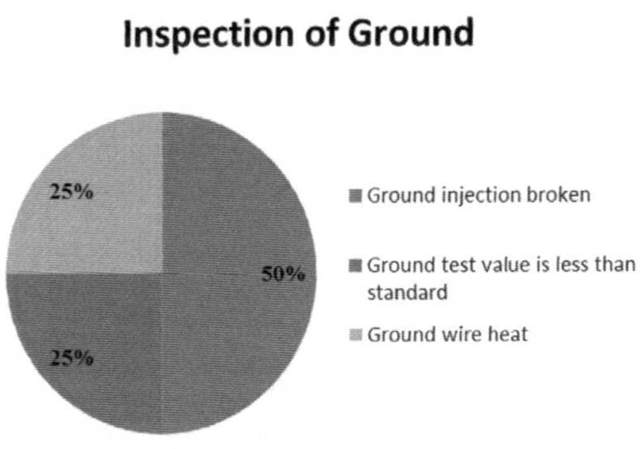

Figure 10. Inspection of ground system.

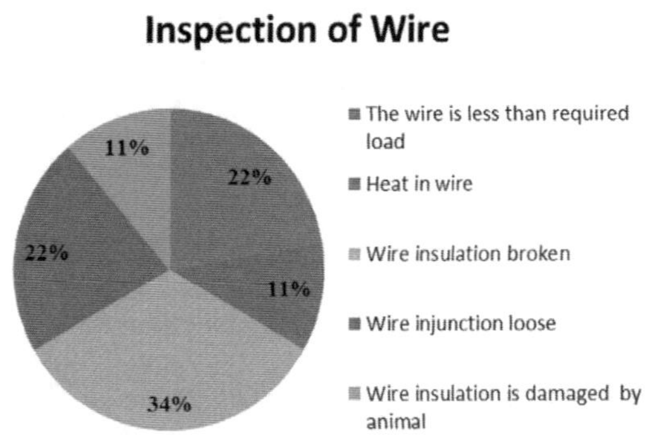

Figure 11. Inspection of electrical wire.

Table 2. Inspection of Transformers.

The List	F1	F2	F3	F4
Transformer tank occur rust	✓	✓		
Oil leak from transformer tank			✓	✓
Dielectric strength of oil is lower than standard	✓	✓	✓	✓
Silica gel is deteriorated	✓	✓	✓	✓
Silica gel is broken	✓			
Rubber shield is broken	✓	✓	✓	✓
Rubber gasket is broken	✓	✓	✓	✓
Bushing broken	✓	✓		

Table 3. Inspection of MDB.

The List	F1	F2	F3	F4
More dust	✓	✓	✓	✓
Nut of air circuit breaker loose		✓	✓	✓
Terminal loose	✓	✓	✓	✓
Case of air circuit breaker heat				✓

Note: F1 is factory of paper with 18,450 kVA electrical.
F2 is factory of paper with 13,450 kVA electrical.
F3 is animal food with 800kVA electrical.
F4 is animal food with 1000kVA electrical.

3.4. Electrical Wire

From the inspection of 4 factories, it is found that all abnormal occurred in wiring system, it shows in table 5. The example of damage is shown in figure 7. From data in figure 11, it is found that the problems which always more found in the ground system are wire insulation broken about 34%. The next are the wire is less than required load and wire injunction loose approximately each of 22%. The finally they are heat in wire and wire insulation is damaged by animal about each of 11%. The examples of problems are shown in figure 7.

Table 4. Inspection of ground system.

The List	F1	F2	F3	F4
Ground injection broken	✓	✓		
Ground test value is less than standard	✓			
Ground wire heat				✓

Table 5. Inspection of Electrical Wire.

The List	**F1**	**F2**	**F3**	**F4**
less than required load			✓	✓
Heat in wire				✓
Insulation broken	✓	✓	✓	
Wire injunction loose			✓	✓
damaged by animal		✓		

4. Discussions

For the problem in transformer; dielectric strength of oil lower than standard, silica gel deteriorated, rubber shield and gasket broken about each of 17%. For ways of preventive maintenance, the oil of transformer should be filtered at lease 6-12 month a year. Silica gel is checked monthly and should be changed when it is deteriorated. The problem of silica deteriorated may be from the rubber shield and gasket broken so they should be changed once a year. The problems in the main distribution board, more dust and terminal loose of 34%. MDB should be clean 6 months a year. The terminal and nut of air circuit breaker should tighten also 6 months a year. The problem in the ground system, 50% is injection broken. It should be checked 6 months a year. It should be changed and connected by mold cast.

5. Conclusion

From experimental results, it can be summarized as following:

1. From 4 systems, the problems are found the most occurring in transformer and the next is in main distribution board. While in wiring and ground system the problems are found quilt less
2. The problems found in the transformer include dielectric strength of oil lower than standard, silica gel deteriorated, rubber shield and gasket broken about each of 17%.

3. The problems found in the main distribution board include more dust and terminal loose about each of 34%. The next is nut of air circuit breaker loose approximately 25%.
4. The problem found in the ground system is ground injection broken at 50%.
5. The problem found in the wire system is wire insulation broken about 34%.

References

[1] Yearbook, G E S 2016 Electricity Domestic Consumption *Available on: https://yearbook. enerdata. net/world-electricity-production-map-graph-and-data. html# electricitydomestic-consumption-data-by-region. html* [Date accessed 20.10. 16].
[2] Energy report, M E T 2018 Power development plan *Available on: http://www.egat.co.th/index.php?option=com_content&view=article&id=325&Itemid=207* [Date accessed 23.08. 18].
[3] Thongchaisuratkrul C 2016 Electrical Energy Conservation *Pissanulok: Ann Production.*
[4] Handbook, W S T 2014. Electrical Safety in the Workplace *Bangkok: The Engineering Institute of Thailand under H.M. The King's Patronage.*
[5] Handbook, W S T 2013 *Electrical Installation Bangkok: The Engineering Institute of Thailand under H.M. The King's Patronage.*
[6] Handbook, W S T 2013 High Voltage Safety *Bangkok: The Engineering Institute of Thailand under H.M. The King's Patronage.*
[7] Report, N S O 2017 Business and Industrial Census. *Bangkok: National Statistical Office.*
[8] Thongchaisuratkrul C 2011 Energy management in buildings *Bangkok: King Mongkut's University of Technology North Bangkok.*
[9] John C 2006 Electrical Safety Handbook **3rd ed**, *NewYork: McGraw-Hill.*
[10] Handbook, N F P A 2004 Standard for Electrical Safety in the Workplace *Massachusetts: National Fire Protection Association.*
[11] Yearly Report, M E A 2016. The Safety Report *Bangkok: Metropolitans Electricity Authority.*

IOP Conf. Series: Materials Science and Engineering **811** (2020) 012040 doi:10.1088/1757-899X/811/1/012040

Analysis of Characteristics and Research on Processing Method of the High Viscosity DW-1 Colloid

Xuyang Liu[1], Haibo Xu[1, *], Li Wang[2], Haiyang Xue[1]

[1]School of Mechanical Engineering, Xi'an Jiao tong University, Xi'an 710049, China
[2]Ke-rui Institute Shanghai Aerospace Equipment's Manufacturer Co, Ltd Shanghai 201100, China

Email： hbxu@mail.xjtu.edu.cn

Abstract. All the variation of the viscosity characteristics of the high-viscosity DW-1 colloid at different temperatures are mainly analysed in the paper. Based on the experimental data, it concludes that the heat treatment method at 45-65°C can keep the colloid with good fluidity within 15 minutes. The inhibitory effect of organic solvents on solid particles in colloids is also studied. By adding an appropriate proportion of Butanone to the DW-1 colloid, a better colloid particle treatment effect is obtained and the colloid curing speed is increased by about 2 times within 24 hours. Finally, the treatment methods and effects of colloids before spraying are summarized.

1. Introduction

High-viscosity DW-1 colloid is often used for spraying the thermal insulation layer of rocket tanks[1]. In order to make the spraying work stable and efficient, and to prevent clogging when spraying DW-1 colloids containing solid particles, appropriate treatment is required. The method can make the colloid maintain good fluidity and inhibit the solid particles in the colloid system. By analysing the viscosity change characteristics of colloids at different temperatures, the heat treatment method to maintain good fluidity was researched. To analyse the influence of organic solvents on the solidification characteristics of colloids, the dissolving and inhibiting effects of organic solvents on the solid particles in the system were studied too. At last, it obtained a suitable and effective treatment method for DW-1 colloid comprehensively.

2. Colloidal Material Composition and Viscosity Analysis

2.1. Composition and properties of colloidal materials

High viscosity DW-1 type colloid is a two-component adhesive. The component A is an isocyanate-terminated polyurethane prepolymer. The component B is a kind of amino benzyl methane. The colloid is made up of two components according to a 5: 1 mass ratio. The main physical parameters at normal temperature are shown in Table 1. The surface tension coefficient of colloids can be calculated by simulation models like continuous surface force [2] and computer-aided experiments[3].

Table 1. Standard value of physical parameters of DW-1 colloid (25°C).

Viscosity /mPa·s	Density/kg·m^{-3}	Surface Tension /N·m^{-1}
4500	1050	0.015

2.2. Analysis of Influencing Factors of the Viscosity of the Colloid

The viscosity of DW-1 colloid is affected by many factors. On the one hand, according to the theory of fluid dynamics, Newtonian hydrodynamic viscosity can be expressed by Newton's formula [4](1). The higher the temperature is, the smaller the internal friction resistance is and the lower the fluid viscosity is. For non-Newtonian fluids [5], the higher the temperature, the lower the fluid viscosity too.

$$\mu = \frac{\tau}{du / dy} \qquad (1)$$

Where, ——Internal frictional resistance per unit area of fluid, ——Velocity gradient.

On the other hand, the viscosity of DW-1 rubber will increase with the curing reaction. The higher the temperature is, the faster the curing reaction speed is, the faster the viscosity growth speed is too. Thus, it shows the rapid decrease of fluidity.

The viscosity change of the thermal insulation layer DW-1 adhesive is related to temperature and time at the same time, which is the result of the comprehensive influence of the friction and the progress of the curing reaction in the system.

3. Research on Viscosity Characteristics and Heating Method the Colloid

3.1. Analysis of the Viscosity Characteristic at Room Temperature

The characteristics of the viscosity of DW-1 colloid at room temperature over time were studied. 25g of component A and 5 of component B were used to prepare 30g of DW-1 colloid. The prepared DW-1 colloid viscosity was measured at 25°C, and the measurement time interval was 1 minute. Take three readings for each measurement and calculate the average value. The changes in colloidal viscosity within 20 minutes are shown in Table 2.

Table 2. Measured value of colloidal viscosity within 20min at room temperature (25°C).

Time/min	1	2	3	4	5	6	7	8	9	10
Viscosity/mPa·s	4380	4461	4602	4770	4932	5124	5340	5553	5784	6012
Time /min	11	12	13	14	15	16	17	18	19	20
Viscosity/mPa·s	6309	6609	6975	7242	7587	8007	8334	8754	9195	9651

According to the experimental data, the curve of the DW-1 type colloid viscosity with time at room temperature is plotted, as shown in figure 1.

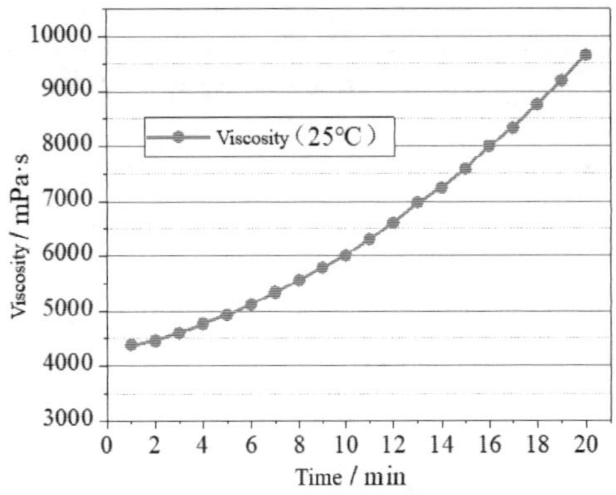

Figure 1. Variation curve of colloidal viscosity with time at room temperature.

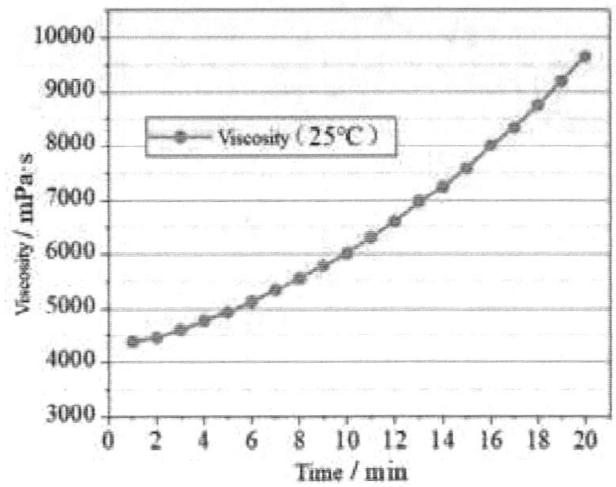

Figure 2. Variation of colloidal viscosity with time at low temperature of 5°C.

It can be seen from the graph that with the progress of the curing reaction, the viscosity of the DW-1 type colloid gradually increased and exceeded 9600 mPa s in 20 minutes, and the viscosity increased by about 120% with a larger increase. In addition, the viscosity increased by about 37% in the first 10 minutes, and increased by about 53% in the last 10 minutes. It can be seen that the rate of increase is gradually accelerating. At normal temperature, the viscosity of DW-1 colloid increases rapidly, and its fluidity becomes worse in a short time, which will affect the normal operation of the colloid spraying system. It is necessary to study the proper temperature range of colloid heat treatment.

3.2. Research on Influences of Temperature on Viscosity Characteristics

Low temperature can reduce the solidification reaction speed of the colloid and delay the viscosity growth rate; while high temperature can reduce the internal friction of the colloid system, thereby improving the fluidity of the fluid, and it can also reduce the viscosity to a certain extent. Therefore, it is necessary to obtain the colloid through research Suitable temperature range for good fluidity. The low temperature 5°C, high 95°C and different heating temperatures were selected to study the viscosity change characteristics of DW-1 type colloid.

(1) Viscosity characteristics of DW-1 rubber at low and high temperatures.

The DW-1 colloid prepared by the same method is used to measure and record the change data of viscosity with time while keeping the temperature at 5°C. The measurement interval is 1min. The measurement results are shown in Table 3. According to the measurement results, draw the curve of the colloidal viscosity change with time at the low temperature of 5 °C, as shown in figure 2.

Table 3. Measured viscosity of colloid at low temperature 5°C.

Time /min	1	2	3	4	5	6	7	8
Viscosity/mPa·s	12152	12571	12850	13094	13378	13588	13815	14081
Time /min	9	10	11	12	13	14	15	16
Viscosity/mPa·s	14271	14404	14580	14768	14974	15198	15443	15661

At low temperature of 5°C, the viscosity of DW-1 colloid basically increased linearly with time. The viscosity value increased by 18.5% within 10 minutes, and the growth rate was significantly slower than that at normal temperature, but the initial value of the colloid at 1 minute The viscosity reached 12152 mPa s. The viscosity characteristics of the DW-1 colloid under the heating condition of 95°C water bath were also studied. The viscosity measurement value of the colloid at 95°C for about 5 minutes exceeded the range. It can be seen that the excessively high temperature greatly improved the gel curing speed. The speed of viscosity increase is also significantly accelerated.

IOP Conf. Series: Materials Science and Engineering **811** (2020) 012040 doi:10.1088/1757-899X/811/1/012040

It concludes that When the temperature is too low, the curing speed of DW-1 type colloid is slowed down, but the friction in the colloid is large, resulting in poor fluidity; when the temperature is too high, the internal friction is reduced It improves the fluidity to a certain extent, but the acceleration of curing speed leads to the excessive increase of viscosity. Therefore, simple low-temperature or high-temperature processing cannot make the DW-1 colloid meet the normal working requirements.

(2) Variation of initial viscosity of DW-1 rubber at different temperatures.

Based on the above analysis and research, the change characteristics of the initial viscosity of DW-1 colloid at different temperatures were studied. Prepare another 7 groups of DW-1 colloids according to the same ratio, and measure the initial viscosity value at 7 different temperatures for 1 minute. All the measurement results are shown in Table 4.

Table 4. Initial colloid viscosity measurements at different heating temperatures (1 minute).

Temperature/°C	5	15	25	35	45	55	65	75	85	95
Viscosity/mPa·s	12152	7975	4380	2670	925	231	1127	3481	4347	4930

According to the measurement results, the curve of the initial viscosity of DW-1 colloid at different temperatures is plotted, as shown in figure 3. The analysis found that the initial viscosity of the DW-1 type rubber was higher when the temperature was lower than normal temperature, which was because the internal friction was large and the fluidity was deteriorated. As the temperature increased, the initial viscosity was obviously obvious within a certain range decrease, then show a slight upward trend.

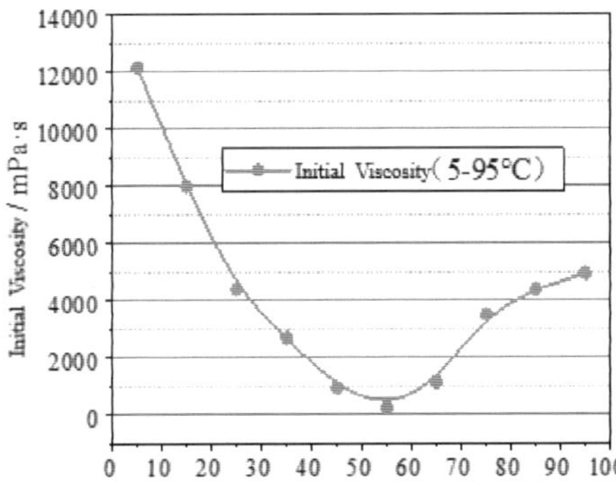

Figure 3. Initial colloid viscosity at different heating temperatures.

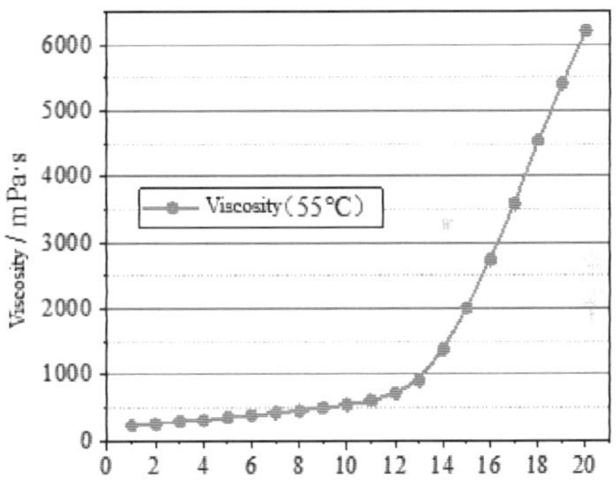

Figure 4. Colloid viscosity change at 55 °C heating temperature.

The reason for the analysis of this change rule is that in the proper heating temperature range, the friction in the colloid is reduced and the fluidity is enhanced. At this time, the acceleration of the curing speed has less effect on the system, and the Colloid shows lower viscosity than normal temperature; When the temperature exceeds this range, the curing reaction speed increases sharply and becomes the dominant factor of the system viscosity, reaching a higher viscosity value in 1 minute. The suitable heating temperature range obtained from the study is about 45-65°C. A heating temperature of 55°C was selected to further determine the change in colloid viscosity with time. The viscosity is continuously measured and recorded. The measurement interval is still 1 minute. The measurement results are shown in table 5. According to the measurement results, the characteristic curve of the viscosity of DW-1 adhesive at 55°C heating time is plotted, as shown in figure 4.

IOP Conf. Series: Materials Science and Engineering **811** (2020) 012040 doi:10.1088/1757-899X/811/1/012040

Table 5. Colloid viscosity measurement value at 55°C heating temperature.

Time /min	1	2	3	4	5	6	7	8	9	10
Viscosity/mPa·s	231	252	294	309	354	390	426	456	501	549
Time /min	11	12	13	14	15	16	17	18	19	20
Viscosity/mPa·s	612	717	903	1380	2013	2742	3579	4545	5409	6201

According to the analysis chart, it can be found that the initial viscosity of the DW-1 colloid is only 231 mPa s at 55°C, and the fluidity is greatly improved compared with that at normal temperature. The heat treatment method at this temperature makes the colloid substantially maintain a lower viscosity of less than 2000 mPa s within 15 minutes, which is significantly lower than that at normal temperature. In general, the colloid obtained by the heat treatment method at 45-65°C has good fluidity, which is more suitable for the stable and efficient colloid spraying of the DW-1 colloid.

4. Research on Processing Method of Colloid Powders by Organic Solvent

4.1. Research on the Effect of Organic Solvent on Colloid Powders

The component B in powder form in DW-1 type colloid determines that it is a solid-containing fluid, and an appropriate treatment method is required to suppress solid particles in the system. Butanone solvent is selected as the research object in this paper. It has strong solubility and good volatility[6]. DW-1 colloid sample A was prepared according to standard methods at room temperature. Sample B was used as a control group, and the mass ratio of component A and component B was still 5:1. During preparation, the component B of DW-1 colloid was prepared into a saturated Butanone solution, and then mixed with component A. The mass ratio of the B component to the Butanone used was 1:2.2. The standard colloidal sample A was opaque milky white, and the sample B containing butanone solvent was transparent light green. Observe the microscopic state of the two colloids through a microscope, as shown in figure 5. Sample A has a lot of opaque particles and bubbles. Because sample B dissolves the solid particles in the system using butanone, the interior of the colloidal system is basically transparent except for a few small bubbles. It can be seen that the treatment method using organic solvents has a better dissolution and inhibition effect on the solid particles in the DW-1 colloid.

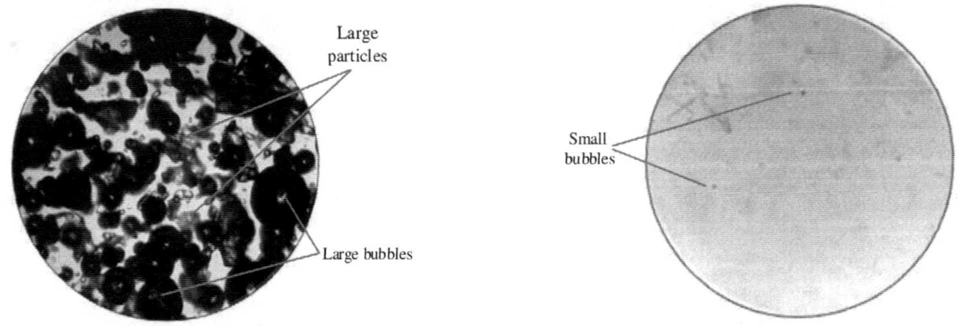

(a) Sample A (b) Sample B (containing butanone)

Figure 5. Microscopic states of two types of DW-1 colloids.

4.2. Research on the Influence of Organic Solvent on the Curing Characteristic of the Colloid

The addition of organic solvents changed the shape of the B component in the colloidal system. In order to study its effect on the solidification characteristics of the colloid, the hardness of the two colloids were measured and analysed at room temperature. Take the average of the three measurement points for each measurement. Continuous measurement of the curing hardness of DW-1 colloid sample A and sample B (containing solvents) was performed at 12 h intervals. The change of curing hardness of colloidal sample A and sample B with curing time is shown in table 6.

Table 6. Cured hardness measurement results of two colloidal samples.

Curing time /h	Sample A hardness /HA	Sample B hardness /HA
24	19.2	57.2
36	56.8	76.2
48	67.2	78.2
60	76	81

According to the measurement results, draw the solidification hardness curve of colloidal sample A and sample B, as shown in figure 6.

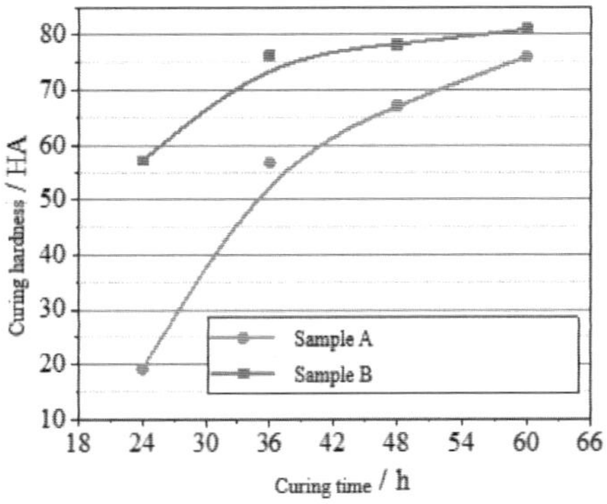

Figure 6. Curves of curing hardness of two colloid samples.

It can be found from the analysis chart that the hardness of the colloidal sample A after curing for 24h is only 19.2HA, and the hardness of sample B has reached 57.2 HA. The curing speed within 24h is about 2 times higher than that of sample A; Curing after 48 hours, the hardness growth rates of A and B both slowed down significantly, and the hardness gradually became consistent. After the two colloidal samples were cured for 60 hours, the surface of the cured layer of Sample B was relatively flat and the overall texture was relatively compact, and there were uneven grain protrusions in the cured layer of Sample A. The two kinds of solidified colloids were sliced and observed, and the obtained microscopic state is shown in figure 7. According to the figure, it can be seen that the size of the solid particles in the system of Sample A is smaller than that at the beginning, but the amount of particulate matter is still large; except for the occurrence of some pores in sample B, the composition in the system is relatively uniform, and it basically contains no particulate matter.

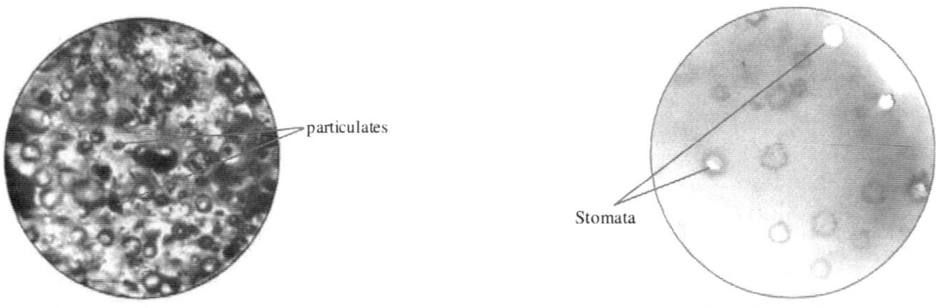

（a）Sample A （b）Sample B

Figure 7. Microscopic state of two colloids cured for 60 hours.

Based on the above analysis, the organic solvent has a better inhibitory effect on the colloidal particles, and at the same time accelerates the solidification speed of the colloid, which can shorten the production cycle of the thermal insulation layer to a certain extent. it is conducive to improving production efficiency too.

5. Conclusions

(1) The viscosity of DW-1 colloid is related to temperature and time at the same time, and its fluidity is the result of the combined influence of internal friction and curing reaction.

(2) Based on the experimental data, the viscosity change characteristics of DW-1 colloid at normal temperature and different heating temperature of 5-95 °C were analysed. The heat treatment method of 45-65 °C can keep the colloid to maintain good fluidity in 15 minutes.

(3) The inhibitory effect of organic solvents on colloidal solid particles was studied with butanone. When the mass ratio of colloid component A: component B: butanone was 5: 1: 2.2, a better colloid particle treatment effect was obtained and The curing speed of the colloid was increased about 2 times within 24 hours.

Acknowledgments

This thesis is supported by the major sub-item of “High-Grade CNC. Machine Tools and Basic Manufacturing Equipment” of the major science and technology project “Aerospace Lightweight Structures Complete Manufacturing Equipment Demonstration Line (2017ZX04005001).”

References

[1] Huang C, Liu D, Wu H, Chang Z 2016 Prospects for the application of composite material storage tanks in China's aerospace vehicles *J. Shenyang University of Aeronautics and Astronautics* **33(2)** 27-35.

[2] Dong L, Zhang H, Chen J 2016 Simulation of fluid surface tension based on IIF model *J. Graphics* **37(3)** 302-7.

[3] Zong Q, Ye W, Hong J 2019 Improvement of computer-assisted liquid surface tension coefficient measurement experiment *Educational Forum* **8** 95-6.

[4] Zhang Z, Cui G 2015 Fluid Mechanics *Tsinghua University Press.*

[5] Mozafarie S. S, Javaherdeh K 2019 Numerical design and heat transfer analysis of a non-Newtonian fluid flow for annulus with helical fins *Engineering Science and Technology, Int. J.*

[6] Almasi M 2019 Thermodynamic study of interactions between 1-alkanol and butanone *Chemical Physics* **527** 110474.

Analysis and Research on the Working Parameters of Colloidal Spraying for the Thermal Insulation Layer of Rocket Tank

Haiyang Xue[1], Haibo Xu[1, *], Yingkun Qian[2]

[1]School of Mechanical Engineering, Xi'an Jiaotong University, Xi'an 710049, China
[2]Ke-rui Institute, Shanghai Aerospace Equipments Manufacturer Co. Ltd, Shanghai 201100, China

Email: hbxu@mail.xjtu.edu.cn

Abstract. In this paper, the working parameters of DW-1 type colloid spray coating for rocket tank insulation are analysed and studied, and the optimal working parameters are determined. Firstly, a flow field simulation analysis model of thermal insulation layer DW-1 colloid spray is established, the spray flow field simulation of different spraying working parameters is performed. Then the relationship between the flow field characteristics such as spray speed and atomization uniformity and the spraying working parameters are initially determined. After that, an analytical model of droplet impact is established to study the influence of spraying speed on the impact dynamics and spreading quality of the droplet, and the optimal working parameters of colloidal spraying are determined.

1. Introduction

The rocket tank is the core component used to store the propellant fuel inside the launch vehicle, and it also plays an important role in the connection and support of the front and rear structures of the rocket. The thermal insulation layer is an important measure to ensure the reliability of the work of the rocket tank. Spray coating using the high-viscosity DW-1 type colloid of the thermal insulation layer is a common method of thermal insulation layer forming. In fact, most of the aerospace accidents caused by the failure of the thermal insulation layer are caused by problems in the coating of the insulation layer. Therefore, the use of appropriate colloid spraying working parameters to obtain better spray coating results is of great significance for ensuring the safety and stability of rocket tanks and launch vehicles.

For the high viscosity DW-1 type colloid of the thermal insulation layer of the rocket tank, it is necessary to obtain a good spray coating effect. In this paper, the flow field simulation analysis of thermal insulation layer DW-1 type colloid spraying is performed. By analysing the flow field characteristics such as spray speed and atomization uniformity, the relationship between colloid spray effect and spray working parameters is studied, and the dynamic characteristics of droplet impact spreading are numerically simulated. After comprehensive analysis, the optimal spraying working parameters of the colloidal spraying system are obtained.

IOP Conf. Series: Materials Science and Engineering **811** (2020) 012041 doi:10.1088/1757-899X/811/1/012041

2. Flow Field Simulation of Colloid Spraying and Analysis of Spraying Effect

This section we will use the computational fluid dynamics (CFD) method to simulate the spraying flow field, analyse the spraying effect, and initially determine the range of spraying working parameters.

2.1. Spray system parameter setting

Material parameters. The insulation layer colloid used in this paper is a DW-1 type colloid containing a butanone solution. The main parameters of the DW-1 type colloid are shown in table 1 [1, [2]:

Table 1. DW-1 type colloid parameters (25°C, containing butanone).

Density (kg·m-3)	Viscosity(mPa·s)	Surface tension ($N\cdot m^{-1}$)
901	950	0.069

Physical parameters. In the spraying system used in this article, a high-pressure airless spray gun is selected as the spray tool, and a wear-resistant nozzle is used in combination with a high-pressure airless spray gun. The models and parameters of the nozzles used in the study are shown in table 2.

Table 2. Three types of nozzles and their diameter parameters.

The types of nozzle	413	523	633
Diameter (mm)	0.330	0.584	0.838

Figure 1 depicts the spray flow field of a fan nozzle and its physical model. The main parameters related to the spraying effect, such as the spray angle α, spray distance S, spray length L, spray width W, etc. In this paper, a fan nozzle with an atomizing angle α of 60° is used to achieve higher spraying efficiency.

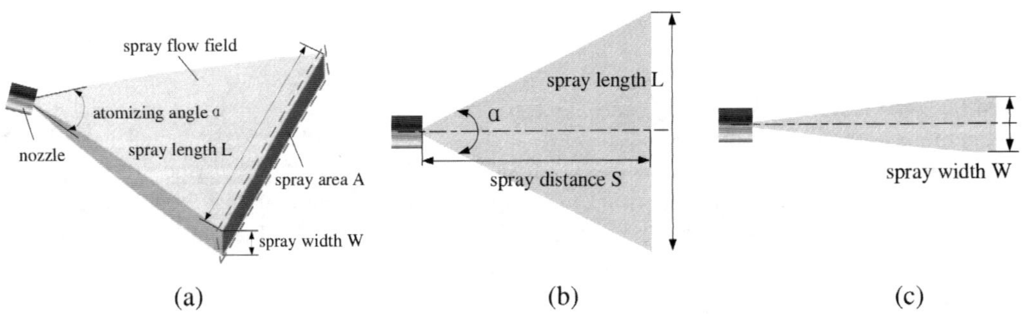

Figure 1. Physical model of fan-shaped flow field.

For the high-viscosity colloid automatic spray coating system of the thermal insulation layer, the working pressure of the plunger type pump is generally within 30MPa. During the research, the spray pressures P_{in} are 5MPa, 10 MPa, 15 MPa, 20 MPa, 25 MPa, and 30 MPa for comparison and analysis.

2.2. Flow field simulation of colloid spraying

2.2.1 Computing domain and grid division. A 500mm×1000mm plane calculation domain with the X axis as the axis of symmetry is established, and the nozzle is located at the origin. The structural unit is used to divide the grid. The minimum size is 0.25mm, the total number of nodes is 49,087, the number of grid units is 48,551, and the grid near the axis is properly encrypted.

2.2.2. Boundary conditions and model settings. In the model, the ambient temperature is set to 298.15K, the gravity is along the -Y direction, the ambient pressure is 101,325Pa, and the air density is 1.225kg·m^{-3}. The turbulence governing equation uses a standard $\kappa-\varepsilon$ two-equation model. The discrete term model (DPM) is enabled, and its governing equation is the equation (1). A new spray model needs to be established under the DPM model, and the spray atomization type is selected as the plane fan atomization model.

$$m_k \frac{dv_k}{dt} = (\sum F)_k \tag{1}$$

As in equation (1): v_k is the speed of the droplet's movement; m_k is the mass of the droplet's movement; $(\sum F)_k$ is the sum of all forces on the droplet.

Set the boundary conditions of the model: The X axis is set to the axis type boundary, the interior area is the interior type, the surrounding wall is the wall boundary, the discrete phase reflection type of the wall is escape, and the other boundary conditions remain the default settings.

Finally, the spray fluid material is set, and the parameters are the DW-1 colloid parameters containing butanone in table 1.

2.3. Study on the relationship between spraying speed and spraying working parameters

With a certain spraying distance *S*, the two most important working parameters in the spraying system are spraying pressure *P* and nozzle diameter *d* [3, 4]. The proper spraying speed is conducive to the adhesion and spreading of colloidal materials, so the spraying speed has an important effect on the colloidal spraying effect [5]. Based on the three nozzle diameters and six spray pressures described above, the CFD simulation of the spray flow field under different spraying parameters of each group is solved. Plot the change curves of spraying speeds v_0 and v_1 under different spraying working parameters, as shown in figure 2, where the spraying speed at the nozzle outlet is v_0 and the spraying speed at 1 meter is v_1.

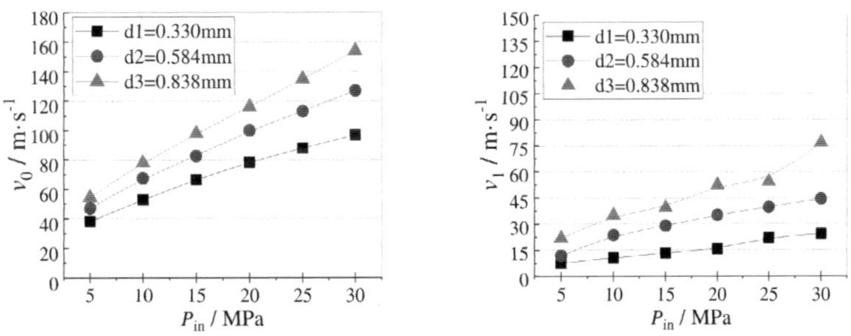

Figure 2. Variation curves of spraying speed under different spraying working parameters.

It can be seen from figure 2 that as the spray pressure increases, both the spray speeds gradually increase, v_1 is greatly reduced than v_0, and v_1 decreases more when the spray pressure is higher, because as the spray pressure increases, the spray coverage area of the flow field expands, and the area of air resistance on the flow field also increases, which accelerates the dissipation of fluid kinetic energy. At the same spray pressure, the larger the nozzle diameter, the greater the spray speeds.

For the nozzle with a diameter of 0.838mm, the entire system is always in a high-speed spray state, which easily causes colloids to splash and bounce against the surface of the rocket tank, so it is not suitable for spraying. In addition, for nozzles with a diameter of 0.330 mm, the width of the spray width at 5MPa and 10MPa starts to decay significantly at a distance of less than 1 meter, and the effective spray distance is short, so it is not suitable for colloid spraying.

For the nozzle with a diameter of 0.330 mm, when the spray pressure is controlled in the range of 15-20 MPa, the spray speed v_1 is basically maintained between 13.3-15.6 m·s^{-1}, and the entire system is in a low-speed spray working state; for a 0.584 mm diameter nozzle, when the spraying pressure is controlled at 15-20 MPa, the variation range of spraying speed v_1 is 28.9-35.1 m·s^{-1}, and the system is in a relatively moderate working state of spraying speed.

2.4. Relationship between atomization uniformity and spraying working parameters

The uniformity of atomization is an important factor affecting the colloid spraying effect. The more uniform the distribution of the material in the spray area, the better the uniformity of the adhesive layer after spraying. Based on the solution results of spraying flow field, the relationship between atomization uniformity and spraying working parameters is studied. In this chapter, the Sauter Mean Diameter is selected as the indicator of the atomization uniformity. Sauter's average diameter is recorded as D_{32} or SMD particle diameter.

Based on the CFD simulation results of three diameter nozzles at different pressures of 5-30MPa, the maximum value Max, minimum value Min, and SMD average value of the atomized droplet size in the spray flow field are calculated, and the atomization under different spraying parameters is plotted in figure 3. The variation curve of the droplet size is shown in figure 3. It can be seen from the figure that the spray pressure has a greater effect on the particle size of the atomized droplets. For the same nozzle diameter, the SMD particle size decreases with increasing spray pressure. Under the same spraying pressure, the smaller the nozzle diameter, the smaller the SMD particle size and the maximum and minimum particle diameters of the atomized droplets.

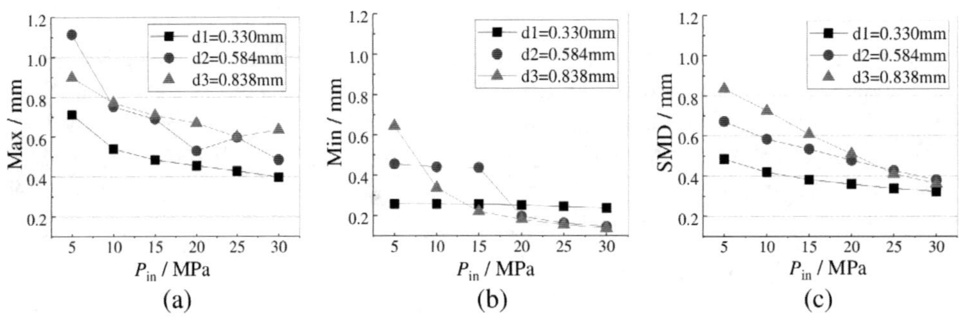

Figure 3. Variation curve of atomized droplet size under different spraying working parameters.

In order to control the spray speed of the flow field at a relatively moderate or low speed, and to achieve a better spray atomization uniformity, in order to obtain a better colloid spraying effect, a suitable spraying working parameter range can be initially determined, as shown in the table 3.

Table 3. Initially determined spraying working parameter range and spraying effect.

Nozzle diameter (mm)	Spraying pressure (MPa)	Spraying speed (m·s^{-1})	Atomized droplet size (mm)
0.330	15-20	13.3-15.6	0.360-0.382
0.584	15-20	28.9-35.1	0.478-0.534

3. Research on Droplet Impact Spreading Dynamics and Optimal Spraying Working Parameters

3.1 Establishment of the droplet ejection impact model

In this section, a CFD analysis model of droplet impact is established, the dynamic characteristics of droplet impact spreading under different ejection speeds are analyzed, and the influence of the variation of ejection speed on the quality of the droplet spread forming is further determined, so as to further determine the optimal colloid spraying working parameters range. As shown in figure 4, a 5mm×5mm CFD simulation calculation domain is established. A DW-1 type colloidal droplet is established at a height of 1mm from the bottom, and the initial diameter D_0 of the colloidal droplet is set to 0.6 mm according to the research results in the previous section.

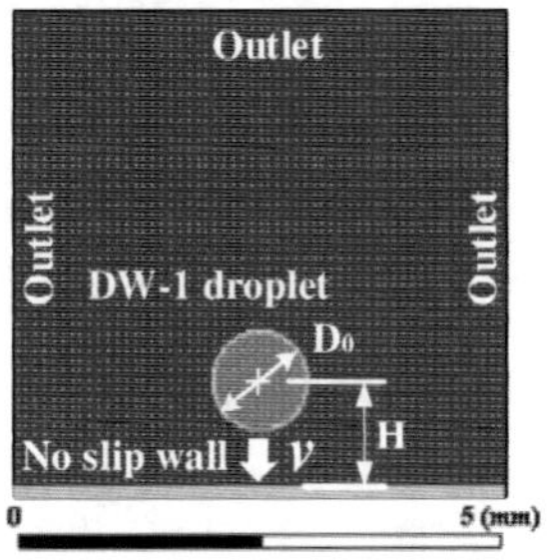

Figure 4. Droplet impact model.

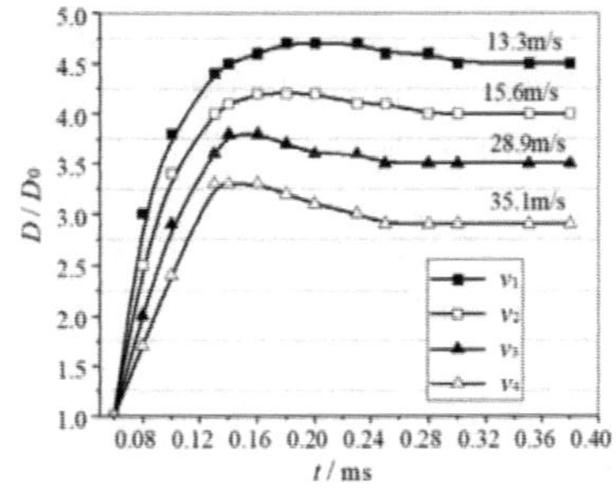

Figure 5. Dynamic changes of droplet spreading coefficients at different spray speeds.

Rectangular structured cells are used for mesh division. The mesh near the bottom region is properly encrypted. The model consists of 62,500 mesh cells. Air is the basic phase and DW-1 type colloid is the second phase. The bottom of the model is the non-slip wall boundary condition, and the wall material is defined as aluminum alloy; the top and the boundary on both sides are atmospheric boundary conditions. The numerical simulation in this section uses the Volume of Fluid (VOF) model.

3.2. Research on droplet impact dynamics and spread forming process

For DW-1 type colloid droplets in the thermal insulation layer, the ejection speed has a great influence on the dynamic process of impact. In this paper, four different spray speeds of v_1, v_2, v_3, and v_4 are selected as 13.3 m·s^{-1}, 15.6 m·s^{-1}, 28.9 m·s^{-1}, and 35.1 m·s^{-1}, respectively, and the numerical simulation of the dynamic process of the impact of the droplet under different spray speeds is carried out. Select the main feature moments from each group of simulation results, as shown in figure 5.

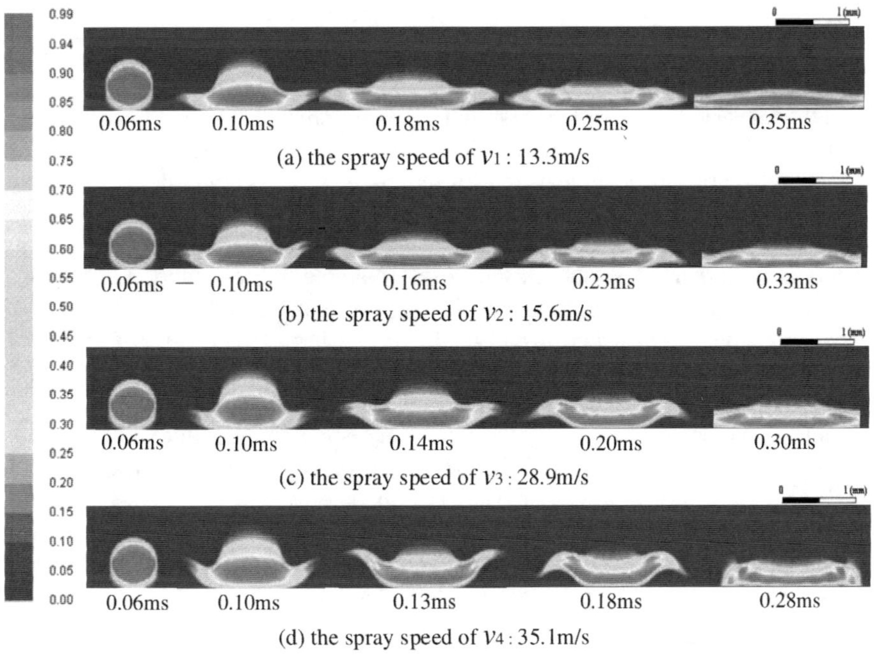

Figure 6. The dynamic process of droplet impact and spread under different spray speeds.

It can be found from figure 5 that as the spray speed increases, the droplets will gain greater impact kinetic energy, and the spreading speed will become faster and faster, that is, the maximum spreading

IOP Conf. Series: Materials Science and Engineering **811** (2020) 012041 doi:10.1088/1757-899X/811/1/012041

state will be reached faster, but the speed is too large. At the same time, there will be splashes and lifts at the edges, the final shrinking and gathering effect of the droplets will become more and more obvious, and the flatness of the spread will also be affected [6].

3.3. Research on the quality of spread spreading

The ratio of dynamic diameter to initial diameter is the spreading coefficient. The change of the spreading coefficient reflects the dynamic characteristics of the spreading of the droplet. The larger the spreading coefficient, the better the overall spreading quality of the colloid droplets. Based on the simulation results of droplet impact spreading at different jet speeds in this section, the dynamic change of the droplet spreading coefficient with time is shown in the graph in figure 6.

It can be seen from figure 6 that after the droplet spreading coefficient reaches the maximum value, it decreases slightly and then stabilizes. The greater the spray speed, the more pronounced the tendency of the droplet to retract. With the increase of the spraying speed, the maximum spreading coefficient of the droplets gradually decreases, and the time to reach the maximum spreading coefficient is relatively shortened, but the overall spreading degree of the droplets is getting lower and lower, and the spreading quality is also worsening.

In general, when the spraying speed v_1 is 13.3 $m \cdot s^{-1}$ and the spraying speed v_2 is 15.6 $m \cdot s^{-1}$, the spreading degree of the atomized droplets after hitting the wall is higher, and the spreading quality of the droplets is higher than that of v_3 and v_4.

Based on the research results of the impact and spreading characteristics of the droplets in this section, when the spray speed of 0.330mm nozzle diameter in the range of 15-20 MPa spray pressure is relatively more suitable for colloid spraying. Finally, the optimal spraying working parameter range of the thermal insulation layer colloid is further determined, as shown in table 4.

Table 4. Range of optimal spraying working parameters and spraying effect.

Nozzle diameter(mm)	Spraying pressure (MPa)	Spraying speed ($m \cdot s^{-1}$)	Atomized droplet size (mm)
0.330	15-20	13.3-15.6	0.360-0.382

4. Conclusions

The research in this paper has obtained the following research results:

A CFD simulation model for DW-1 type colloid spraying of adiabatic layers was established, and the flow field simulation results were analyzed. Based on the CFD simulation results of the spray flow field, the spray velocity of the flow field was studied separately. Based on the relationship between atomization uniformity and spraying working parameters, a suitable spraying working parameter range is initially determined after analysis: when the nozzle diameter is between 0.330-0.584 mm and the spraying pressure is within the range of 15-20 MPa, the spraying speed can be set to v_1 controlled at 13.3-35.1 $m \cdot s^{-1}$, while the SMD particle size of the atomized droplets is less than 0.6 mm.

A spray impact analysis model for atomized droplets was established, and the DW-1 type droplet impact spreading dynamics at different spraying speeds were numerically simulated, and the influence of spraying speed on the droplet spreading coefficient and spreading quality was analysed. It is found that when the spraying speed is relatively low, the spreading quality after the impact of the droplet is better. Optimal spraying working parameters were obtained.

Based on the analysis of the characteristics of the flow field and the research on the impact of the droplets, the optimal spraying parameters were obtained: the nozzle diameter is 0.330 mm, and the spraying pressure is in the range of 15-20 MPa.

Acknowledgments

This thesis is supported by the major sub-item of "High-Grade CNC. Machine Tools and Basic Manufacturing Equipment" of the major science and technology project "Aerospace Lightweight Structures Complete Manufacturing Equipment Demonstration Line (2017ZX04005001)."

References

[1] Teng X Y, Min G H, Liu H L, Shi Z, Wang H, Ye Y 2001 Correlations between viscosity and surface tension with the structure of liquid pure iron *Mater. Sci. Technol.* **9** 383-6.

[2] Belda R 2009 A proposed equation of correlation for the study of thermodynamic properties (density, viscosity, surface tension and refractive index) of liquid binary mixtures *Fluid Phase Equilibria* **282(2)** 88-99.

[3] Cai Y K, Luo X 2018 Simulation Study of Dynamic Behaviour of Water Droplet on Laser Machined Surface *WCMNM*.

[4] Han X, Qin J, Feng M 2014 A Numerical Study on Dynamics of Drop Impact on Solid Hot Surface in Microgravity *Manned Spaceflight* **20** 538-42.

[5] Liang C Wang H, Zhu X, Chen R, Ding Y, Liao Q 2013 Numerical simulation of droplet impact on surface with different wettabilities *CIESC J.* **64(8)** 2745-51.

[6] Mehdi-Nejad V, Mostaghimi J, Chandra S 2003 Air bubble entrapment under an impacting droplet *Phys. Fluids* **15(1)** 173-83.

IOP Conf. Series: Materials Science and Engineering **811** (2020) 012042 doi:10.1088/1757-899X/811/1/012042

Experimental Research on Ferro Foam Concrete I Beams with Different Number of Wire Mesh Layer

M Afifuddin[1], M Isya[1], Mahlil[1], M Dwihandayani[2]

[1]Civil Engineering Department, Engineering Faculty, Syiah Kuala University, Banda Aceh, Indonesia
[2]Magister Student of Civil Engineering Department, Engineering Faculty, Syiah Kuala University, Banda Aceh, Indonesia

Abstract. The paper presents the experimental study on Ferro foam concrete I beam with different number of wire mesh layers. Ferro foam concrete is alternative materials, which consist of a combination between wire mesh and foam concrete. This material is lightweight and easy to be produced. There are 3 specimens has been tested. Specimens' configuration was 300 mm height, 150 mm width of upper and lower flange, and 60 mm web with 2200 mm span. The different between specimens was number of wire mesh layers, 2 layers, 3 layers, and 4 layers. Foam concrete used in this research consists of cement, water, foam agent, and 10 % Pozzolan. Compression strength of concrete was 23.23 MPa, and Tensile strength of reinforcement D8 and wire mesh G12.7 were 421 MPa. The specimens were tested using four points loading. Three units of LVDT were placed at a distance of 250 mm and 1000 mm from the support. Based on the results, it was found that the maximum load that can be carried out by specimen was 8.48 Tonf, 8.00 Tonf, and 9.50 Tonf, for 2 layer, 3 layers, and 4 layers, respectively. Maximum deflection for each specimen was 14.01 mm, 13.53 mm, and 21.37 mm, for 2 layers, 3 layers, and 4 layers, respectively. Number of layer in I beam 300.60.60 has no significant effect for the beam in their ability to carried the load.

1. Introduction

One reason for the isolation of a region is the unavailability of the bridge infrastructure. With the isolation of an area of economic growth in the region will move very slowly. Many products from that area cannot be sold, and the people in the area to be so far behind. Many simple bridge build in the village made of wood. Wood is a material which can be easy to be found, and available in the village. Unfortunately, since moratorium logging announced by Aceh Government through governor instruction No. 05/INSTR/2007 on 6 June 2007, wood material is very difficult to be find. To solve this problem, this research gives an alternative material, which is called Ferro foam concrete. Structural and Material Laboratory of Civil Engineering Syiah Kuala University, since 2006 has developed research on Foam Concrete. Foam concrete is a type of Lightweight concrete, which can be produced with different quality [1]. Abdullah [2], pozzolan with certain percentage can give positive influence on compression strength of foam concrete. Based on the research, it is possible to create an alternative material, which is a modification of Ferro cement. The objective of this study is to study experimentally Ferro foam concrete I beam with different number of wire mesh layers. Three Ferro foam concrete I beam with 300 mm height with 2, 3, and 4 wire mesh layers were tested in Structural Laboratory of Syiah Kuala University. This profile has a span of 2000 mm, and 100 mm wingspan.

IOP Conf. Series: Materials Science and Engineering **811** (2020) 012042 doi:10.1088/1757-899X/811/1/012042

Foam concrete used in this research consists of cement, water, foam agent, and 10% pozzolan. Average concrete compression strength was 24.23 MPa, and 4217.14 kgf/cm^2 (421 MPa) for yield strength of reinforcement with D8, and 4200 kgf/cm^2 (420 MPa) for wire mesh G 12.7. The specimens were tested using four-point loading. Three units of LVDT placed at a distance of 250 mm and 1000 mm from the support. Deflection readings every 100 kg of load increase.

2. Literature Review

2.1. Foam Concrete
Foamed concrete is defined as a light cellular concrete, which can be classified as a lightweight concrete (density of 400–1850 kg/m^3) with random air-voids created from the mixture of foam agents in mortar. Foamed concrete is recognized for its high flowability, low cement content, low aggregate usage [1-3], and excellent thermal insulation [4]. Furthermore, the foamed concrete is considered as an economical solution in fabrication of large scale lightweight construction materials and components such as structural members, partitions, filling grades, and road embankment infills due to its easy production process from manufacturing plants to final position of the applications [2, 5-7]. In practice, foamed concrete has been commonly used in construction applications in different countries such as Germany, UK, Philippines, Turkey, and Thailand [8, 9].

2.2. Foam Agent
Foam agents control the concrete density through a rate of air bubbles created in the cement paste mixture. Foam bubbles are defined as enclosed air-voids formed due to the addition of foam agent. The foam agents are commonly synthetic, protein-based, detergents, glue resins, hydrolysed protein, resin soap, and saponin, [13, 15, 16]. The most common foam agents are synthetic and protein based. The protein based foam agents result in a stronger and a more closed-cell bubble structure which permits the inclusion of greater amounts of air and also provides a more stable air void network while the synthetic ones yield greater expansion and thus lower density [7, 10, 17]. The content of the foam agent has a considerable effect on properties of both fresh and the hardened concrete [11, 18]. It is reported that the excessive foam volume results in a drop in flow [14, 19]. However, the flow is significantly affected by mixing time. As reported, the greater the mixing time, the more the entrained air, albeit, prolonged mixing may cause the loss of entrained air by dropping the air content [10, 12].

2.3. Ferro Foam Concrete
Ferro foam concrete is an alternative material, which consists of a combination between wire mesh and foam concrete. Ferro foam concrete is a type of thin composite material made of foam concrete reinforced with uniformly distributed layers of continuous relatively small diameter, wire meshes. The design and construction of ferro foam concrete almost similar with Ferro cement which is described in two state-of-the-art reports published by the ACI committee 549 [20]. This material will give highly innovative solutions, since the strengths of foam concrete can be achieved more than 25 MPa [21, 22].

3. Methodology

3.1. Material
The material used in this study were cement, pozzolan, water, deform steel reinforcement, foam agent, wire mesh and admixture. Cement used was Portland cement type I production from PT. Semen Padang, Banda Aceh, Indonesia. Laboratory examination of semen is not done because the cement has been considered to meet the Indonesian National Standard (SNI) No. 15-2049-1994 and ASTM C.150-81. Tests done only by visual observation and examination of the wrapping bag friability, as well as the colour of the cement fineness. Natural pozzolan sand used should be cleaned of organic waste and filtered using a 4.76 mm sieve. Sand natural pozzolan is extracted from Ujong Bate sub district, Aceh Besar. Natural pozzolan also examined his physical density, absorption inspection, and fineness modulus. There was also tested in the chemical properties of Research and Testing Laboratories Industry Standards in Banda Aceh The water used in the concrete mix foam is water available at the

Laboratory of Building Construction and Materials Engineering Faculty Syiah Kuala University derived from wells. Foam agent used in this study derived from synthetic foam, which has been treated with chemicals to produce a similar foam lather that can be used as filler mixed concrete. Reinforcement used for tensile reinforcement is reinforcement steel screw with a diameter of 8 mm. Wire mesh used in this study were obtained from the nearest hardware store. Wire mesh, which will be used in this research diameter of 1 mm and 12.71 mm spacing as reinforcement. The rectangular wire meshes and conforms to ASTM A-185. Mix proportion for foam concrete used based on previous research done by Azzani [23]. Table 1 shows mix design for this research.

Table 1 Mix Proportion of Foam Concrete

Specific Gravity	Cement (kg)	Pozzolan (kg)	Water (kg)	Foam Agent (l)
1.6	**1028.57**	**160**	**411.43**	**178.71**

3.2. Detail Specimens

Detail of specimen shown in figure 1. Figure 1(a) shows detail profile I with number of reinforcement, and wire mesh layer. Figure 1(b) shows length of specimen which was 100 mm from the pin and roll support.

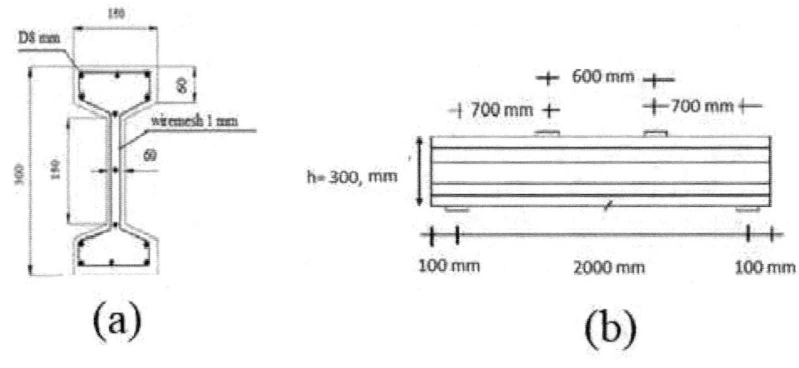

Figure 1 Detail Specimen

3.3. Experimental Setup

The three beams were simply supported and were subjected to two, symmetrically placed, point loads. The distance between the two loading points was kept at 600 mm and 700 mm from the support. The beams were suitably instrumented for measuring deflections at 250 mm and 1000 mm from the support. All the readings from the measurements were recorded using data logging system. Throughout the test, photos were taken for cracking behaviour, failure mode and any other features that were observed. Details of experimental Setup is shown in figure 2.

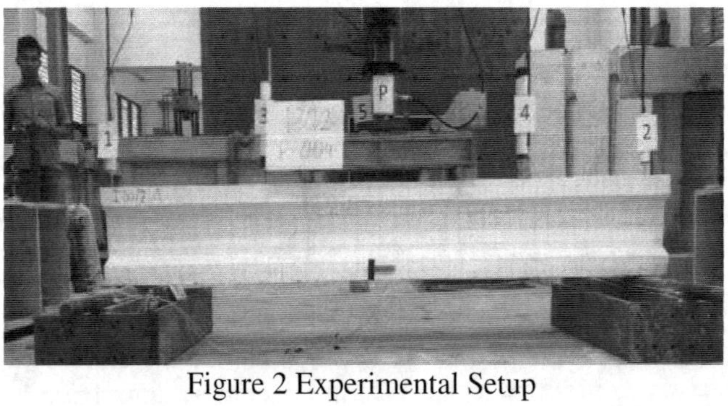

Figure 2 Experimental Setup

4. Results and Discussion

4.1. Material Properties

Table 2 shows concrete compression strength of results. Three-cylinder specimen has been tested before I beam specimen tested. Tensile strength result for reinforcement was 4217.136 kg/cm^2 (421 MPa), and for wire mesh 4200kg/cm^2 (420 MPa).

Table 2 Concrete Compression Strength

Specimen name	f'c (kg/cm^2)	Average f'c(kg/cm^2)
C1	**24.97**	
C2	**23.86**	**24.23**
C3	**23.86**	

4.2. Load-Deflection Relationship

Load-Deflection relationship shows in figure 3. It can be seen from the figure that the rigidity for all beam almost the same. First crack occurs for I beam 300.60.60 with 2-layer wire mesh when the load was 1.8 Tonf, and almost same for 3 layers and 4 layers. Maximum load that can be carried by the beam 2, 3 and 4 layer is 8.48 Tonf; 8 Tonf; 9.5 Tonf, respectively. After maximum load is reach on each specimen, the load continues to decrease until all specimens' collapse. The load that can be carried by each beam only slightly different. It can be said that number of wire mesh layer was not so significant for the beam in their ability to carry the load. Maximum deflection for each specimen was 14.01 mm, 13.53 mm, and 21.37 mm, for 2 layers, 3 layers, and 4 layers, respectively.

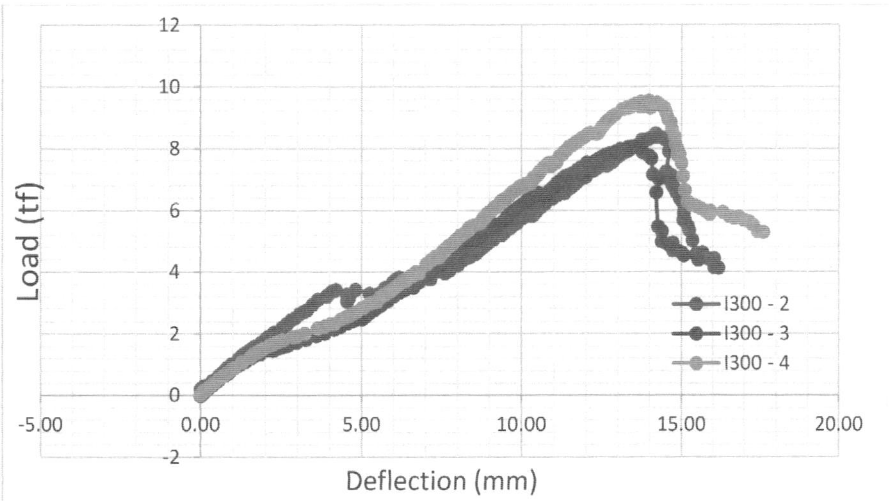

Figure 3 Load-Deflection Relationship for I Beam 300.60.60 with different number of layer

4.3. Crack Pattern

Crack pattern for each beam shown in figure 4. It shown from the figures that crack pattern and maximum load for each beam. It has the same pattern and all the failure due to crush of concrete from the support. It can be seen from the figure that all the beam has shear failure. Figure 4 Crack Pattern for each Beam.

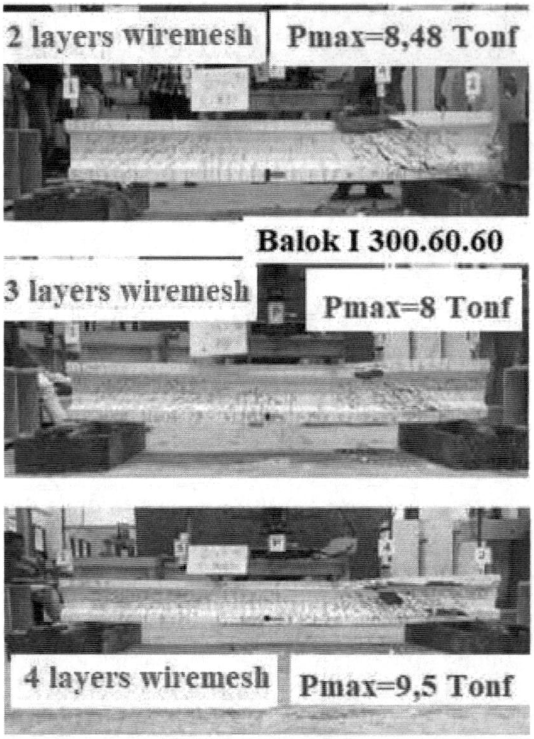

Figure 4 Crack Pattern for each Beam

5. Conclusions

1. Maximum load which can be carried by I beam 300.60.60 with 2, 3 and 4-layer wire mesh is 8.48 Tonf; 8 Tonf; 9.5 Tonf, respectively.
2. Number of layer in I beam 300.60.60 has no significant effect for the beam in their ability to carried the load.
3. All the final failure due to crush of the concrete in the support, all the beam has shear failure.

Acknowledgments

Authors wishing to acknowledge financial support from LPPM Unsyiah for research loan on skim PCP (Candidate Professor Loan).

References

[1] Valore R C 1954 Cellular concretes Part 2 physical properties *J. Proc.* **50(6)** 817-36.
[2] Ramamurthy K, Nambiar E K, Ranjani G I S 2009 A classification of studies on properties of foam concrete *Cement Concr. Compos.* **31(6)** 388-96.
[3] Jones M R, McCarthy A 2005 Preliminary views on the potential of foamed concrete as a structural material *Mag. Concr. Res.* **57(1)** 21-31.
[4] Richard T, Dobogai J, Gerhardt T, Young W 1975 Cellular concrete–a potential load-bearing insulation for cryogenic applications? *IEEE Trans Magnet.* **11(2)** 500-3.
[5] Uddin N, Fouad F, Vaidya U K, Khotpal A, Serrano-Perez J C 2006 Structural characterization of hybrid fiber reinforced polymer (FRP)-autoclave aerated concrete (AAC) panels *J. Reinf. Plast. Compos.* **25(9)** 981-99.
[6] Tarasov A S, Kearsley E P, Kolomatskiy A S, Mostert H F 2010 Heat evolution due to cement hydration in foamed concrete *Mag. Concr. Res.* **62(12)** 895-906.
[7] Tikalsky P J, Pospisil J, MacDonald W 2004 A method for assessment of the freeze–thaw resistance of preformed foam cellular concrete *Cement Concr. Res.* **34(5)** 889-93.
[8] Weigler H, Karl S 1980 Structural lightweight aggregate concrete with reduced density-lightweight aggregate foamed concrete *Int. J. Cem. Compos. Lightweight Concr.* **2(2)** 101-4.

[9] Mydin M A O, Wang Y C 2011 Structural performance of lightweight steel-foamed concrete-steel composite walling system under compression *Thin-walled Struct.* **49(1)** 66-76.
[10] Beningfield N, Gaimster R, Griffin P 2005 Investigation into the air void characteristics of foamed concrete *Use of Foamed Concrete in Construction: Proceedings of the International Conference held at the University of Dundee, Scotland, UK* 51-60.
[11] Jones M R, McCarthy A 2005 Behaviour and Assessment of Foamed Concrete for Construction Applications *Use of Foamed Concrete in Construction: Proceedings of the International Conference held at the University of Dundee, Scotland, UK* Thomas Telford Publishing, 61-88.
[12] Mindess S. (Ed.) 2019 Developments in the Formulation and Reinforcement of Concrete Woodhead Publishing.
[13] De Rose L, Morris J 1999 The Influence of Mix Design on the Properties of Microcellular Concrete *Sp. Tech. Mater. Constr.* 185-97.
[14] Nambiar E K, Ramamurthy K 2006 Influence of filler type on the properties of foam concrete *Cem. Concr. Compos.* **28(5)** 475-80.
[15] Van Deijk S 1919 Foam Concrete *Concrete* **25(5)**.
[16] Bing C, Zhen W, Ning L 2011 Experimental research on properties of high-strength foamed concrete *J. Mater. Civ. Eng.* **24(1)** 113-8.
[17] ACI523.2R-96 1996 Guide for Precast Cellular Concrete Floor, Roof, and Wall Units *ACI Committee 523.*
[18] Aldridge D 2005 Introduction to foamed concrete: what, why, how? *Use of Foamed Concrete in Construction: Proceedings of the International Conference held at the University of Dundee, Scotland, UK* Thomas Telford Publishing 1-14.
[19] Jones M R, McCarthy A 2006 Heat of hydration in foamed concrete: effect of mix constituents and plastic density *Cement Concr. Res.* **36(6)** 1032-41.
[20] Batson G B, Castro J O, Guerra A J, Iorns M E, Johnston C D, Naaman A E, Tilsen B L 1988 Guide for the design, construction, and repair of ferrocement *ACI Struc. J.* **85(3)** 325-51.
[22] Afifuddin M 2016 Performance of Ferro Foam Concrete Girder Beam Subjected to Static Load *Appl. Mechan. Mater.* **845** 214-9.
[23] Azzani 2010 Effect of Substitution Pozzolan on Mechanical Behavior of Foamed Concrete *Under Graduate Tesis, Engineering Faculty, Syiah Kuala University, Banda Aceh.*

IOP Conf. Series: Materials Science and Engineering **811** (2020) 012043 doi:10.1088/1757-899X/811/1/012043

Design and Analysis of Flow Field in Electrochemical Machining of Micro Grooves on Gear Tooth Surface

Yuanlong Chen[1, a, *], Tengfei Zhang[1, b], Hua Lin[1], Qi Chen[1]

[1]School of Mechanical Engineering, Hefei University of Technology, Hefei 230009, China

[a]chenyuanlong@.hfut.edu.cn, [b]1831766842@qq.com

Abstract. The gear with solid lubricating material embedded in micro grooves on gear tooth surface can better adapt to severe lubrication conditions. Electrochemical machining with the mask has shown good feasibility in machining micro grooves for filling solid self-lubricating materials on the tooth surface of gear. The electrolyte flow field is a vital contributor to the accuracy of electrochemical machining, therefore, it is important to design a suitable electrolyte flow mode. In this paper, based on the distribution characteristics of micro grooves on gear tooth surface, two kinds of shaped cathodes are designed, one with electrolyte flow slot and the other without. Each cathode can form two electrolyte flow modes respectively. The effects of four electrolyte flow modes on the stability of the machining gap flow field are analyzed by using computational fluid dynamics. The simulated results show that the cathode without electrolyte flow slot is beneficial to electrolyte flow during electrochemical machining with the mask, notably when the electrolyte flows into the machining gap from the face of the gear.

1. Introduction

In view of the deficiency of present self-lubricating gear technology, Chen et al. proposed a new type of self-lubricating gear [1]. The salient feature of the gear is that staggered micro-grooves for filling solid self-lubricating materials are machined on the gear tooth surface. Depending on the excellent lubrication characteristics of filling materials, this new type of gear can better adapt to harsh conditions of insufficient lubrication.

At present, various methods have been researched for the fabrication of micro grooves, such as laser processing, electrical discharge machining and electrochemical machining (ECM) [2 3]. Among them, ECM is a metallic anodic dissolution process in which the metallic workpiece and tool are separately the anode and cathode [4]. The major advantages of ECM are high removal rate in combination with the damage-free of workpiece surface [5]. Liu et al. proposed a method of ECM using multifunctional cathode to machine micro grooves on workpiece with the width of 0.7 mm and the depth of 0.46 mm [6]. In ECM with the mask, metal is dissolved from unprotected areas of a mask-patterned workpiece that acts as the anode in an electrolytic cell [7]. Xu et al. fabricated micro grooves on SS304 metal surface with the width of 0.25 mm and the depth of 0.1 mm by employing ECM with the mask [8]. However, research on electrochemical machining of micro grooves on involute tooth surface of gear is still deficiency.

In this paper, according to the distribution features of micro grooves on gear tooth surface, two kinds of shaped cathodes are designed, one with electrolyte flow slot and the other without. Each cathode can form two electrolyte flow modes separately. In order to obtain the most reasonable electrolyte flow mode for ECM with the mask. The flow field distribution in the machining gap of those four flow modes are analyzed by computational fluid dynamics.

2. The Principle of ECM with the Mask Using Shaped Cathode

The research object of this paper is involute spur gear with the module is 2mm and the tooth thickness is 15mm. The width and depth of micro grooves can be calculated to be 0.38 mm and 0.16 mm respectively. The grooves are distributed on the involute tooth surface. In ECM with the mask, micro-patterns can be produced on photoresist-coated workpiece using photolithography, the metal is then selectively dissolved from the unprotected regions, so the method has high feasibility.

The paper adopted the shaped cathode of single-tooth method for the tool cathode design. The shape of cathode is designed based on the shape of gear tooth space, but the overall reduction is 0.3 mm compared with the spacewidth, this gap is used for electrolyte flow [9]. Figure 1 shows a schematic diagram of ECM with the mask using shaped cathode. The workpiece is covered by the photoresist mask with micro grooves pattern. The area of the workpiece exposed to the electrolyte can be dissolved when in an energized state. The dissolved metal, the Joule heat produced and the generated gas bubbles are carried away by flowing electrolyte.

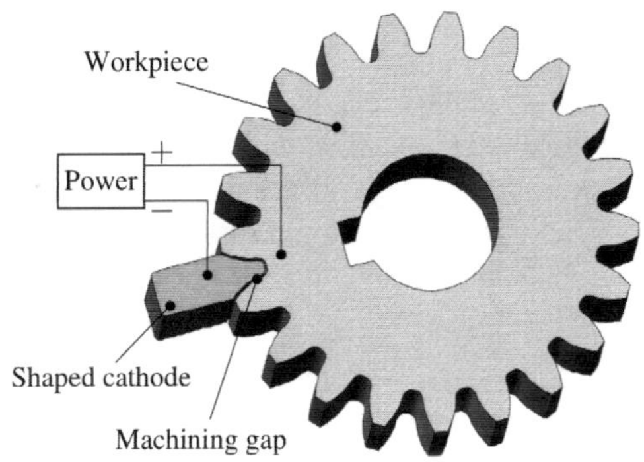

Figure 1. Schematic diagram of the ECM with the mask using shaped cathode.

3. The Design of Cathode and Flow Mode

The micro grooves studied in this paper are mainly distributed on the involute tooth surface of the gear, and there is no distribution at the bottom of the gear tooth space. Based on the feature, two kinds of shaped cathodes are designed, one with electrolyte flow slot and the other without. As shown in figure 2(a), the kind of cathode is called the shaped cathode with electrolyte flow slot. In the middle of the cathode, there is a channel for the electrolyte to flow into the machining gap. The shaped cathode with electrolyte flow slot as shown in figure 2(b). The cathode does not directly participate in the supply of electrolyte, and the flow mode of electrolyte is lateral flow.

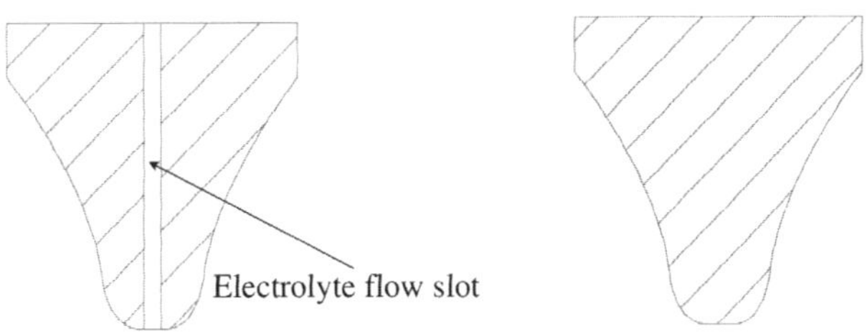

(a) Cathode with electrolyte flow slot (b) Cathode without electrolyte flow slot

Figure 2. Shaped cathode model.

The flow mode of electrolyte depends on the design of the cathode and the shape of the workpiece. Four electrolyte flow modes were proposed. Among them, Mode-I and Mode-II are composed of cathode with electrolyte flow slot and workpiece, Mode-III and Mode-IV are composed of cathode without electrolyte flow slot and workpiece. As shown in figure 3(a), the kind of flow mode is called Mode-I in which the electrolyte flows from the inlet on the cathode into the machining gap, and flows out of the machining gap from the electrolyte outlet. As shown in figure 3(b), the electrolyte flow path of Mode-II is opposite to that in Mode-I. As shown in figure 3(c), the electrolyte in Mode-III flows from the inlet at the gear tooth tip into the machining gap and then flows out from the outlet. From figure 3(d), it is clear that the electrolyte in Mode-IV flows into the machining gap from the face of the gear. To obtain the most suitable flow mode, a computational fluid dynamics (CFD) will be carried.

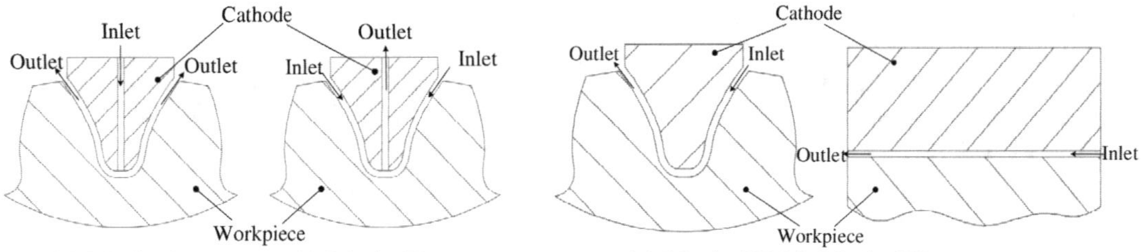

(a) Mode-I (b) Mode-II (c) Mode-III (d) Mode-IV

Figure 3. Schematic diagram of the electrolyte flow mode.

4. Numerical Simulation of Electrolyte Flow Field

The flow field is the interspace between the cathode and the workpiece, which is determined by the design of the electrolyte flow mode and the shape of the workpiece. Since the object of the study is the flow field in the machining gap, the micro grooves and mask are not considered during modeling. The flow field model of Mode-I and Mode-II is show in figure 4(a). In the model, the plane A in the middle of the cathode is the inlet or outlet of the electrolyte. Figure 4(b) shows the flow field model of Mode-III and Mode-IV.

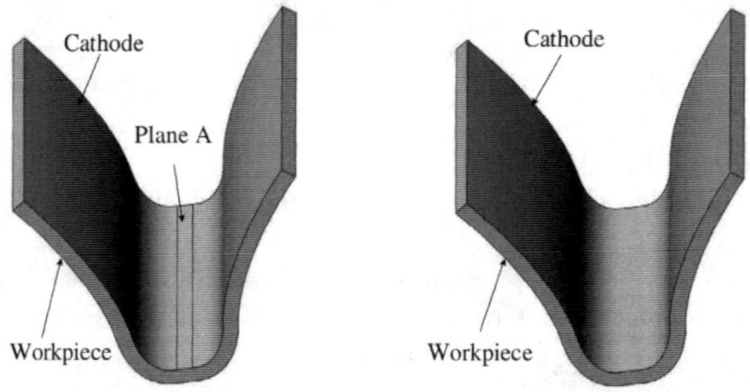

(a) Model of Mode-I and Mode-II (b) Model of Mode-III and Mode-IV

Figure 4. Geometric model of flow field simulation.

4.1. Mathematical Model

The simulation is based on following assumptions: (1) the fluid is an incompressible, continuous, and viscous Newtonian fluid. (2) The change of the electrolyte temperature and the energy dissipation caused by the temperature differences are ignored, and the flow motion obeys the mass and momentum conservation equations which can be written as

$$\begin{cases} \rho\dfrac{\partial\vec{u}}{\partial t}+\rho(\vec{u}\cdot\nabla)\vec{u}=\nabla\cdot\left[-PI+(\mu+\mu_T)(\nabla\vec{u}+(\nabla\vec{u})^T)\right] \\ \rho\nabla\cdot(\vec{u})=0 \end{cases} \quad (1)$$

Where u is the velocity vector along the x-axis, P is the mean electrolyte pressure, μ is the dynamic viscosity, ρ is the density of the fluid, and μ_T is the turbulent viscosity coefficient.
The obvious improvements of the RNG $k-\varepsilon$ model compared with the standard model is due to the

better treatment of near-wall turbulence effects [10]. In the ECM process, the electrolyte is sealed in the fixture, the flow in the machining gap is complex, and the electrolyte streamlines are curved. Thus, the RNG $k-\varepsilon$ model is suitable for the current situation and is adopted in this paper to simulate the flow field. The equations are as follows:

$$\frac{\partial(\rho\kappa)}{\partial t}+\frac{\partial(\rho\kappa u_i)}{\partial x_i}=\frac{\partial}{\partial x_j}[\alpha_k\mu_{eff}\frac{\partial k}{\partial x_j}]+G_k+\rho\varepsilon \tag{2}$$

$$\frac{\partial(\rho\varepsilon)}{\partial t}+\frac{\partial(\rho\varepsilon u_i)}{\partial x_i}=\frac{\partial}{\partial x_j}[\alpha_\varepsilon\mu_{eff}\frac{\partial\varepsilon}{\partial x_j}]+\frac{C^*_{1\varepsilon}\varepsilon}{k}G_k-C_{2\varepsilon}\rho\frac{\varepsilon^2}{k} \tag{3}$$

Where κ is turbulent kinetic energy, ε is dissipation rate, G_k is turbulent kinetic energy caused by gradient change, the model constants are given as $C_\mu=0.0845$, $C_{1\varepsilon}=1.42$, $C_{2\varepsilon}=1.68$,and $\alpha_k=\alpha_\varepsilon=1.39$.

4.2. Numerical Simulation Results and Analysis

Based on the RNG $k-\varepsilon$ turbulent model, the flow field simulation in the machining gap is carried out employing the ANSYS software. The boundary conditions determined by initial work are that the electrolyte inlet pressure is 0.4 MPa and the outlet pressure is 0.1 MPa. After simulation, the pressure and velocity distribution of electrolyte in the machining gap of Mode-I and Mode-II can be obtained, as shown in figure 5. Mode-I, the simulation results show that the electrolyte pressure in the machining gap is low as a whole, and there are negative pressure areas near the bottom of the cathode. The electrolyte flow forms a vortex in the negative pressure area, which will lead to defects such as flow marks or grooves on the machined surface. In Mode-II, the electrolyte pressure is relatively uniform as a whole. However, the electrolyte pressure in the flow channel is high, resulting in a low electrolyte flow velocity on the machining surface. The electrolytic products will be hard to remove because of the low velocity. Based on the above description and the simulation results, the flow field uniformity of Mode-I and Mode-II is poor.

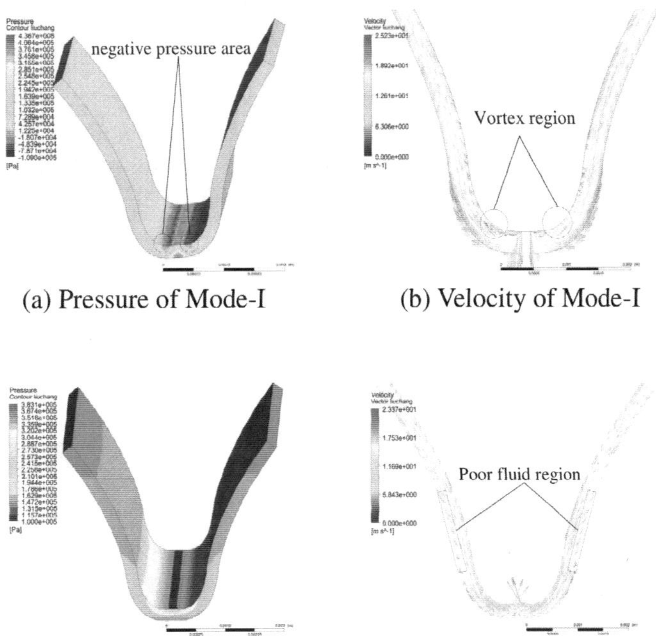

(a) Pressure of Mode-I (b) Velocity of Mode-I

(c) Pressure of Mode-II (d) Velocity of Mode-II

Figure 5. Electrolyte pressure and velocity distribution in the machining gap.

When the electrolyte flows in Mode-III and Mode-IV, the velocity and pressure distribution of the flow field in the machining gap are shown in figure 6. In Mode-III, gradient change of electrolyte pressure in the machining gap, the electrolyte flow velocity is relatively uniform as a whole. However, because the flow path is curved, the electrolyte flow at the bottom of the machining gap is confused. In Mode-IV, the electrolyte pressure changes in gradient and the flow velocity distribution is uniform.

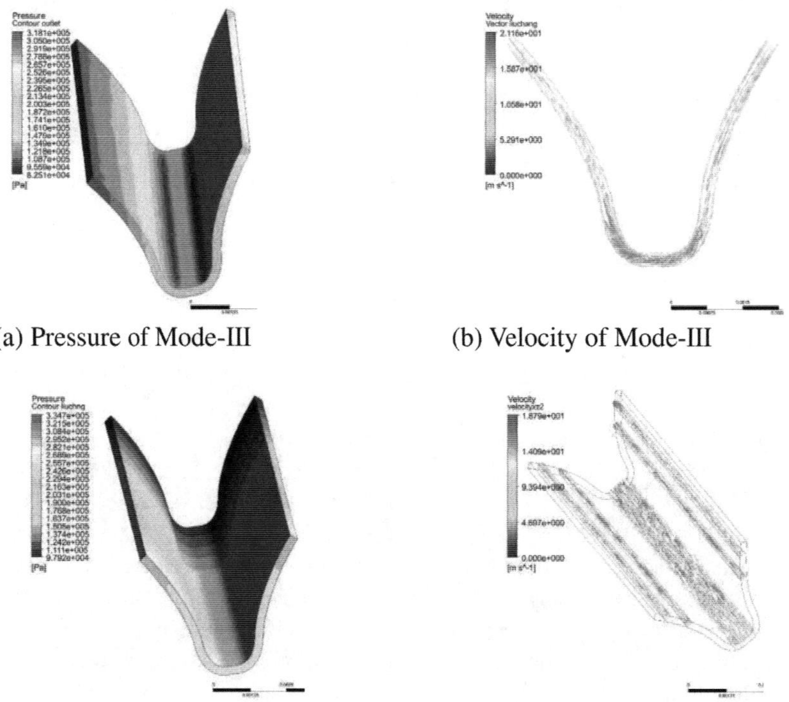

(a) Pressure of Mode-III (b) Velocity of Mode-III

(c) Pressure of Mode-IV (d) Velocity of Mode-IV

Figure 6. Electrolyte pressure and velocity distribution in the machining gap.

In order to get the best flow mode of the electrolyte. The velocity curves of Mode-III and Mode-IV are obtained as shown in figure 7. As figure 7(a) shows, In Mode-III, the flow velocity of the electrolyte fluctuates greatly, and the uniformity of the flow field is poor, which will seriously affect the stability of the ECM process. For Mode-IV, as described in figure 7(b), the electrolyte flow velocity is relatively uniform without large fluctuations. It indicates that the flow field distribution of this mode is uniform.

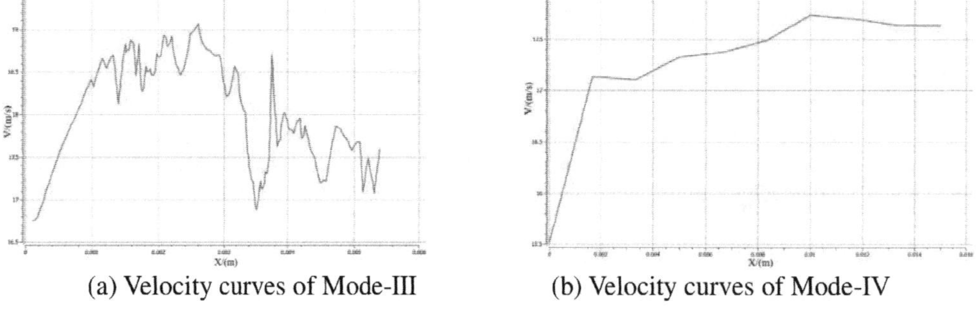

(a) Velocity curves of Mode-III (b) Velocity curves of Mode-IV

Figure 7. Velocity curves of electrolyte under Mode-III and Mode-IV.

5. Conclusions

The electrolyte flow field is a critical factor that affects processing stability and surface quality during ECM with the mask. In this paper, the ECM with the mask using shaped cathode has been adopted to fabricate micro grooves on the gear tooth surface. Four electrolyte flow modes consisting of the shaped cathode have been simulated by means of a finite element fluid analysis method. The simulated results have indicated that the cathode without electrolyte flow slot was beneficial to electrolyte flow during ECM with the mask, especially when electrolyte flows into the machining gap from the face of the gear.

Acknowledgement

This work is financially supported by the National Natural Science Foundation of China (51775158).

References

[1] Chen Q, Yao Z G, Xu F, Li H 2017 CN Patent 106678347 A.

[2] Schreck S, Gahr K H Z 2005 Laser-assisted structuring of ceramic and steel surfaces for improving tribological properties, *J. Appl. Surf. Sci.* **247** 616-22.

[3] Uhlmann E, Piltz S, Doll U 2005 Machining of micro/miniature dies and moulds by electrical discharge machining—Recent development, *J. Mater. Process. Technol.* **167** 488-93.

[4] Bhattacharyya B, Munda J, Malapati M 2004 Advancement in electrochemical micro-machining, *J. Int. Mach. Tool. Manuf.* **44** 1577-89.

[5] Lohrengel M M, Rataj K P, Münninghoff T 2016 Electrochemical Machining - Mechanisms of anodic dissolution, *J. Electrochim. Acta.* **201** 348-53.

[6] Liu G X, Zhang Y J 2017 The tool design and experiments on PECM of micro channel arrays on metallic bipolar plate using multifunctional cathode, *Int. J. Adv. Manuf. Technol.* **89** 407-16.

[7] Kern P, Veh J, Michler J 2007 New developments in through-mask electrochemical micromachining of titanium, *J. Micromech. Microeng,* **17** 1168-77.

[8] Xu W J 2015 Experimental Study on ECM of Microgrooves, *Electromachining. Mould,* **1** 3-6.

[9] Zhou J J, Pang G B 2009 Design of Flow Field for Finishig Spiral Bevel Gear in Pulse Electrochemica Finishing, *China Mech. Eng.* **20** 1425-32.

[10] Wang F J 2004 Computational fluid dynamics analysis, Tsinghua University Press, Beijing.

Thermal Analysis of the Urban Canyon Based on the Variables: Orientation, Wind and Vehicular Anthropogenic Heat

Ruth María Grajeda Rosado[1, a], Elia Mercedes Alonso Guzmán[2, b], Wilfrido Martínez Molina[2, c *], Juan Alberto Bedolla Arroyo[1, d], Carlos Javier Esparza López[3, e], Hugo Luis Chávez García[2, f]

[1]Professor of the Universidad Veracruzana, School of Architecture and student of the Inter-Institutional Doctorate Program at the Michoacan University of San Nicolas de Hidalgo, Morelia, Michoacan, Mexico, 5804
[2]Materials Department, Civil Engineering Faculty, Universidad Michoacana de San Nicolas de Hidalgo, Morelia, Michoacan, Mexico, 58040
[3]Research-Lecturer, Head of Master in Bioclimatic Architecture UCO-CA25 Arquitectura y medio ambiente / Architecture and environment Research Unit, Facultad de Arquitectura y Diseño, Universidad de Colima

[a]ruthgrajeda10@yahoo.com.mx, [b]eliamercedesalonso@gmail.com, [c]wilfridomartinezmolina@gmail.com, [d]bedollaalberto@gmail.com, [e]cesparza@ucol.mx, [f]hchavezenator@gmail.com

Abstract. The city and its behavior in the transfer of heat, generates lines of research that study phenomena such as the Island of Urban Heat (ICU); Within its formation factors, the heat expelled by the vehicle is the least analyzed in the literature. Thanks to the Computational Fluid Dynamics (CFD) software and its validation with measurements in the field, data is acquired to examine the thermal behavior in an urban canyon based on the number of cars passing through it and the heat expelled by its engines, the speed of the wind and orientation. The results show that the temperature increases between 26% to 10% more, and the highest values are in the simulated cases with a north-south orientation, affecting the passer-by as to the energy efficiency of the buildings.

1. Introduction

The knowledge of urban climate today, is an issue that concerns all specialists involved in the planning of cities. Situation that, when not being considered, leads to the increase of the phenomenon of the Island of Urban Heat (rise of internal temperatures in comparison to its surroundings) [1]. And, therefore, to an excessive use of electrical energy to solve the thermal conditioning of the space.

All the factors involved such as: a) the decrease of the green area, b) the low wind speed due to the high density of the building, c) the lining materials of the construction surface and d) the anthropogenic heat (vehicular combustion, of energy and human metabolic heat), have generated this problem in the outer spaces, so they are issues that must be addressed in depth to be handled properly and thus achieve the balance that satisfies both the population and the integration of cities in sustainability so demanded today.

Considering that the research uses different urban scales to determine the weighting of each of these factors. In this work, the urban canyon is used, which according to Stromann-Andersen & Sattrup [2] is a simple basic urban structure that offers, in field measurements, data with repetitive patterns and his form is defined as a microclimate [3].

In the present work, the variable to investigate is the anthropogenic heat (Q_F) that occurs in the urban environment due to vehicles (Q_{FV}). According to Sailor [4], the inclusion of Q_F has significant implications for urban climate, air quality and energy demand. A proposed way has been the estimation associated with the densities of the urban population, coupled with a homogeneous spatial distribution in magnitude and location equal to natural insolation; now it is known that the magnitudes are locally different in Wm^{-2} and it depends on the location and time.

According to Lin et al. [5], within mesoscale urban modeling, software uses established parameters such as 90, 50, and 20 Wm^{-2}, respectively, for high, medium, and low density urban categories [6]. Currently, research such as Dong et al. [7] seek the estimation of these values in more detail in several cities, giving results: Hong Kong (493), Singapore (353), New York (297), Portland (188), Macau (185), Seoul (175), Pusan (147), Montreal (147), Chicago (141), and Tokyo (141) (unit: Wm^{-2}).

The analysis of Sailor and Lu [8] suggests that, in cities such as Houston, the total anthropogenic warming of the vehicle sector is 300 Wm^{-2} during the afternoon peak hour. In Beijing a study, has identified a correlation with Q_{FV} (53.9Wm^{-2}) and the amplifying of the temperature from 0.91°C at 8:00 am, as well as the conclusion that the average heat intensity of the vehicle during the day it is 2 to 10 times the intensity during the night [9].

To estimate QFV and its weighting in the temperature patterns of an urban canyon, this document uses a three-dimensional idealized configuration model, to be analyzed by the CFD software; the results are validated against measurements taken in the field.

2. Description of the Simulation Model

In this investigation, the heat transfer calculations are solved through the unstructured finite volume solver ANSYS FLUENT 14.5, in general, according to Mirzaei [10], the urban environment can be investigated using CFD software.

An air volume of 190x180x60 meters was idealized, from which two rectangular 100x50x10 meters are subtracted to define two block blocks with a separation of 10m, resulting in an aspect ratio H/W=1 (height of buildings between widths of the street). The separation of the two blocks simulates the urban canyon, from which the volumes of the cars are extracted. The maximum number of cars is 20 units, followed by simulations with 15, 10, 5 and 0 cars to determine the thermal behavior of the canyon.

Another variant in the simulation is the orientation of the canyon, obtaining two more variables, North-South and East-West. The materials used in the simulation are specified in table 1, as well as their thermal and radiation properties, all based on [11].

Table 1. List of thermal properties, materials used in the walls of buildings, bearing surface and car body, data based on [11].

	Materials	Thickness (m)	Density (Kg/m^3)	Specific Heat (j/kg-k)	Conductivity (W/m-k)	Emissivity E
Ground	Concrete	0.20	2,300	840	1.00	0.90
Wall	Flattened	0.15	1,570	1,000	0.53	0.90
Bodywork	Steel	0.01	7,850	456	50	0.50

The model assumes that the flow of heat has as properties to be constant, isotropic and without viscous dissipation. The Reynolds Avaraged Navier-Stokers (RANS) equations are used, and for the turbulence model K-epsilon Realizable Model is applied, together with the option of Scalable Wall Functions for the treatment in the vicinity of the walls. The Radiation model, Surface to Surface, is also used, and using the Solar calculator, the radiation values in the model are estimated with the coordinates where the field data were taken for the validation.

The border conditions used were: the symmetry on the sides of the volume; on the back the condition of constant pressure was fixed; for the front face, a wind speed of 0.0, 1.2 and 2.2m/s is set; the buildings, the pavement and the car were established as walls that intervene in the radiation. To simulate the heat that the cars issue, a constant temperature of 60°C was established in the hood.

3. Validation

The values obtained by the simulation are compared with measurements taken in the city and port of Veracruz, Mexico, with geographic location of 19°10'51" north latitude and 96°08'34" west longitude, and altitude above sea level of 15 m. The average temperature in the city is 32 ° C with 70% relative humidity. For more information on the methodology refer to [12]. As shown in the graph, the behaviors are very similar, however, there are variations of 0.1 to 1°C (figure 1). The explanation to the variation can be given to the entry of the wind creating vortices in the ends of the canyon and therefore altering the values obtained in the field measurements.

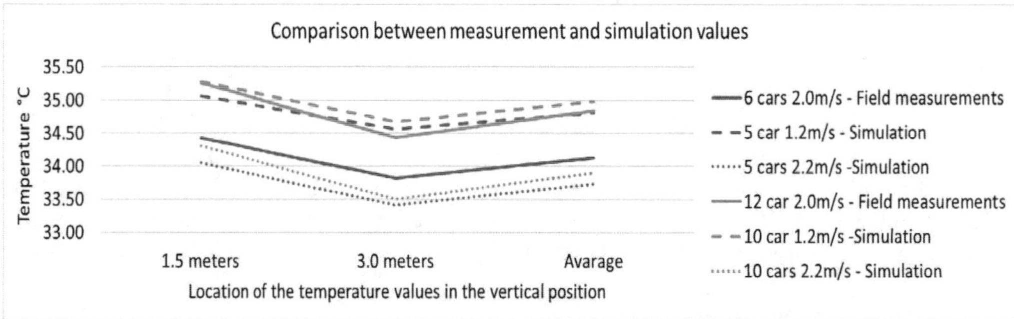

Figure 1. Comparison between field measurements and simulation averages.

4. Results and Simulation

The results show, in relation to the number of cars an expected rise in temperature, however, the increases, although they are progressive, were not proportional. According to table 2, if the initial value of the canyon is taken with zero cars as a base and a comparison is made with the 5,10,15 and 20 canyons, the following can be observed: a) the East-West oriented canyons with five automobiles, at any wind speed the average thermal gain is 0.70%, b) on streets with E-O orientation, with 10 units the increase is 1.7%, however, there is a clear difference of almost 2.5°C with the North-South orientation, c) the E-O orientation values, with 15 and 20 units become stable with 2.6% to 2.9% more gain and d) a progressive increase of 2% in the N-S orientation is observed for every 5 units that are added.

Table 2. Percentage table according to the number of cars.

Initial Value-	0 cars	5 cars	10 cars	15 cars	20 cars
0.0m/s E-O	32.12	0.8%	2.4%	3.5%	3.3%
1.2m/s E-O	38.39	0.7%	1.5%	2.5%	3.0%
2.2m/s E-O	30.79	0.5%	1.2%	2.0%	2.5%
0.0m/s N-S	35.12	2.0%	3.7%	5.4%	9.1%
1.2m/s N-S	30.61	2.2%	4.5%	6.5%	8.6%
2.2m/s N-S	34.18	2.1%	4.5%	6.5%	9.1%

Regarding the wind speed, the E-O oriented canyons between a wind of 1.2 and 2.2 do not vary much with a general loss of 1.6°C, but the N-S, as shown in table 3, with a wind of 1.2 m/s, the loss is 3.3°C in general and when the value increases to 2.2m/s the loss is a difference of 1°C more.

Table 3. Table of losses due to the effect of wind in °C.

	# Cars	0	5	10	15	20
0.0m/s E-W	Base value in °C	32.12	32.37	32.88	33.22	33.18
1.2m/s E-O		-1.5	-1.6	-1.8	-1.8	-1.6
2.2m/s E-O		-1.3	-1.4	-1.7	-1.8	-1.6
0.0m/s N-S	Base value in °C	38.39	39.16	39.82	40.48	41.87
1.2m/s N-S		-3.3	-3.3	-3.1	-3.1	-3.7
2.2m/s N-S		-4.2	-4.2	-4.1	-4.1	-4.6

The smallest amount of increase of the thermal profile in the urban canyons, are those oriented East-West, that is, with façade facing north and south, taking them as a base, when there is no wind, from 0 to 15 cars the gain is on average, 20%, however, rises to 26% when 20 units or cars are reached (table 4).

Table 4. Percentage table according to the increase in °C due to orientation.

		0 cars	5 cars	10 cars	15 cars	20 cars
0.0m/s E-W	Base value ºC	32.12	32.37	32.88	33.22	33.18
0.0m/s N-S		19.54%	20.95%	21.09%	21.84%	26.20%
1.2m/s E-W	Base value ºC	30.79	30.95	31.15	31.39	31.55
1.2m/s N-S		14.09%	15.95%	17.77%	19.13%	20.92%
2.2m/s E-W	Base value ºC	30.61	30.82	31.08	31.39	31.53
2.2m/s N-S		11.65%	13.27%	14.97%	15.93%	18.31%

5. Conclusions

Figure 2 shows the average of the air temperatures in each simulation, considering that the incoming wind has a constant of 30ºC. As you can see the average temperature of the air (T_{aw}) higher, is presented in the canyon with 20 cars with N-S orientation and no wind. The lowest T_{aw} in the canyon with East-West orientation and wind of 2.2 m/s, however, there is not much difference between the temperature that is generated with wind speeds of 1.2 and 2.2 m / s for both orientations.

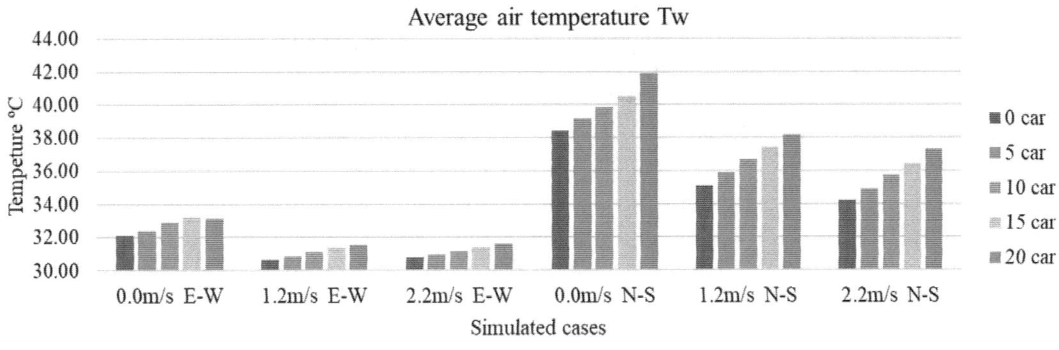

Figure 2. Summary of Tw of the 30 simulated cases.

The North-South orientation of the canyon (facades towards East-West) are those that represent higher elevation in the thermal profile, because the solar path crosses perpendicularly; causing that both the horizontal area (street) and the façade facing east, add twice as much area heated by solar radiation.

We can conclude with base to the results that the investigation threw that a) the majors increases are located in the North-South orientation (facades East-West), b) the wind speed is a very important factor because in conditions of zero speed the data is greater but the difference between 1.2 and 2.2m/s is not relevant and; c) that the number of cars is indistinct in relation to the wind speed and the temperature decrease, where the E-O orientation loses from 1.5 to 1.8 ° C and N-S from 3.3 to 4.6 ° C.

An objective to follow is to determine how the temperature is modified taking the width and height of the blocks as a variant. The present information seeks the awareness of having updated data of the Q_{FV} for the simulation of buildings that seek their energy efficiency, in relation to their environment.

References

[1] Santamouris M, Haddad S, Saliari M, Vasilakopoulou K, Synnefa A, Paolini R, Ulpiani G, Garshasbi S, Fiorito F 2018 On the energy impact of urban heat island in Sydney: Climate and energy potential of mitigation technologies, *Energy Build.* **166** 154-64.

[2] Stromann-Andersen J, Sattrup P A 2011 The urban canyon and building energy use: Urban density versus daylight and passive solar gains, *Energy Build.* **43(8)** 2011-20.

[3] Santamouris M 2013 Energy and Climate in the Urban Built Environment, *Routledge*, 102-38.

[4] Sailor D J 2011 A review of methods for estimating anthropogenic heat and moisture emissions in the urban environment, *Int. J. Climat.* **31(2)** 189-99.

[5] Lin C Y, Chen F, Huang J C, Chen W C, Liou Y A, Chen W N, Liu S C 2008 Urban heat island effect and its impact on boundary layer development and land-sea circulation over northern Taiwan, *Atmosp. Enviro.* **42(22)** 5635-49.

[6] Sailor D J, Georgescu M, Milne J M, Hart M A 2015 Development of a national anthropogenic heating database with an extrapolation for international cities, *Atmosp. Enviro.* **118** 7-18.

[7] Dong Y, Varquez A C G, Kanda M 2017 Global anthropogenic heat flux database with high spatial resolution, *Atmosp. Enviro.* **150** 276-94.

[8] Sailor D J, Lu L 2004 A top-down methodology for developing diurnal and seasonal anthropogenic heating profiles for urban areas, *Atmosp. Enviro.* **38(17)** 2737-48.

[9] Wang Y, Sun R, Chen L 2017 The impact of vehicle emissions on microclimate in Beijing, *metropolis. Acta. Ecolog. Sinica.* **37(3)** 953-9.

[10] Mirzaei P A 2015 Recent challenges in modeling of urban heat island, *Susta. Cities Soci.* **19** 200-6.

[11] Cengel Y, Ghajar A 2011 Heat and mass transfer. Fundamentals and applications (4th. ed.). Barcelona: McGrawHill.

[12] Grajeda R, Alonso E, Esparza C 2018 Vehicular Anthropogenic Heat in the Physical Parameters of an Urban Canyon for Warm Humid Climate, *34th Int. Conf. Passive Low Energ. Archite.* **1** 225-30.

IOP Conf. Series: Materials Science and Engineering **811** (2020) 012045 doi:10.1088/1757-899X/811/1/012045

Mapping the LST (Land Surface Temperature) with Satellite Information and Software ArcGis

Ruth Maria Grajeda Rosado[1, a], Elia Mercedes Alonso Guzmán[2, b], Carlos Javier Esparza Lopez[3, c], Wilfrido Martínez Molina[2, d*], Hugo Luis Chávez García[2, e], Eduardo Limón Yedra[4, f]

[1]Professor of the Universidad Veracruzana, School of Architecture and student of the Inter-Institutional Doctorate Program at the Michoacán University of San Nicolas de Hidalgo, Morelia, Michoacan, Mexico, 5804
[2]Materials Department, Civil Engineering Faculty, Universidad Michoacana de San Nicolas de Hidalgo, Morelia, Michoacan, Mexico, 58040
[3]Research-Lecturer, Head of Master in Bioclimatic Architecture UCO-CA25 Arquitectura y medio ambiente / Architecture and environment Research Unit, Facultad de Arquitectura y Diseño, Universidad de Colima
[4]Student of the Veracruz University of the Faculty of Construction Engineering and Habitat.

[a]ruthgrajeda10@yahoo.com.mx, [b]eliamercedesalonso@gmail.com, [c]cesparza@ucol.mx, [c]wilfridomartinezmolina@gmail.com, [e]hchavezenator@gmail.com, [f]yedra314@gmail.com

Abstract. This document describes the procedure to calculate the LST (Land Surface Temperature). The series of steps and equations is described using the thermal bands offered by the LandSat 8 satellite and the ArcGIS program tools. Although there are other methods and tools this procedure explicitly breaks down the use of information.

1. Introduction

The first official reports of urban climatology made in the city of London by Luke Howard, a century ago, have shown an increase in the urban area with respect to the rural area [1], at that time this increase was attributed to the intense fuel consumption in the city. Nowadays, the inhabitants look for new ways of monitoring the climatic, environmental, economic and social conditions within the cities, to obtain accurate data that suits their needs.

However, the growth of cities has increased exponentially in the last three decades. Faced with this, warnings about climate change emerge due to the alarming report of the United Nations Organization regarding the demographic behavior in the cities, because "54 percent of the world's current population lives in urban areas and it is predicted that by 2050 will reach 66 percent ... ", [2]. Intellectual work in technical areas such as climatology, architecture, urban planning or remote sensing should be applied with an ecological conscience aimed at predicting and solving the problems they cause in order to solve the urban-consumer environment. Currently 75% of the planet's energy resources are consumed in large cities, causing the highest percentage of environmental deterioration in the form of pollution and gases that induce climate change [3].

The phenomenon of the Island of Urban Heat (IUH) has motivated architects, urban planners and engineers to analyze how heat transfer is generated within urban and suburban areas. Factors such as wind, population density, construction materials, vegetation, land use, are some of the typical ones that researchers have considered. A method of much use to determine the IUH isotherms has been calculating the Superficial Island of Urban Heat (SIUH) through the Land Surface Temperatura (LST).

The EarthExplorer platform of the United States Geological Survey (USGS) provides free access for the download of Landsat satellite images from Mexican territory. The OLI (Operational Land Imager) and TIRS (Thermal Infrared Sensor) sensors are instruments on board the Landsat 8 satellite, which was launched in February 2013. From the page https://earthexplorer.usgs.gov/ the images taken by the satellite are discarded specifying the name of the region or area of analysis as well as the date, it should be mentioned that generally the satellite makes two shots per month.

2. Description of the Method

1) Using the ArcGIS program loads the images (thermal bands) Lansat 8 and the metadata, pressing "add data" selects the bands with the ending "B2, B3, B4, B5, B10 and B11" and the metadata "MTL", the last one in ".txt" format (figure 1).

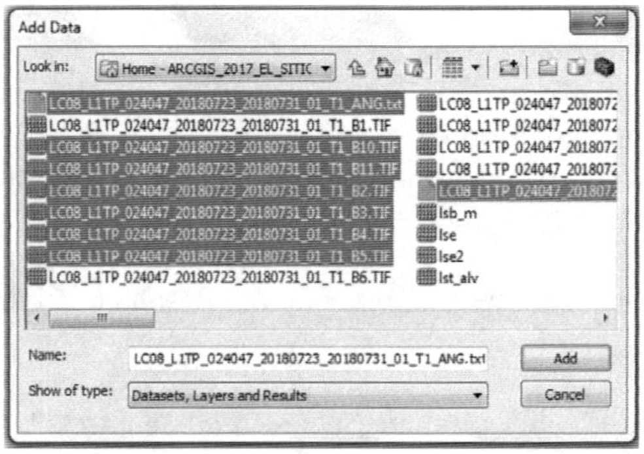

Figure 1. Add data from ArcGis.

For practical purposes, a clipping of the analysis area is done, with the "clip" tool, which will cut what is inside the work area, the tool is in the top menu, (windows-image analysis) will be displayed a window in which the bands should be selected and use the trimming tool (image analaysis - processing - trimming). This creates a series of new layers in the table of contents, indicated by the word "clip" at the beginning of the name. The following new files will be used in the future, replacing the originals (figure 2).

Figure 2. Using the tool clip and image analysis.

To create an RGB image (network, green, blue) from the combination of bands 2, 3 and 4, to visualize the analysis area. In order that the constructed image has the right color, the bands must be arranged in a specific order (4,3,2). This is provided directly with the right mouse button over the composition (layer-function properties-composite layer functions) (figure 3a). In the tab "composite bads" the order of the bands is exchanged using the arrows (figure 3b).

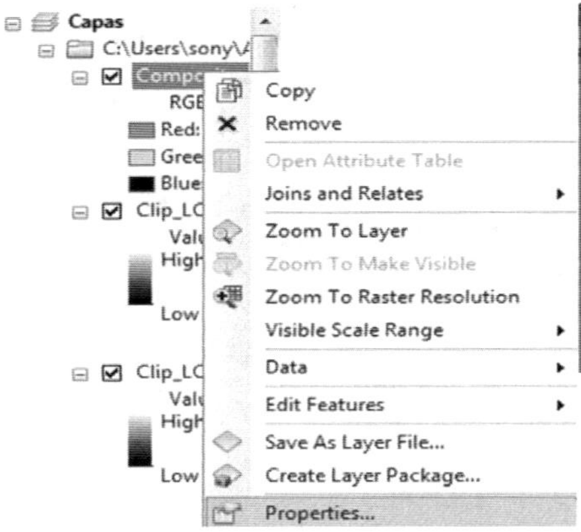

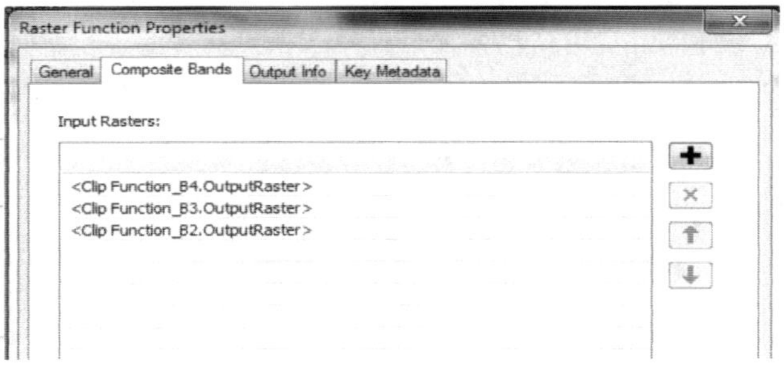

Figure 3a. Using the function properties tool (left).
Figure 3b. Using composite bands tool (right).

2) The calculation of the radiance of the thermal bands 10 and 11 (TIR bands) is achieved using the "raster calculator" (figure 4a), located in the menu (ArcToolBook - Spatial Analyst tools - Map algebra - Raster Calculator). Where the equation applies:

$$TOA\ (L) = M_L * Q_{cal} + A_L \qquad (1)$$

Where:

- ML = Specific multiplicative scaling factor of each band. Value obtained from the MTL metadata file under the name of "RADIANCE_MULT_BAND_X".
- Q_{cal} = Is the band or the cut of it.
- A_L = Value included in the MTL metadata "Radiance_Add_Band_X", where X corresponds to the number of the band (figure 4a).

This equation must be applied to bands 10 and 11, transferring this information to the raster calculator (figure 4b) like this: Radiance_Mult_Band_X * BAND_TERMICA_X + Radiance_Add_Band_X, as a result of this equation a new "radiance" image is generated.

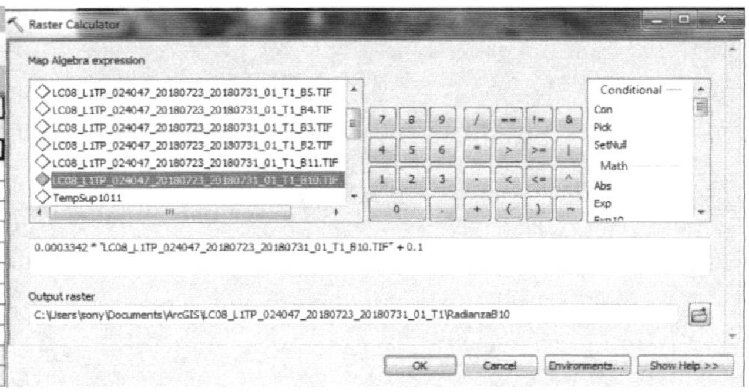

Figure 4. Information of the radiance (left) and **Figure 4b.** Using calculator raster tool (right).

3) The brightness temperature (BT) is the next step. The equation used is:

$$BT = (K_2 / (\ln (K_1 / L) + 1)) - 273.15 \qquad (2)$$

Where:

- K_1 and K_2 = Conversion constants, included in the metadata (K_1_CONSTANT_BAND_x and K_2_CONSTANT_BAND_x) apply to each band, 10 and 11.
- ln (natural logarithm) = Function in the raster calculator.

The results are given in the unit of Kelvin degrees, if the user wishes to convert it to Celsius degrees, the formula must incorporate the value 273.15. This equation must be done by substituting the values corresponding to bands 10 and 11. As in the previous case, the result of this operation will generate an output raster for each thermal band (figure 5).

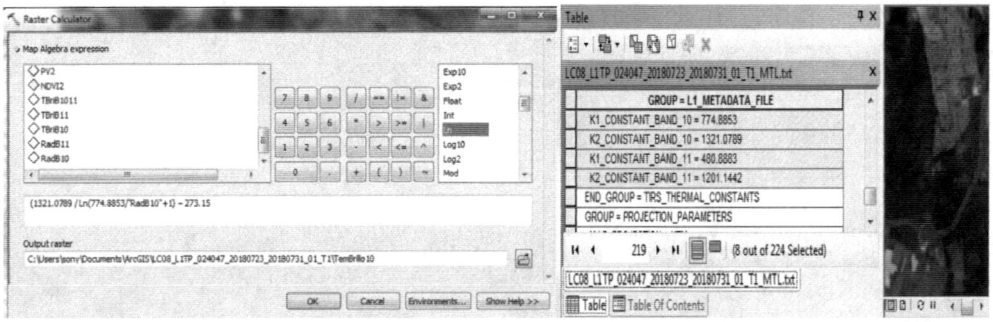

Figure 5. Raster calculator and using values of the metadata files.

To combine the temperature layers with the thermal bads (TemBrillo 10 and TemBrillo11) and obtain only one, we use the cell statistics tool located in ArcToolBox- Spatial Analyst tools- Local- Cell Statistics. It displays a window in which the images obtained as a result of the use of the raster calculator (TemBrillo10 and TemBrillo11) must be dragged. The name of the output raster must be specified (figure 6).

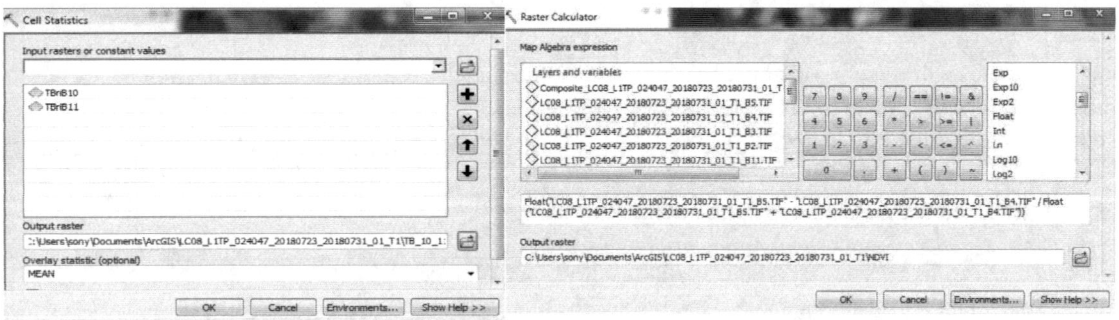

Figure 6. Tool Cell Statistics to combine thermal bands.

4) The following is to calculate the NDVI (Normal Difference Vegetation Index), for this bands 4 and 5 are used, applying the raster calculator, the equation is as follows:

$$NDVI = (Band\ 5 - Band\ 4) / (Band\ 5 + Band\ 4) \quad (3)$$

The data obtained is required to obtain the proportion of vegetation (P_v), of the next step. The degree of coverage or proportion of vegetation is calculated by the equation:

$$Pv = Square\ ((NDVI - NDVI_{min}) / (NDVI_{max} - NDVI_{min})) \quad (4)$$

Where:

- Square = Corresponds to squaring the formula
- $NDVI_{max}$ = Maximum values visible in the "table of contents"

- NDVImin = Minimum values visible in the "table of contents"

5) The value of the emissivity, as a last step, is obtained with the raster calculator, introducing the following equation:

$$e = m\,Pv + n \qquad (5)$$

Where:

- m = value of emissivity of vegetation, in this case 0.004 was used
- Pv = corresponds to the percentage of vegetation
- n = Soil emissivity value, in this case 0.986 was used
- Substituting remains:
- LSE = 0.004 * P_v + 0.986

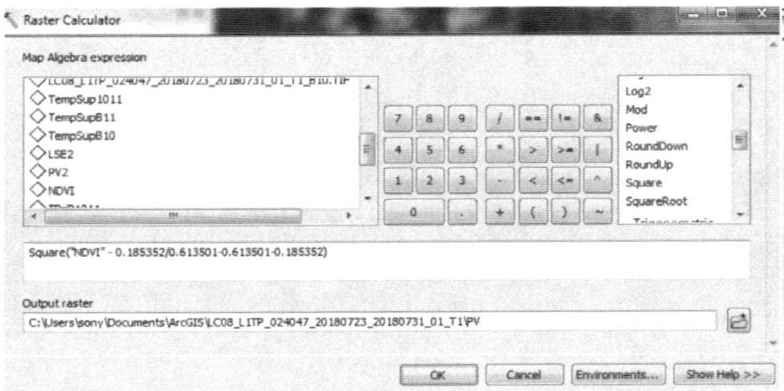

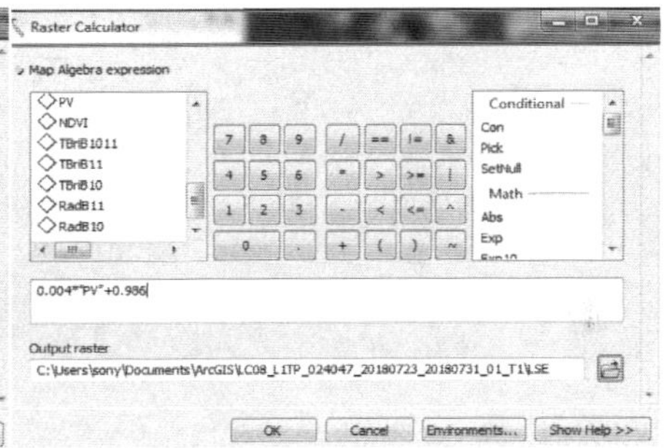

Figure 7a. Raster calculator tool for calculate the NDVI (left) and **Figure 7b.** for calculate the emissivity (right).

6) Finally, the Land Surface Temperature (LST), is applied with the equation:

$$LST = BT / 1 + w\,(BT / p) * Ln\,(\varepsilon) \qquad (6)$$

Where:

- BT = Brightness temperature (band 10 or 11 depending on the case)
- w = Length of the emitted radiation (band 10 or 11 as the case may be)
- p = Constant value obtained by the formula h * c / s that when substituting the values is 1.438 * 10^ -34Js and results in 14,380 and ε is LSE obtained previously, this equation should be applied in bands 10 and 11 separately for later unify the results with the tool (cell statistics)

Each equation that is made in the raster calculator generates an output raster which must be named and can be in a specific folder (figure 8a, 8b).

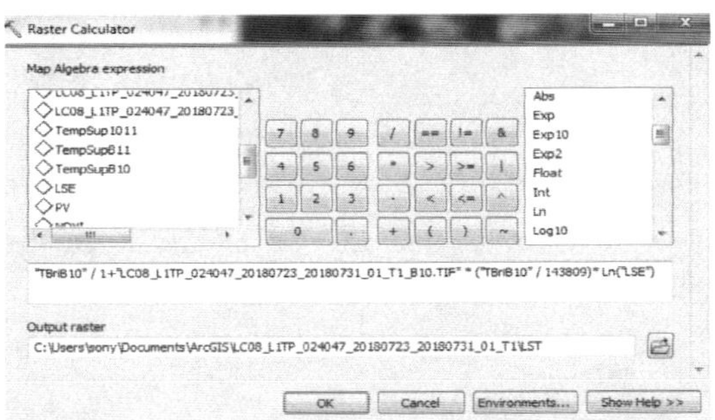

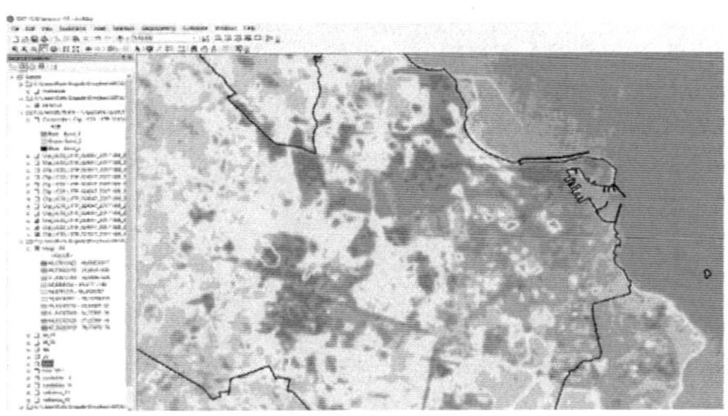

Figure 8a. Raster calculator tool for calculate LST (left) and **Figure 8b.** Corresponding land surface map of Veracruz, Mexico registered on May 21, 2018 (right).

3. Applications

The detection of the Land Surface Temperature (LST) in a city, besides determining the isotherms of the Superficial Island of Urban Heat (SIUH), allows to focus mitigation measures such as the reforestation of the zones to improve the evapotranspiration, the adequate distribution of the land use and the balanced of physical-thermal characteristics (emittance and albedo) of building materials. Examples of this are the works of [4-6].

Reference

[1] Gartland L M 2008 Heat Islands-Understanding and mitigation heat in urban areas, London: Routledge.
[2] ONU. (July 10, 2014). Department of Economic and Social Affairs. http://www.un.org/es/development/desa/news/population/world-urbanization-prospects-2014.html
[3] Yeang K, Richards I 2007 Eco Skyscrapers I, Images Pub. 1.
[4] Tomlinson C J, Chapman L, Thornes J E, Baker C 2011 Remote sensing land surface temperature for meteorology and climatology: A review, *Meteoro. App.* **18(3)** 296-306.
[5] Anderson M C, Allen R G, Morse A, Kustas W P 2012 Use of Landsat thermal imagery in monitoring evapotranspiration and managing water resources, *Remo. Sens. Enviro.* **122** 50-65.
[6] Mia Md B, Bromley C J, Fujimitsu Y 2012 Monitoring heat flux using Landsat TM/ETM+ thermal infrared data—A case study at Karapiti ('Craters of the Moon') thermal area New Zealand, *J. Volcano. Geother. Res.* **235-236** 1-10.

IOP Conf. Series: Materials Science and Engineering **811** (2020) 012046 doi:10.1088/1757-899X/811/1/012046

Prediction of the Fatigue Life of the AISI 1513 Carbon Steel Lower Arm Based on Strain-Life Approach

HUSAINI[1, a], TEUKU Edisah Putra[1, b*], MUHAMMAD Reza Rizky[1, c]

[1]Department of Mechanical Engineering, Universitas Syiah Kuala, Darussalam 23111, Banda Aceh, Indonesia

[a]husainiftm@unsyiah.ac.id, [b]edi@unsyiah.ac.id, [c]mrezarizky@mhs.unsyiah.ac.id

Abstract. The purpose of this study is to predict the fatigue life of a lower arm as a vehicle is driven on a straight and turn roads. The measurement of strain signals was carried out by attaching a strain gauge to the left lower arm and driving it with an acceleration of 30 km/h. According to the results, the lowest fatigue life was obtained just as the vehicle turned clockwise, which was 2.6E+6 cycles to failure. This value was lower by 123 % and 1,862 % than it was driven straight and counter-clockwise. Similarity, when the vehicle turned to the right, the lower arm on the left side received a higher strain, contributing to a shorter fatigue life.

1. Introduction

The stability and control of the vehicle entirely depends on the friction between the tire and road surface [1]. This dynamic interaction results in vibrations detremental to vehicle components. Vibration contributes to the mechanical damage of a component due to the dynamic load. Therefore, the suspension system dampens the vibrations in the wheel due to uneven road surfaces, therefore, it is not felt directly by the driver or passengers [2-7].

Several components have varieties components in a suspension system, such as the lower arm, which functions to connect the suspension system to the main frame of the vehicle and as a wheel drive controller, both back and forth. There are several studies that predict the fatigue life of a lower arm [8, 9], however, these studies utilized strain signals measured on a straight road. Since a lower arm functions to stabilize the vehicle when turning, therefore, the main purpose of this study is to determine its fatigue life when driven on straight and bent roads based on the strain-life approach.

2. Materials and Method

In order to determine the material, a chemical composition test was performed. The tested specimen was injected with argon gas electrodes. Furthermore, the lower arm was designed using computer aided design (CAD) and was analyzed using finite element analysis (FEA). FEA is a numerical method for solving technical problems through finite element modeling, in the form of meshes with boundary conditions and certain physical quantities. It can also be used to analyze non-structures, by modeling several types of materials and variations in the size of elements at the location of the structure. In the simulation process, it is necessary to review certain things that impacts hugely on the results, such as meshing, force, and support.

The mesh process was carried out to divide parts of geometry into small parts or elements. In this research, the mesh used was a tetrahedron [10] with a size of 5 mm. Its size will greatly affect the level of accuracy of the simulation results because the smaller its size, the more thorough results obtained. It produced high-quality meshing used for the solid boundary representation model imported from the

CAD system, with 32,812 nodes and 16,887 elements. In this research, the load given was the weight of the vehicle, passenger(s) and goods which was 1,045 kg, and 420 kg culminating to 1,465 kg or 14,366.74 N. The total force was divided by four which depicts the number of wheels that hold the vehicle. Therefore, the maximum load received by the lower arm was 3,591.68 N with a displecement support and force in the bushing and ball joint sections, as shown in figure 1.

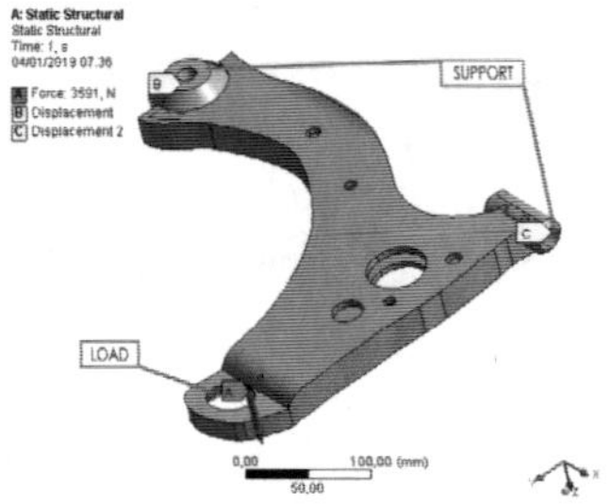

Figure 1. Load and boundary conditions.

Equivalent stress is often referred to as von Mises stress [11] which is usually obtained from hoop, longitudinal and radial directions. These are represented by σ_1, σ_2, σ_3. The equivalent stress σ_e is mathematically represented as follows:

$$\sigma_e = \frac{1}{\sqrt{2}}\left[\left(\sigma_x - \sigma_y\right)^2 + \left(\sigma_y - \sigma_z\right)^2 + \left(\sigma_z - \sigma_x\right)^2 + 6\left(\tau_{xy}^2 + \tau_{yz}^2 + \tau_{zx}^2\right)\right]^{1/2} \quad (1)$$

where σ_z, σ_y, σ_z are the normal stresses in the direction of the *x*, *y* and *z*-axes, while τ_{xy}, τ_{yz}, τ_{zx} are the shear stresses.

Strain signals were measured by attaching a strain gauge at the lower arm, with the installation location chosen based on the stress distribution. The frequency for measuring the strain signals was 500 Hz [12]. After mounting the measuring equipments, the vehicle was then ran on a straight, clockwise and counter-clockwise roads, at the speed of 30 km/h.

Generally, fatigue signals are characterized by statistics, therefore, there are a number of statistical parameters used for monitoring patterns and classification of random signals. The parameters usually used in observing the behavior of strain signals include mean, standard deviation (SD), root-mean square (r.m.s.), and kurtosis. For a signal F_j with the number of data *n*, the mean value $\bar{x}$ is estimated by:

$$\bar{x} = \frac{1}{n}\sum_{j=1}^{n} Fj \quad (2)$$

SD is used to determine how data are distributed in a sample and the number of average variability in a data set. SD can be stated as follows:

$$SD = \sqrt{\frac{1}{n}\sum_{j=1}^{n}(F_j - \bar{x})^2} \quad (3)$$

r.m.s. is used to estimate the total amount of energy contained in a discrete data F_j. This is expressed by:

$$r.m.s = \sqrt{\frac{1}{n}\sum_{j=1}^{n}F_j^2} \quad (4)$$

Kurtosis is a statistical parameter that is sensitive to spikes in the data with a discrete equation as follows:

$$K=\frac{1}{n(SD)^4}\sum_{j=1}^{n}(F_j-\bar{x})^4 \tag{5}$$

In general, it is estimated that 50 - 90 % of structure failures are caused by fatigue, thereby, making it a serious problem [13-15]. An accurate fatigue life assessment can be performed by the strain-life approach while considering the plastic form in the local area. Its analysis is used to estimate the risk of damage caused by repetitive loads and determine the life of a structure or component. By predicting the fatigue life, the risk of total damage can be minimized and the structure used to predetermine design targets. This approach is generally used for flexible materials.

The Coffin-Manson model [16, 17] can be used to obtain the relationship between stress amplitude and a number of cycles. This model is stated by:

$$\varepsilon=\frac{\sigma'_f}{E}\left(2N_f\right)^b+\varepsilon'_f\left(2N_f\right)^c \tag{6}$$

where ε is the strain amplitude, σ'_f is the fatigue strength coefficient, E is the material modulus of elasticity, N_f is the number of cycles to failure for a particular stress range and mean, b is the fatigue strength exponent, ε'_f is the fatigue ductility coefficient and c is the fatigue ductility exponent.

According to the Morrow model [18], the mean stress can be enhanced by adjusting the elastic strain-life. The model is defined by:

$$\varepsilon=\frac{\sigma'_f-\sigma_{mean}}{E}(2N_f)^b+\varepsilon'_f(2N_f)^c \tag{7}$$

where σ_{mean} is the normal mean stress. The SWT parameter [19] was proposed by K.N. Smith, P. Watson and T.H. Topper and mathematically expressed by:

$$\sigma_{max}\varepsilon=\frac{\sigma'^2_f}{E}(2N_f)^{2b}+\sigma'_f\varepsilon'_f(2N_f)^{b+c} \tag{8}$$

where σ_{max} is the maximum stress.

Fatigue damage for each loading cycle D_i is represented by:

$$D_i=\frac{1}{N_f} \tag{9}$$

The Palmgren-Miner rule [20, 21] was used to determine the cumulative fatigue damage, which is:

$$D=\sum\left(\frac{n_i}{N_f}\right) \tag{10}$$

3. Results and Discussion

The results of the chemical composition test show that the material consists of 0.08 % carbon (C), 1.31 % manganese (Mn), 0.002 % silicon (Si) and of 0.005 % phosphor (P). From the compositions of manganese and carbon, it was concluded that this material belongs to the medium alloy steel type. According to ASTM A29/A29M-05 [22], the material used to fabricate the lower arm was the AISI 1513 carbon steel. Figure 2 illustrates the stress distribution of the lower arm, with a maximum value of 358.82 MPa, which occurred on the upper side near the bushings. This value was smaller than the yield

IOP Conf. Series: Materials Science and Engineering **811** (2020) 012046 doi:10.1088/1757-899X/811/1/012046

strength of the AISI 1513 carbon steel, at 450 MPa [23]. This result was similar to study by Kim et al. [9].

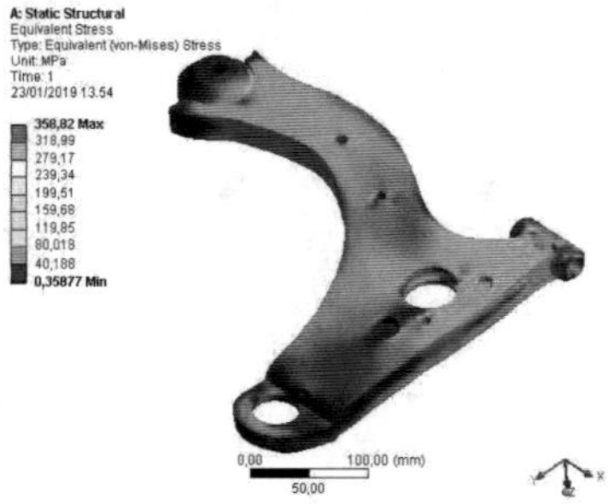

Figure 2. Stress distributions of the lower arm.

The measured strain signals are shown in figure 3, with the mean values of -12 με, -9 με, and -0.5 με for the straight, clockwise, and counter-clockwise roads, which shows that the lower arm accepted a compressive load. The road with the highest strain amplitude range was the clockwise, which was 8 με to -33 με with the strain gauge affixed at the left lower arm. Therefore, when the vehicle was turning to the right (clockwise), the component received a higher stress. However, on the counter-clockwise road, the stress received by the component was lower, at 8 με to -13 με, with the straight road providing a strain amplitude range between 2 με to -28 με. The clockwise road provided the highest strain amplitude ranges, at 3 με and an r.m.s. value of 10 με. While the straight and counter-clockwise roads gave an SD of 2 με, with an r.m.s. value of 12 με. All the strain signals gave the kurtosis value above 3, which means that they were non-stationary [24, 25].

According to the Coffin-Manson model, the fatigue damage for the straight, clockwise, and counter-clockwise roads were 5.6E-6 damage per block, 1.9E-5 damage per block, and 9.8E-7 damage per block, respectively. The Morrow model gave the fatigue damage of 3.1E-6 damage per block, 1.3E-5 damage per block, and 9.3E-7 damage per block. The SWT model gave the lowest fatigue damage, as 3.8E-11 damage per block, 8.6E-7 damage per block, and 6.8E-7 damage per block. The clockwise road gave the highest value, followed by the straight and counter-clockwise. These values were proportional to the SD, where the clockwise road also gave the highest SD. The high fatigue damage of a component lowers its fatigue life. Therefore, the clockwise road with the highest fatigue damage value gave the lowest fatigue life, of 2.6E+6 cycles to failure. The straight road gave the fatigue life of 5.8E+6 cycles to failure and the counter-clockwise road gave 5.1E+7 cycles to failure.

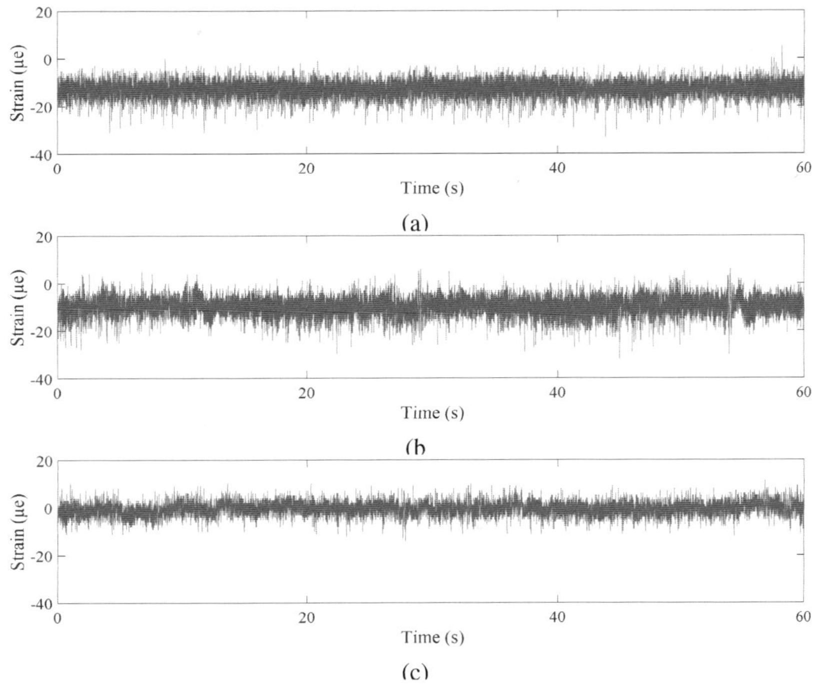

Figure 3. Strain signals: (a) straight road, (b) clockwise road, (c) counter-clockwise road.

4. Conclusion

This work aims to predict the fatigue life of a lower arm just as a vehicle was driven on a straight and turn roads. The results show that when the vehicle turned in a direction, the lower arm located on the opposite received a higher strain, thereby, giving a lower fatigue life of 1,862 %. In this case, when the vehicle turned to the right, the lower arm on the left side received a higher strain, contributing to a shorter fatigue life. This corresponds to the primary function of the component that is to stabilize a turning vehicle. With this function, a lower arm works more active just as the vehicle turns, and tend to be passive provided the vehicle runs on a straight road.

Acknowledgement

The authors would like to express their gratitude to Universitas Syiah Kuala for financial support for research through the Professor's research grant No. 13/UN11.2/PP/PNBP/SP3/2019 and also some funds assisted by the Laboratory research grant No. 313/UN11.2/PP/PNBP/SP3/2019.

References

[1] Jin L Q, Ling M, Yue W 2017 Tire-road friction estimation and traction control strategy for motorized electric vehicle, *PLoS One.* **12(6)**.

[2] Tandel A, Deshpande A R, Deshmukh S P, Jagtap K R 2014 Modeling, analysis and PID controller implementation on double wishbone suspension using SimMechanics and Simulink, *Proced. Eng.* **97** 1274-81.

[3] Dehkordi H B 2014 Vibration and force analysis of lower arm of suspension system, Uni. of Quebec.

[4] Anubi O M, Crane C 2014 A new semiactive variable stiffness suspension system using combined skyhook and nonlinear energy sink-based controllers, *IEEE Trans. Contr. Sys. Technol.* **23(3)** 937-47.

[5] Wang J, Zhang J 2015 Model-free tracking control for vehicle active suspension systems, *Proceed. 34th Chinese Contr. Conf.* 8067-72.

[6] Huihui P, Weichao S, Huijun G, Jinyong Y 2015 Finite-time stabilization for vehicle active suspension systems with hard constraints, *IEEE Trans. Intel. Transp. Sys.* **16(5)** 2663-72.
[7] Ka'ka S, Himran S, Renreng I, Sutresman O 2018 The pneumatic actuators as vertical dynamic load simulators on medium weighted wheel suspension mechanism, *IOP Conf. Series: Journal of Physics: Conf. Series,* **962**.
[8] Abdullah S, Al-Asady N A, Ariffin A K, Rahman M M, Nopiah Z M 2008 FEA based fatigue life assessment of an automobile lower suspension arm using various strain-life models, *Int. Conf. Appl. Theo. Mechan.* 81-8.
[9] Kim D H, Choi D H, Kim H S 2014 Design optimization of a carbon fiber refinfoard composite automotive lower arm, *Compo. Part B: Eng.* **58** 400-7.
[10] Mulia T M, Kadam S J, Kengar V S 2012 Finite element analysis of helical coil compression spring for three wheeler automotive front suspension, *Int. J. Mech. Ind. Eng.* **2(3)** 74-7.
[11] von Mises R 1913 Mechanik der festen Körper im plastisch-deformablen Zustand, Nachrichten von der Gesellschaft der Wissenschaften zu Göttingen, Mathematisch-Physikalische Klasse 1, 582-92.
[12] Ilic S 2006 Methodology of Evaluation of In-Service Load Applied to the Output Shafts of Automatic Transmissions, Ph.D. Thesis, Uni. of New South Wales.
[13] Puff R, Barbieri R 2014 Effect of non-metallic inclusions on the fatigue strength of helical spring wire, *Eng/ Fail. Analy.* **44** 441-54.
[14] Tsubouchi T, Takashi K, Kuboki T 2014 Development of coiled springs with high rectangular ratio in cross-section, *Proced. Eng.* **81** 574-9.
[15] Husaini, Kikuo K, Hanji M, Mitsuo N 2016 Investigations of the mixes mode crack growth behavior of an aluminium alloy, *J. Eng. Appl. Sci.* 11 885-90.
[16] Coffin L F Jr. 1954 A study of the effects of cyclic thermal stresses on a ductile metal, *Trans. ASME.* **76** 931-50.
[17] Manson S S 1965 Fatigue: a complex subject-some simple approximation, *Exp. Mech.* **5(4)** 193-226.
[18] Morrow J 1968 Fatigue Properties of Metal Fatigue Design Handbook, *Soci. Autom. Eng.*
[19] Smith K N, Watson P, Topper T H 1970 A stress-strain function for the fatigue of materials, *JMLSA.* **5** 767-78.
[20] Palmgren A 1924 Die Lebensdauer von Kugellagern, *Zeitschrift VDI,* **68** 339-41.
[21] Miner M A 1945 Cumulative damage in fatigue, *J. Appl. Mech.* **67** A159-A164.
[22] ASTM A29/A29M-05 2006 Standard Specification for Steel Bars, Carbon and Alloy, Hot-Wrought, General Requirements for. ASTM Int.
[23] nCode, GlyphWorks, nCode International, Ltd., Sheffield, 2018.
[24] Chen X 2014 Analysis of crosswind fatigue of wind-excited structures with nonlinear aerodynamic damping, *Eng. Struc.* **74** 145-56.
[25] Kihm F, Ferguson N S, Antonic J 2015 Fatigue life from kurtosis controlled excitations, *Proced. Eng.* **133** 698-713.

IOP Conf. Series: Materials Science and Engineering **811** (2020) 012047 doi:10.1088/1757-899X/811/1/012047

Higher Modes and Superposed Mode Behavior for Flexible Pavement Layer System

NORFARAH NADIA Ismail[1, a*], NUR IZZI Md Yusoff[2, b], WAN NUR AIFA Wan Azahar[1, c], AHMAD BUKHARI Ramli[1, d], NADIAH Md Husain[1, e], NORHIDAYU Kassim[1, f]

[1]Department of Civil Engineering, Faculty of Engineering, International Islamic University Malaysia, Jalan Gombak, Selangor
[2]Smart and Sustainable Township Research Centre (SUTRA), Faculty of Engineering and Built Environment, Universiti Kebangsaan Malaysia, 43600 UKM Bangi, Selangor

[a]norfarahnadia@iium.edu.my, [b]izzi@ukm.edu.my, [c]aifa@iium.edu.my, [d]bukhari@iium.edu.my, [e]drnadiah@iium.edu.my, [f]ayukasim@iium.edu.my

Abstract. This paper presents the study of modes behavior in pavement layer system using Common Array Profiling Spectral Analysis of Surface Waves (CAP-SASW) testing method. Method for evaluation and testing the materials at pavement sites are traditionally destructive. The utilization of non-destructive method such as CAP-SASW is a very useful advantage, however deeper study on its mode behavior for pavement layer is not proven. Comparing both theoretical and experimental generated modes, this paper aims to provide precise configuration measurement in sampling pavement materials besides understanding its modes behavior.

1. Introduction

Roads and highways are crucial part of today's infrastructure and economy. Flexible pavement is widely used in all hierarchy of roads and streets. Therefore, maintenance of the road network is a significant consideration in highway and transportation engineering researches [1]. Due to the budget constraint and with the purpose of minimizing the effects on traffic, a complete renovation or major restoration of the roads would not be practical [2]. Pavement assessment conventionally uses destructive methods such as coring and CBR. But a good non-destructive method would be such a useful advantage, as it is proved that suitable pavement rehabilitation design is feasible to be developed to provide a practical solution to overhaul and maintain a nation's road network [2].

In surface wave propagation, general mode and higher modes are the unique characteristics in multilayered system. When the surface waves travelled through this multilayered system, the stiffness of each layer will affect the surface wave propagation [3, 4]. The difference in layer stiffness could refrain the stress waves in some layers, or causes some refractions and reflections, which leads to different ray paths, and finally producing different propagation velocities even at the same frequency point. To identify the velocities of these general mode and higher modes, transfer matrix method [5, 6] and dynamic stiffness matrix method [7] could be used.

The superposed mode in surface wave propagation is also an essential characteristic, because this mode is the actual mode that is generated during the measurement. The superposed mode is suitable for 3-dimensional wave propagation, which is in cylindrical pattern. This propagation is always observed when the location of the impact source is near to the receiver location, and when the wave still possesses the cylindrical shape. Superposed mode does not fall into any modes in specific, but is superposed in between normal modes. Normal modes consisted of general mode and higher modes. The superposed mode is also called apparent velocity or effective velocity. The apparent phase velocities that are calculated or measured will be depending on the actual location of impact source and the receivers [8, 9]. In the measurement of surface wave, there are two approaches to the usage of these surface waves modes. This paper (CAPSASW method) uses the superposed mode, due to the fact that for impact source which is located near to the receivers' location, the modes separation for the waves measured are not practical. The other approach used by other techniques such as MASW and f-k method is utilizing the general mode and higher modes, because in that approach, the location of impact source is far from the receivers, and this allows better modes separation for the waves measured. However, the later approach also needs larger number of receivers to help with the mode separation [8, 9].

2. Methodology

Spectral Analysis of Surface Waves (SASW) utilizes mechanical surface wave, called Rayleigh wave to evaluate subsurface material of a medium. The surface wave penetrates the media from the surface until the depth of at least one wavelength. This means, we are able to evaluate media with different depth by utilizing different wavelength. One blow from the impact source could produce combination of Rayleigh surface waves which represented a large range of frequency. With the aid of recording device i.e dynamic signal analyzer and high-powered computer, all information in that frequency range could be recorded for particular time.

Typical configuration of Common Array Profiling- Spectral Analysis of Surface Waves (CAP-SASW) at site is to use two number of receivers in an array and an impact source (seismic source) is acted at the end of the array, as shown in figure 1. The same receiver distance is used throughout the measurement. Adequate wavelength range is needed to sample materials in a pavement layered system.

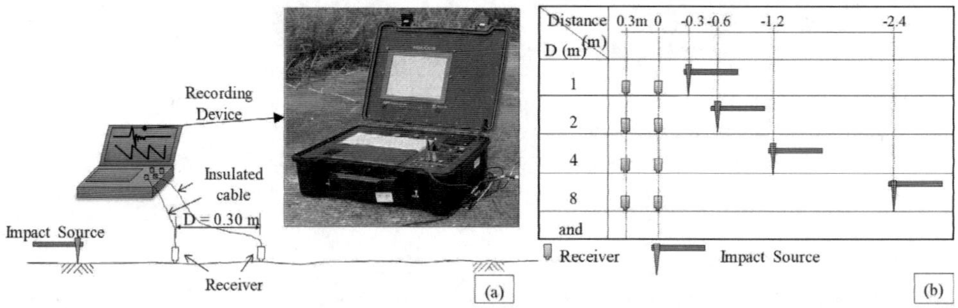

Figure 1. (a) Hardware configuration for field measurement of CAP-SASW test and (b) CAP-SASW measurement configuration.

3. Result and Discussion

The objective of this paper is to study the wave propagation behavior for pavement layer system. The pavement layer system has irregular dispersive profile, in which the top layer is stiffer-has higher shear wave velocity- compared to the underlying layer. This shows that the velocity of each layer is decreasing with the depth. This feature is opposed to the dispersion profiles of other geological site such as the normally dispersive site or the shallow bedrock site. A site with sandwiched layered system has the nearest uniqueness to pavement layered system. However, for this research of modes behavior in the pavement layered system, one specific profile as shown in table 1, which is similar to the actual pavement layered system as used by Malaysia's Public Work Department Guideline 2013 [10] on local road has been used. The material characteristics for pavement layered systems allows the presence of inverse velocity.

This research is divided into two studies, which are 1) using receiver distance, D = 0.15 m, and 2) using receiver distance, D = 0.30 m. Other than that variable, the frequency range for pavement layer system is found to be relatively higher than the geological sites, thus for both receiver distance, this research kept the frequency range at 10 kHz. The sample interval taken is 0.0001 s, which gave Nyquist frequency of 5000 Hz, adequately used for profiling the subsurface material as studied in this paper. To get reasonably good resolution, record length of 2048 samples is chosen, with time history record of 0.2047 s. The time signal record is shown in figure 2(a) and figure 2(b).

Table 1. Pavement Layer System.

Layer	Depth (m)	Shear Wave Velocity, Vs (m/s)	Poisson's Ratio, υ	Damping Ratio, D (%)	Density, ρ (kg/m³)
1	0.10	1200	0.350	0.02	2300
2	0.50	300	0.400	0.02	2100
3	0.25	180	0.400	0.02	2100
Half-space	∞	160	0.333	0.02	1900

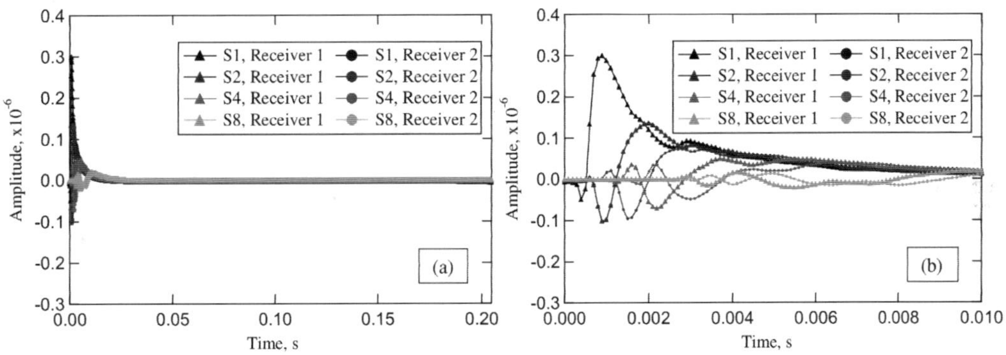

Figure 2. Synthetic seismograms of pavement layer system in time domain, generated theoretically with source distance, S1 = 0.30 m, S2 = 0.60 m, S4 = 1.2 m and S8 = 2.4 m: (a) full signal and (b) expanded signal.

Figure 3 shows phase spectrum plot for theoretical simulation of CAP-SASW, for pavement layer system, for receiver distance, D = 0.15 m while figure 4 shows the same plot when receiver distance, D is set to 0.30 m. Early evaluation shows that there is a discrepancy in number of *jump* at frequency range of 9 kHz – 11 kHz for phase spectrum with D = 0.15 m. In this case, it is assumed that the real phase cycle is only up until 9000 Hz. Furthermore, if compared with figure 4, figure 3 is found to be incomplete as one full cycle. One complete cycle starts with value 0° at the phase angle and ended when the cycle goes back to 0°. The initial description for this figure is the measurement configuration for receiver distance, D = 0.15 m might not be adequate to sample the whole layered system as needed. This initial description, however, does not always represent the final outcome, which is the dispersion curve. Therefore, to prove the initial observation, further study is continued to get dispersion curve as needed.

Utilizing CAP-SASW method, apparent phase velocity dispersion curve has been obtained. The phase velocity dispersion curve is essential because it is used in the inversion process to obtain stiffness profile of the subsurface material. The theoretical phase velocities use forward modelling process [7]. Therefore, it is significant for each profile model studied to give correct information of material for each layer in the system. In the calculation of the theoretical dispersion curve, the *array* inversion process uses all information regarding the actual location of impact source and receivers during CAP-SASW measurement. On the other hand, the *global* inversion process does not take into account the actual location of impact source and receivers.

Figure 5 shows the behavior of modes obtained from the CAP-SASW calculation, and normal modes for the same profile, for two receiver distances; i.e 0.15 m and 0.30 m. For theoretical superposed mode, it is calculated using *global* inversion process. Hence, if the superposed modes are plotted together for all source distance S1 until S8, it can be seen that the dispersion for each source distance is similar. The experimental superposed modes obtained in this figure though, are not the ones obtained from field measurement, but rather theoretical superposed modes, which takes into account the actual location of impact source and receivers, hence represents the actual location on the field. In other words, experimental superposed modes in this context are the theoretical superposed modes calculated using *array* inversion process.

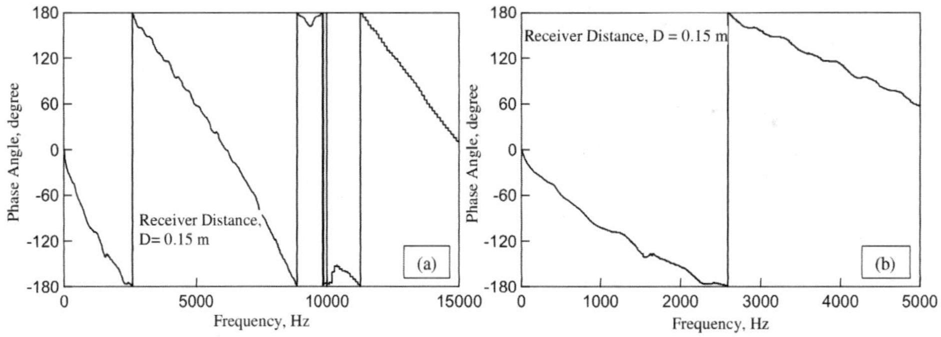

Figure 3. Phase spectrum for theoretical simulation CAP-SASW for pavement layer system, for receiver distance, D = 0.15 m for frequency range of: (a) 15 kHz and (b) focused to the first 5 kHz.

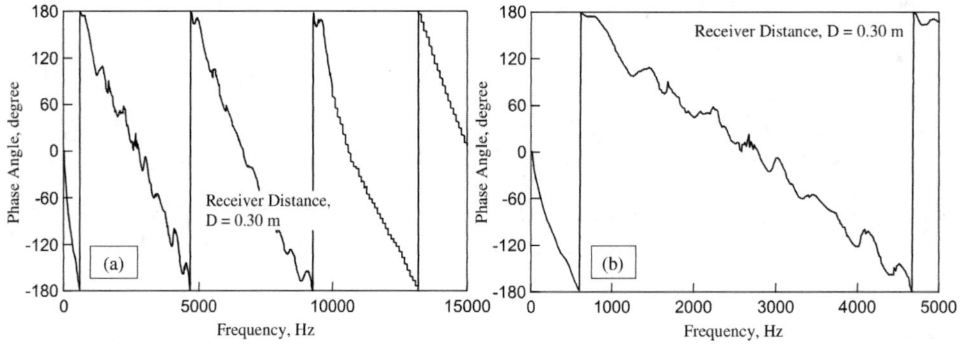

Figure 4. Phase spectrum for theoretical simulation CAP-SASW for pavement layer system, for receiver distance, D = 0.30 m for frequency range of: (a) 15 kHz and (b) focused to the first 5 kHz.

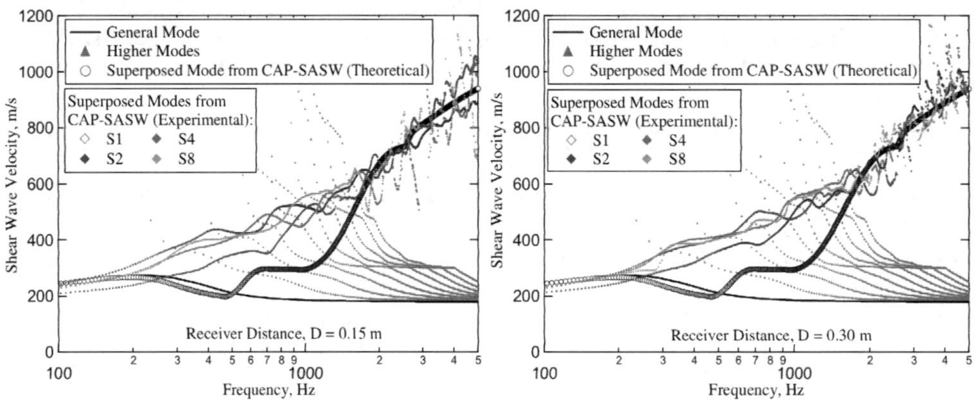

Figure 5. Normal modes of Rayleigh wave and superposed modes from CAP-SASW calculation for pavement layer system with receiver distance of 0.15 m and 0.30 m.

Based on both plots of shear wave velocity, some observations could be made. Firstly, the media profile of pavement layered system has inverse velocity (velocity is decreasing with depth) and totally opposite to the normally dispersive media. Seismic measurement testing on irregular dispersive materials needs scrutinized attention so that the final outcome obtained later won't be misleading and contradictory.

Secondly, the apparent velocity for pavement layer system profile does not follow the governing (general) mode. As similar as the sandwiched case, the apparent velocity involved superposition of general mode and higher modes. At frequency range lower than 200 Hz (or wavelength longer than 2m), the apparent velocities obtained from superposed mode follows the trend of the general mode. However, from the frequency of 200 Hz, the velocity steadily increases until frequency of 2000 Hz (wavelength around 0.35m). For all of the plots, the apparent velocity in theoretical superposed mode has steeper graph gradient as compared to the experimental superposed mode. Again, this happened due to the inversion process which is done to obtain the theoretical superposed mode is not taking into account the actual location of impact source and receivers into its calculations.

Thirdly, as the array inversion process is used to obtained the experimental superposed mode, therefore information regarding the location of impact source and receivers are crucial during the measurement. In both of the graphs, it is observed that for the smaller receiver distance, D = 0.15 m, the experimental superposed modes obtained from all source distance (S1-S8) are more scattered as compare to the receiver distance, D = 0.30 m. This shows that for smaller receiver distance, the modes obtained are less precise, although they are all accurate. The outcomes also conclude that the smaller receiver distance is not suitable or adequate enough to be used to sample or characterized subsurface material of pavement layer system with the required depth and thickness.

4. Conclusion

The usage of superposed modes is significant for CAP-SASW measurement of flexible pavement layered system as it alleviates the chances of misleading result of normal and higher modes on complex site such as pavement site. Receiver distance of 0.30 m is proven useful for better characterizations of subsurface materials on pavement layered system.

Acknowledgement

Authors would like to thank IIUM for financing the project under IIUM Research Acculturation Grant Scheme, IRAGS18-022-0023.

References

[1] Ivanova E, Masanova J 2013 Importance of road infrastructure in the economic development and competitiveness, *Eco Manage.* **18(2).**

[2] Chao C, et al. 2017 Non-destructive evaluation of a city roadway for pavement rehabilitation: A case study, *Int J. Pave. Res. Techno.*

[3] Al-Hunaidi M O, Rainer J H 1995 Analysis of multi-mode signals of the SASW method, *Soil Dynam. Earth Eng. VII.* 259-66.

[4] Gucunski N, Woods R D 1992 Numerical simulation of the SASW test, *Soil Dynam. Earth Eng.* **11(4)** 213-27.

[5] Haskell N A 1953 The dispersion of surface waves on multilayered media, *Bulle. Seism. Soc. America,* **43(1)** 17-34.

[6] Thomson W T 1950 Transmission of Elastic Waves through a Stratified Solid Medium, *J. Appl. Phy.* **21(2)** 89-93.

[7] Kausel E, Roesset J M 1981 Stiffness matrices for layered soils, *Bulle. Seism. Soc. America,* **71(6)** 1743-61.

[8] Joh, Ho S 2006 Optimal Use of Stress Waves in Non-Intrusive Seismic Techniques for Geotechnical Applications, *Kor. Sci.* 434-78.

[9] Stokoe K H, Joh S H, Woods R D 2004 Some Contributions of in Situ Geophysical Measurements to Solving Geotechnical Engineering Problems, *Pro Geot. Geop. Site Charac.* 97-132.

[10] Technical Guideline (Road) 5/85. 2013 Manual for the Structural Design of Flexible Pavement, Public Work Department, Malaysia.

Effect of Hinge Properties on Nonlinear Analysis of Eccentrically Steel Braced Frames

Dwinita Apritasari[1, a], Josia Irwan Rastandi[1, b], M Orientilize[1, c*], B O B Sentosa[1, d]

[1]Departement of Civil Engineering, Universitas Indonesia, Depok, 16424, Indonesia

[a]dwinita26@gmail.com, [b]jrastandi@eng.ui.ac.id, [c]mulia@eng.ui.ac.id, [d]bastian.sentosa@ui.ac.id

Abstract. Eccentrically Braced Frame (EBF) is one of seismic resistance of steel structures. Dissipation of seismic energies is designed by allowing plastic hinges to form in link element. Design of link elements allow plastic hinge to form due to shear or flexural failure. Structural analysis using finite element software require shear plastic hinges to be modelled. It is found that hinge properties proposed by two references, FEMA 356 and Richards & Uang are different. Study of effect of hinge properties on non-linier analysis of EBF was performed. The research is limited to hinge properties of shear link of EBFs. Numerical study was conducted on one and three storey of EBFs in two and three dimensional geometric of EBFs. Two different hinge properties from FEMA 356 and Richard-Uang were compared each other. Results show that hinge properties affect the structural response. It is found that EBFs modelled by hinge properties according to Richards-Uang has close result to experimental result than modelled built by FEMA 356. The difference between FEMA 356 and Richard-Uang to experiment are 34,54% and 15,55%, respectively.

1. Introduction

Eccentrically Braced Frame (EBF) is one of earthquake resistant structural systems for steel structure. It is a hybrid system, offering lateral stiffness from its concentrically brace, and ductility from moment frame system. EBF has a link that serves as an element of energy dissipation as well as the place of plastic joints. The link is intended to provide weaker section to allow plastic deformation so that energy from earthquake can be absorbed. Longer link will experience flexural yielding while shorter link will experience shear yielding. According to Bruneau [1], shear yielding allows the development of plastic deformation without an excessive amount of local strain, a phenomenon that appears with flexural yielding. According to Kasai [2], shear yielding provides larger dissipative energy capacity than flexural yielding. Hence, brace configuration, link length and link section properties are three important factors in EBF design. Shear link EBFs are more stable and more ductile than flexural link.

An analytic model of the shear link was proposed by Ramadan and Ghobarah [3] in 1995. It is modelled by three translational and three rotational sub-spring in each ends of the link. The stiffness and yielding of the link are determined by the force-deformation relationship of the sub-springs. Revised model was proposed by Richards and Uang in 2006 [4].

An earlier research by Inel [5] found that different parameters of hinge property could affect the structural response to earthquakes. Study conducted to RC structures showed that displacement capacity of structure was affected by plastic hinge length and the amount of transversal reinforcements in the hinge region. Therefore, bear in mind to conduct similar studies on EBF structure. This research aims to compare the effects of the two hinge properties used in FEMA 356 and Richard-Uang. The study conducted by using OpenSEES software.

OpenSEES (Open System for Earthquake Engineering Simulation) is an open source software that is largely used by researchers. OpenSEES provides various materials, elements and algorithm. Each behaves differently and affects the results of the analyses. Results of the analyses conducted by OpenSEES can closely resembles the results of an advanced analysis. Thus, the results can be considered as advanced analysis. However, structural analysis using OpenSEES are considered costly since it is difficult to use. For that purpose, commercial software such as ETABS remain to be favoured instead of OpenSEES. However, there needs to be an adjustment in ETABS so that the analysis results can closely resemble advanced analysis. Hence, this research also aims to find out the best way to modelled EBFs with shear link in ETABS so that the result can be identical as the OpenSEES.

2. Numerical Model of EBF and Validation

As mention, the research objectives are to study the differences of hinge properties modelling on EBFs using the guidelines in FEMA 356 [6] and in Richard-Uang [4]. Then it aims to find out the strategy to model EBFs frame in ETABS so that the results are closed to OpenSEES. Pushover analysis was performed on EBFs using both software. Prior to this research, model validations need to be done to ensure that numerical model can represent the real condition. The reference are EBFs used by Shujun [7] who conducted experimental and numerical study of EBFs. Figure 1 show the EBFs used in Shujun's study. It is Inverted-V EBF with a shear link as shown in in figure 1. The frame used H350x350x10x16 for columns, H300x150x6.5x9 for beams and ⌀150x8 for braces. Steel has yield stress as 235 MPa.

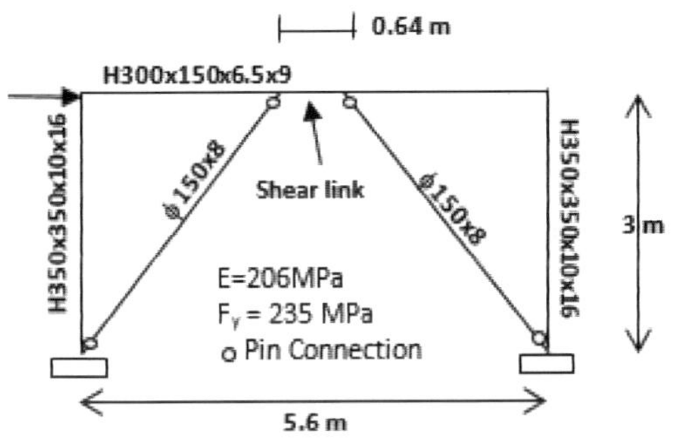

Figure 1. EBF as reference in validation stage.

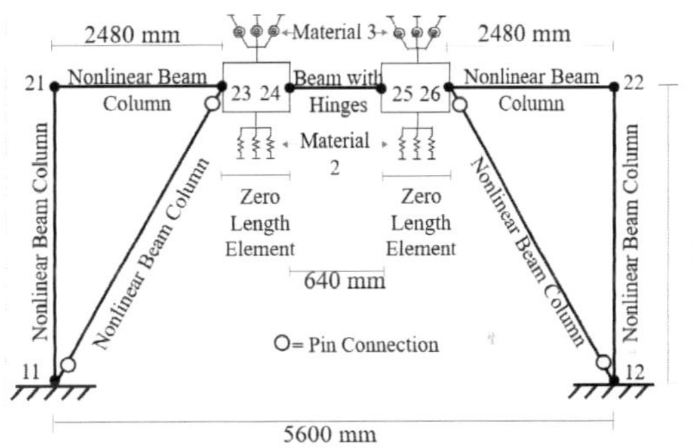

Figure 2. EBF Model in OpenSEES.

Figure 2 shows EBF model used in openSEES. The shear link refers to Richards-Uang hinge property where link is presented as 3 rotational and 3 translational springs where each has different stiffness presented as multi-linear graph as shown in figure 3. Stiffness of each linier line can be determined based on Richards-Uang theory [4]. The equations are as follows:

$$K_{V1} = \frac{2GA_{shear}}{e} \tag{1}$$

$$K_{V2} = 0.030\ K_{V1}. \tag{2}$$

$$K_{V3} = 0.015\ K_{V2}. \tag{3}$$

$$K_{V4} = 0 \tag{4}$$

$$V_p = 0.6F_yA_{shear}. \tag{5}$$

Steel02 is selected for steel material which is a uniaxial Giuffre-Menegotto-Pinto model with isotropic strain hardening. Truss element is assigned for bracing since it only receives axial loads. Beam and column use the NonlinearBeamColumn element, an element that assumes plastic deformation to be distributed along the element. Link is defined with BeamWithHinges command, where the plasticity is concentrated at the end of link. Each uses the ElasticPP command, which represents an elastic-perfectly plastic material. This allows the three springs are parallel to each other.

Push over analysis with P-delta effect was carried to represent loading stage conducted in the experiment and to allow the non-linearity in geometric. Results are presented on base shear and lateral displacement as shown in figure 3. As can be seen, result of openSEES agrees well with Shujun's. It can be concluded that model built by openSEES is valid since it can represent the real structure, thus the openSEES model can be used for further study.

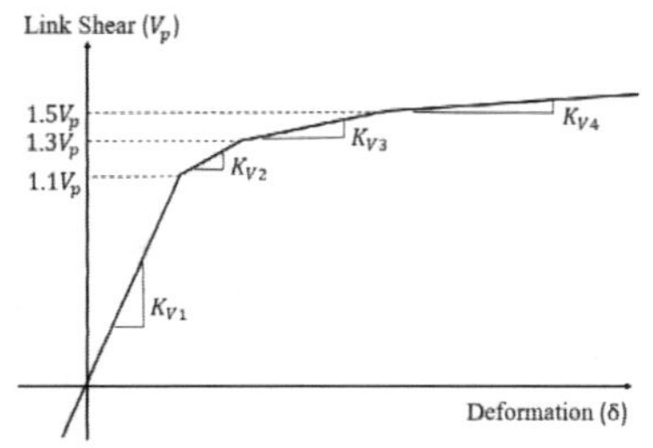

Figure 3. Multi-linear Link Model.

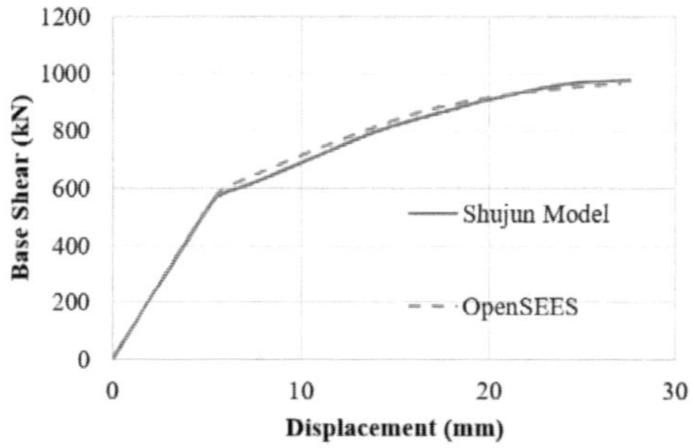

Figure 4. Validation of Numerical Model of One Storey EBF.

3. Research Methodology

Since the numerical model built by openSEES is valid, further study was continued to three-storey EBF. The frames were also built in three dimensional (3D) to see the geometric effect. Hence there are four EBF models investigate in this study; one-storey EBF in 2D geometric (1S-EBF2D) as presented in figure 1, one-storey EBF in 3D geometric (1S-EBF3D), three storey EBFs in 2D (3S-EBF2D) and 3D model (3S-EBF3D) as shown in figurer 5. The 3D model is an extension of the 2D model, thus having the same dimensions and specifications. The 3S-EBF3D is a single system where EBFs are only at grid A and D and frame on grid B and C have pin connection on each joint.

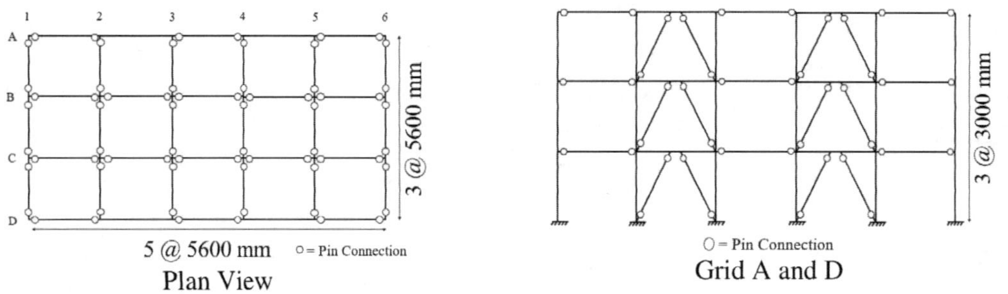

Figure 5. Three Storey of EBFs.

The different of FEMA 356 and Richard-Uang in modelling the hinge on link element is described in figure 6. As shown, the main difference is hinge location. FEMA assume that hinges are located on both link-end whereas Richard-Uang put hinges not exactly on link-end with an eccentricity as 5% of link length. This assumption affects force-deformation curve to represent the plastic hinge properties as shown in figure 7. As shown, point A, B, C, D, and E depends on value of a and b. The parameters are required when defining hinge properties in ETABS and modelling springs in OpenSEES. The different value of those parameters used in numerical model is presented on Table 1.

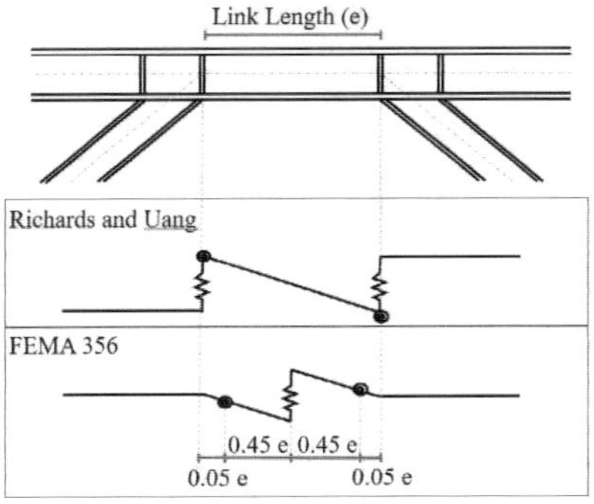

Figure 6. Link Model Based on Richards-Uang and FEMA 356.

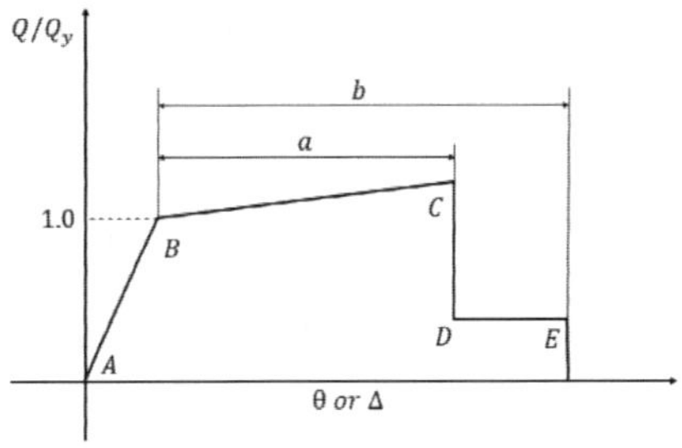

Figure 7. Force-Deformation Curve on FEMA 356.

Table 1. Hinge Property of shear spring in OpenSEES and ETABS model.

Point	OpenSEES model				ETABS model			
	FEMA 356		Richard-Uang		FEMA 356		Richard-Uang	
	Force (N)	Displacement (mm)	Force (N)	Displacement (mm)	Force/SF	Displacement/SF	Force/SF	Displacement/SF
A	0	0	0	0	0	0	0	0
B	258453	0.795493682	284298.3	0.62	1	0	0.916	0
C	284298.3	25.71428571	335988.9	4.38	1.27	25.714	1.3	4.0153
D	206762.4	25.71428571	387679.5	11.9	0.8	25.714	1.5	10.9
E	206762.4	29.14285714	387679.5	27.43	0.8	29.143	1.5	25.002

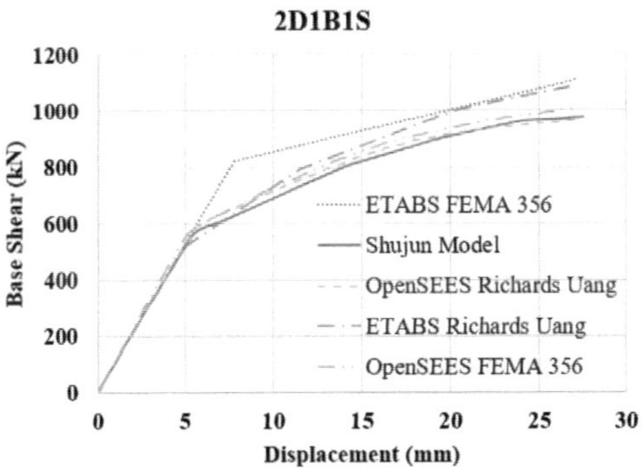

Figure 8. Comparison Results of 2D 1 Bay 1 Storey EBF.

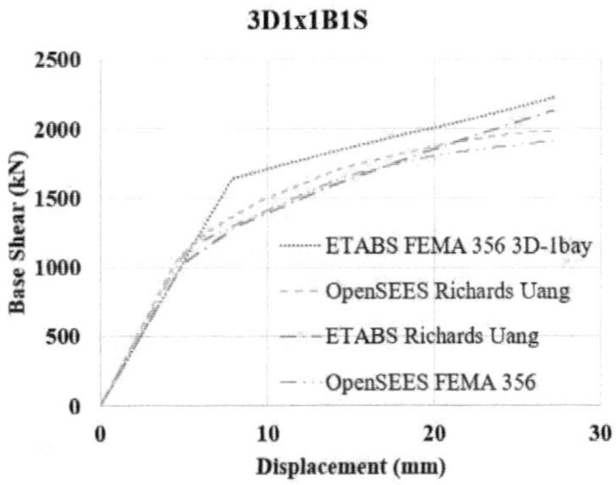

Figure 9. Comparison Results of 3D 1x1 Bay 1 Storey EBF.

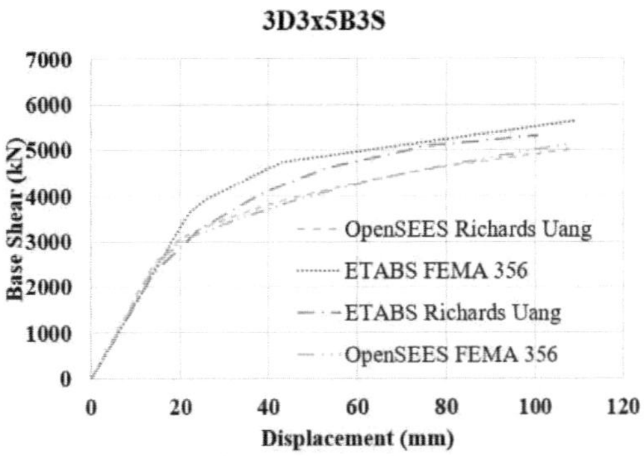

Figure 10. Comparison Results of 3D 3x5 Bay 3 Storey EBF.

4. Results and Discussions

Results of comparative study of hinge properties proposed by Richard-Uang and FEMA 356 are shown in figure 8 to figure 10. Figure 8 present analysis results based on OpenSEES and ETABS software. Using the same hinge property proposed by Richard-Uang has slightly different results when analysis conducted by OpenSEES and ETABS. As shown, results of openSEES has closer results to Shujun's experiment than ETABS. Results based on Richards-Uang's model is 15,55% higher than Shujun's whereas FEMA 356 are 34,54% higher. The difference is explained by figure 5. The Richards-Uang has rotational and linear springs at the ends of the link while the FEMA 356 has rotational springs with eccentricity as 5% from link-ends and put linear spring at the midpoint of the link. Modelling EBFs with ETABS using Richards-Uang hinge model results close agreement with advanced analysis. It can be concluded that the Richards-Uang shear link model is more accurate in modelling shear links in EBF steel structures.

Based on above results it can be concluded that hinge property greatly affects the analysis results. Hinge property use in FEMA 356 results result higher structural response than those model use with Richards-Uang's hinge property. Therefore, it is suggested to adopt Richard-Uang model in defining hinge properties for EBF in order to get closer results to real structure.

5. Summary

Hinge properties proposed by Richards-Uang result a Point B displacement as 33% smaller than that of FEMA 356, thus earlier yielding occur. During the elastic stage, model with FEMA and Richard-Uang show the same behaviour since hinge properties only affect post-yield behaviour. Using OpenSEES software, hinge properties based on Richards-Uang and FEMA 356 have almost similar results. However, it is suggested to use Richard-Uang hinge since its closer to experiment results than FEMA 356.

Acknowledgement

This works is supported by Hibah PITTA 2019 funded by DRPM Universitas Indonesia No. 5000/UN2.R3.1/ HKP.05.00/2019.

References

[1] Berman J W, Bruneau M 2006 Experimental and Analytical Investigation of Tubular Links for Eccentrically Braced Frames, *Eng. Struc.* **29(8)** 1929-38.
[2] Kasai K, Popov E 1986 Cyclic Web Buckling Control for Shear Link Beams, *J. Struc. Eng.* **112(3).**
[3] Ghobarah A, Ramadan T 1995 Analytical Model for Shear-Link Behaviour, *J. Struc. Eng.* **121(11)**.
[4] Richards P W, Uang C 2006 Testing Protocol for Short Links in Eccentrically Braced Frames, *J. Struc. Eng.* **132(8)**.
[5] Inel M, Ozmen H B 2006 Effects of Plastic Hinge Properties in Nonlinear Analysis of Reinforced Concrete Buildings, *Eng. Struc.* **28(11)** 1494-502.
[6] FEMA 356 2000 Pre-standard and Commentary for the Seismic Rehabilitation of Buildings, *American Soc. Civ. Eng.*
[7] Hu S 2015 Practical Advanced Analysis for Eccentrically Braced Frames, *Adv. Ste. Const.* **11(1)** 95-110.
[8] OpenSEES version 1.7.0 2017 OpenSEES Manual, Berkeley.
[9] ETABS version 16 2016 Analysis Reference Manual, Comp. Stuct. Inc.
[10] AISC and Sructural Steel Educational Council, AISC Seismic Design Manual, AISC Inc. 2006.

IOP Conf. Series: Materials Science and Engineering **811** (2020) 012049 doi:10.1088/1757-899X/811/1/012049

Dynamic Behavior Analysis of Feed Drive System Using Linear Guideways with Different Ball Diameters

Kazuma Tanaka[1, a*], Takanori Yamazaki[2, b]

[1]Graduate School of Science and Engineering, Tokyo Denki University, Ishizaka, Hatoyama, Hiki, Saitama, Japan
[2]School of Science and Engineering, Tokyo Denki University, Ishizaka, Hatoyama, Hiki, Saitama, Japan

[a]18rmt18@ms.dendai.ac.jp, [b]yamazaki@mali.dendai.ac.jp

Abstract. The feed drive system for machine tool consist of rolling elements such as ball screw and linear guides. It is well known that in the positioning of linear guides using steel ball rolling, the contact surface has nonlinear spring characteristics. This phenomenon is deteriorating the accuracy of the machine tool. In this study, we have manufactured a feed drive system in which the physical contacts are only linear guides by using a linear motor as actuator. At first, we measured the responses to some fine feed amount for feed drive system. Next, the same experiment was performed by changing the linear guideways. This paper presents the transition of nonlinear spring characteristic region by considering the response to linear guides under different conditions.

1. Introduction

In recent years, positioning systems consisting of mechanically non-contact elements are increasing. There elements include the drive mechanism using the linear motor, and hydrostatic guideways using air or oil pressure. However, it is still dominant the positioning systems have contact element such as ball screw, guideway using rolling and sliding for the cost and usage conditions [1]. The increase in the physical elements of the system means the increase in the elements to be considered in improving the accuracy of the system. The friction and spring behavior caused by the mechanical contact of the elements have nonlinear characteristics, and the responses of positioning system are changed by element size and displacement amount. For the reason, several studies improve the positioning accuracy by tuning the input force or control parameters, and compensating for friction, respectively. [2, 3].

In rolling guides, which are often used as guideways for machine tools, it is well known to have three different spring characteristics depending on the displacement [4]. According to a study by Futami et al. decreasing the displacement, the hysteresis loop of displacement and friction force transit from steady rolling to nonlinear spring characteristics. If the displacement is further reduced, the relationship becomes not the hysteresis loop but linear. Similar analysis of rolling guideways using sinusoidal displacement is well done [5-8].

Since these studies are considering characteristics (hysteresis loop) for one linear guide, the spring characteristics when changing the linear guide have hardly studied. In addition, it is also reported that the nonlinear friction zones appear frequently when switching the forward and reverse rotation in sinusoidal displacement [9]. The positioning accuracy is evaluated for step input; it is considered necessary to clarify the behavior for microscopic displacement.

In this study, we will compare and consider the transition of spring characteristics caused by changing the size of the linear guideways (ball diameter) for the step input. Specifically, entering different steps of displacement for the two guideways, NS15 (preload 294 N) and NS25 (preload 98 N) made in NSK, the transition of nonlinear spring characteristics is discussed from their dynamic behavior.

2. Experimental Device

2.1. Feed Drive Mechanism

The drive device used in this paper is shown in figure 1. This device consists of a table, a linear motor (made by GMC Hillston, S200Q), a guide rail, and a base. This drives the table one axis. The guideways details used in this paper shown in table 1. The device table is attached with an encoder scale (made by HEIDENHAIN, LIP201, grid scale pitch is 512 nm) for displacement measurement, and an encoder head (made by HEIDENHAIN, AK LIP 21) at the base using a jig. Guideways on height adjustment rails, encoders are mounted to achieve specified torque and parallelism.

2.2. Control Device

The experimental device configuration is shown in figure 2. In this experiment, two feed drive devices using different guideways are put on the same surface plate. These two feed drive devices use a common control configuration system by switching the connection with the linear motor and the encoder from the servo amplifier. For control, Programmable Multi Axis Controller (made by OMRON, CKE3E-1210, after this it called PMAC) is used as a host device, and PMAC is operated by the software of the PC. PMAC is operated by PC and torque commands from PMAC are transmitted to the servo amplifier, the servo amplifier supplies power to the linear motor to drive the table. The encoder reads the displacement, and the signals passing through the multiplier (made by HEIDENHAIN, maximum 16384 divisions) are fed back to the servo amplifier and PMAC.

2.3. System

The control system installed in PMAC is shown in figure 3. PMAC can select either velocity or torque control. In this experiment, position loop and velocity loop are performed by PMAC, it is torque control that to transmit current signals to the servo amplifier. The current loop is performed by the servo amplifier, which supplies power to the linear motor.

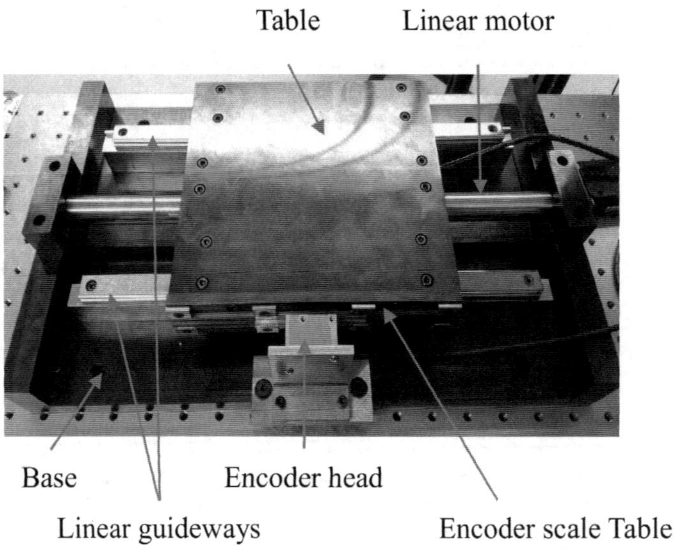

Figure 1. Drive device.

Table 1. The guideways details.

	NS15	NS25
Rail width[mm]	15	25
Preload[N]	294(middle)	98(fine)
Steel ball diameter [mm]([inch])	2.7781(7/64)	3.9688(5/32)

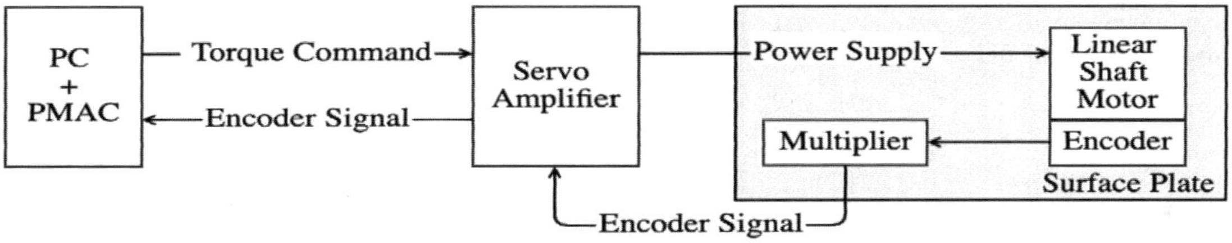

Figure 2. Experimental device configuration.

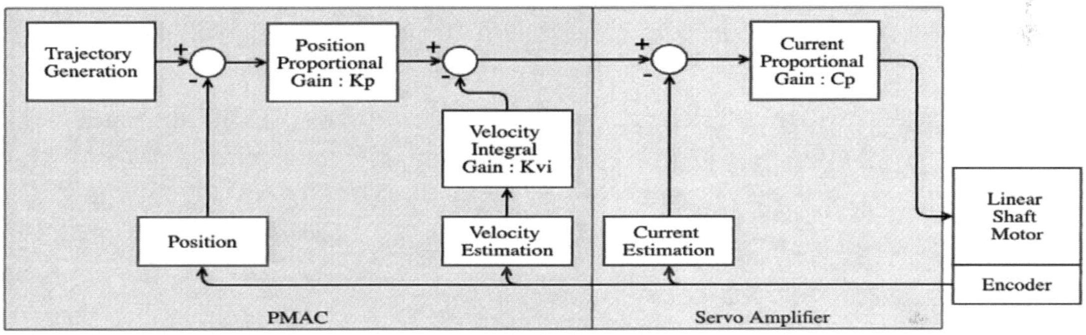

Figure 3. Control configuration diagram.

3. Feed Experiments

3.1. Experimental Conditions

The linear scale used in the experiment has a grid scale pitch of 512 nm, and the minimum resolution of 0.03125 nm can be obtained by electrically multiplying this by 2^{14} (=16,384). The servo amplifier does not support to this resolution. For this reason, the signals from the multiplier were divided by 16(ignored 4 bits) and the encoder resolution was set to 0.5 nm.

$$512 \times \frac{\text{Division(16)}}{16384} = 0.5 \text{ nm}. \quad (1)$$

The servo amplifier and PMAC gains were set to that position loop proportional gain Kp is 0, velocity loop integral gain Kvi is 0.013, current loop proportional gain Cp is 1000, and current loop integral gain Ci is 0. The displacement of the step inputs is 10, 5, 1, 0.5, 0.1, 0.05 and 0.1 μm. In this paper, 30 steps of input were performed using the displacement of each step input.

3.2. Comparison of Actual Force for Each Guideway

First, we compared the actual force at each linear guideway for different displacements. Here, the several displacements are extracted, and figure 4(a)-(d) show actual forces at displacement of 10, 1, 0.1 and 0.01μm of one step, respectively. In these figures, the red line is the actual force, the magenta is the average for each step of the actual force, the black is the step input command, the solid line is NS15 and the dotted line is NS25. The left and right axes represent the displacement and the actual force.

The actual force for both guideways are almost constant after the first step input at a displacement of 10μm in figure 4(a). This shows that the spring characteristics are constant because the steel ball is in the steady rolling area at this displacement. In figure 4(b), the actual force for both linear guideways show that the nonlinear spring characteristics up to around 9 μm (9th step). More than this displacement, the actual force for NS15 is increase, but the actual force for NS25 is almost constant. From that, it is considered that range of spring characteristics by linear guideway is different.

The large difference is appeared depending on the spring characteristics by the guideways in the area shown in figure 4(c). In the solid line NS15, the increase in force up to around 500 nm (5th step) is different from after that displacement. On the other hand, the increase in force after the first step is almost constant in the dotted line NS25. This shows that the area of the spring characteristics of NS15 is transitioning around this displacement. In figure 4(d), there is almost no difference between the two actual force, and the increase is almost constant. This indicates that both are linear spring characteristic area in this displacement.

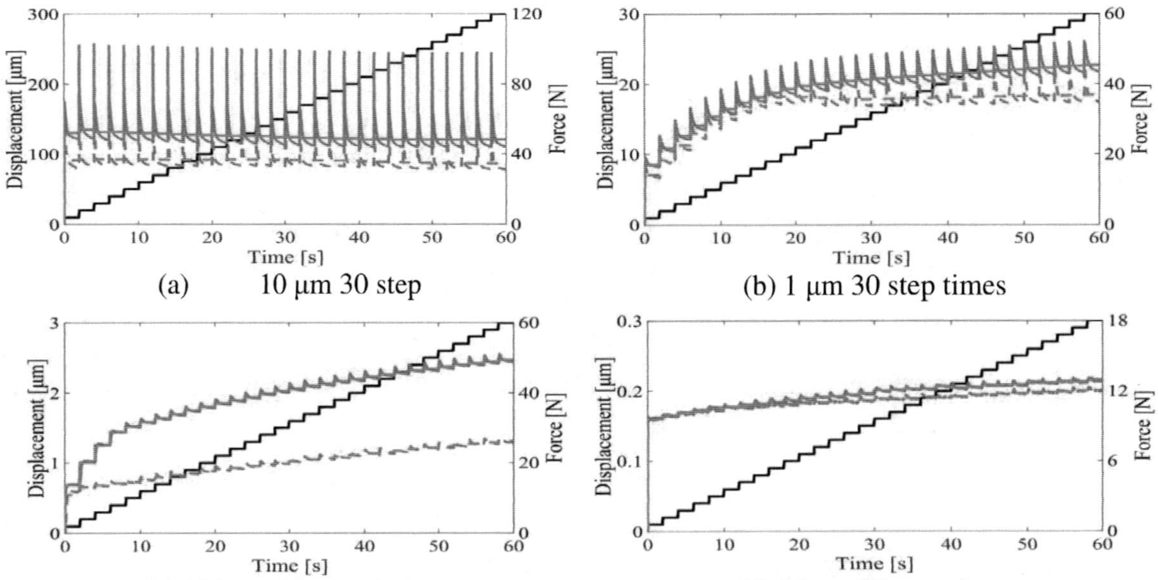

(a) 10 μm 30 step (b) 1 μm 30 step times

(c) 100 nm 30 step times (d) 10 nm 30 step times

Figure 4. Comparison of actual force for each guideways.

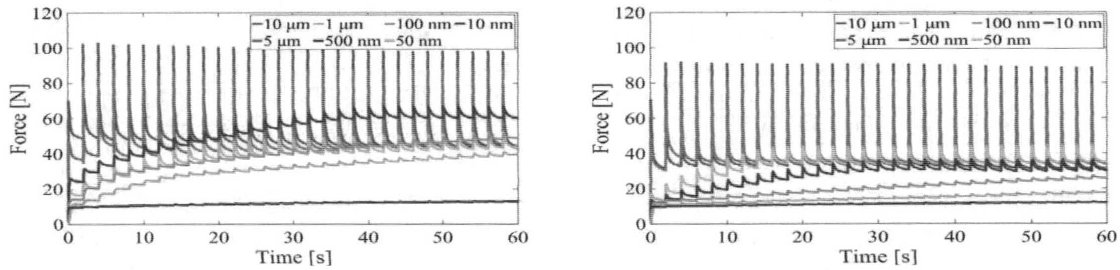

(a) N (b) N

Figure 5. Comparison of actual force for each displacement.

Table 2. Range of spring characteristics area.

	Linear area [μm]	Nonlinear area [μm]	Steady rolling area [μm]
NS15 (294 N)	~0.3	0.3~10	10~
NS25 (98 N)	~3	3~5	5~

IOP Conf. Series: Materials Science and Engineering **811** (2020) 012049 doi:10.1088/1757-899X/811/1/012049

3.3. Comparison of Actual Force for Each Displacement

Second, we compared the actual force at different displacements for each linear guideway. Figure 5(a) and (b) show actual forces at NS15 and NS25 for displacement of 10 μm-10 nm of one step. From figure 5(a), the actual force at the steps input is approximately the same at a displacement of 10 μm. At a displacement of 5 μm, 1 μm and100 nm, the forces converge to around 40 N. This indicates that the drive force becomes constant when the step input is performed multiple times, and it is considered that steel balls are in the steady rolling area for the more than those displacements.

At a displacement of 50 nm, the transition of the spring characteristic area is observed and does not reach the steady rolling area. At a displacement of 10 nm, linear spring characteristics can be observed. From figure 5(b), the actual force at the steps input is approximately the same at a displacement of 10 and 5 μm. At a displacement of 1 μm and, the forces converge to around 35 N, the transition of the spring characteristic area is observed. At a displacement of 100, 50 and 10 nm, linear spring characteristics can be observed.

4. Summary

In this paper, we performed steps input experiments with different displacements using two types of linear guideways. The experimental results were classified by guideways (ball diameter) and displacements, and the actual torque was compared. As a result, each area is summarized as shown in table 2. The linear spring area is smaller in NS15 according to table 2. It is considered that the NS15 which the diameter is small have small elastic deformation until the steel balls roll or slide. On the other hand, NS25 transitions to a steady rolling area at a displacement smaller than NS15. The rolling friction is assumed to be smaller than NS15 because NS25 has a small preload. In the future, we will change the preload of the guideways and to study the transition of spring characteristics in more detail.

References

[1] Oiwa T, Katsuki M 2015 Survey of Questionnaire on Ultra-Precision Positioning, *J. Jap. Soc. Preci. Eng.* **81(10)** 904-10.

[2] Iwasaki M, Maeda Y, Kawafuku M, Hirai H 2006 Improvement of Precise Positioning Performance by Modeling and Compensation for Nonlinear Friction, *IEEJ Trans.* **126(6)** 732-40.

[3] Otsuka J, Takahashi M, Usuda T, Tofuku M, Aoki Y 1989 Study on Precision Positioning by Friction Drive (2nd report) –Trapezoidal Velocity Control, *JSPE.* **55(1)** 123-8.

[4] Futami S, Furutani A 1991 Nanometer Positioning Using AC Linear Motor and Rolling Guide (2nd Report) –Tribology of the Rolling Guide, *JSPE.* **57(10)** 1808-13.

[5] Otsuka J, Aoki I, Ishikawa T 2000 A study on Nonlinear Spring Behavior of Rolling Elements (1st Report)-Two Simple Measuring Methods, *J. Jap. Soc. Preci. Eng.* **66(6)** 944-9.

[6] Tanaka T, Oiwa T, Otsuka J, Onda H 2007 Study on Linear Ball Guideway for Precision Positioning–Method of Estimating Dynamic Characteristic from Static Behavior in Nonlinear Spring Area, *J. Jap. Soc. Preci. Eng.* **73(3)** 350-4.

[7] Sato R, Tsutsumi M, Imaki D 2007 Experimental Evaluation on the Friction Characteristics of Linear Ball Guide, *Trans of the JSME.* **73(734)** 173-81.

[8] Ohashi T, Shibata H, Futami S 2017 Effects of preload change and with/without ball retainers for infinitesimal motions of a linear motion ball guide, *JSPE.* 319-20.

[9] Tsuruta K, Murakami T, Futami S 2003 Nonlinear Friction Behavior of Discontinuity at Stroke End in a Ball Guide Way, *J. Jap. Soc. Preci. Eng.* **69(12)** 1759-63.

IOP Conf. Series: Materials Science and Engineering **811** (2020) 012050 doi:10.1088/1757-899X/811/1/012050

Vibration Compensation of Checkweigher Using Acceleration Sensor

Akinori Tsuzura[1, a *], Yuji Yamakawa[2, b], Takanori Yamazaki[3, c]

[1]Graduate School of Science and Engineering, Tokyo Denki University, Ishizaka, Hatoyama, Hiki, Saitama, Japan
[2]Institute of Industrial Science, The University of Tokyo, 4-6-1 Komaba, Meguro, Tokyo, Japan
[3]School of Science and Engineering, Tokyo Denki University, Ishizaka, Hatoyama, Hiki, Saitama, Japan

[a]18rmt19@ms.dendai.ac.jp, [b]y-ymkw@iis.u-tokyo.ac.jp, [c]yamazaki@mali.dendai.ac.jp

Abstract. The checkweigher is a device that measures mass continuously and it carries out mass measurement while conveying an object on a belt conveyor. It is a device that operated an active role in various fields such as food, medicine, and industrial products. Although the load cell scale is often used to measure measurement in the checkweigher, due to the structure of load cell, the accuracy is susceptible to environmental vibration. It is difficult to generate low frequency input such as floor vibration with the general vibration exciter. We first manufactured a device that can reproduce environmental vibration. In this study, we present the response of the load cell to excited by the device, moreover consider the vibration compensation using an acceleration sensor.

1. Introduction

The checkweigher is a device that measures mass continuously and it carries out mass measurement while conveying an object on a conveyor belt. This device is operated in various fields such as food, medicine, and industrial products. Although the load cell scale is often used to measure measurement in the checkweigher, the measurement accuracy is susceptible to environmental vibration due to the structure of load cell.

The several studies have been conducted to improve the accuracy of mass measurement using a load cell up to the present date. Kamimura et al. reported that the accuracy of mass measurement has improved by designing the FIR filter that removes the environmental vibration for the load cell [1]. Umemoto et al. derived a notch filter that can remove only signals of specific frequency components, and the transfer speed is improved by applying it to the checkweigher [2]. Sun et al. proposed a method to remove environmental vibration using signal processing by newly installing a load cell for vibration detection separately from the measurement load cell. In this method, the natural frequencies of the two load cells do not match (there is a difference in dynamic characteristics), which may adversely affect the measurement accuracy. Hence, by applying the principle of relative compensation, it has been shown to improve the accuracy of mass measurement [3]. In addition, it takes 40 ms to identify on-line in this method, and high-speed mass measurement is difficult. Therefore, the shortening of the identification time was considered by the compensation of the different in dynamic characteristic using a polynomial approximation model [4]. However, it seems that there are not many reports on compensation using an accelerometer.

General vibration exciters use hydraulic pressure; it is difficult to reproduce low frequency range. We first manufactured a vibration exciter using an electric cylinder combining a servomotor and a ball screw. It has been confirmed that this exciter can reproduce vibration of 0.1 to 10 Hz. On this basis, we measured the response of the load cell when applying vibration with the manufactured exciter and propose vibration compensation of checkweigher using an acceleration sensor.

2. Vibration Device Configuration

The scale of the checkweigher used in this study is a load cell type, the specifications of this device are as follows: measuring range 12-3000 g, scale 0.1 g, selection accuracy ± 0.2 g. The vibration exciter is manufactured using an electric actuator (SMC, LEY32DNZC-100) and a servomotor (Yaskawa Electric, Σ-II), it is possible to control the motor by sending a position command from the host device (PC+LabVIEW) to the servo amplifier and to vibrate in the vertical direction. The apparatus of the vibration exciter is shown in figure 1.

The actual measuring conveyor consists of multiple parts, but a dummy conveyor was manufactured to simplify the vibration phenomenon in this experiment. The dummy conveyor is shown in figure 2. The dummy conveyor reproduces the same mass, size, and center of gravity position as an actual conveyor. The checkweigher is attached with acceleration sensors for the purpose of vibration compensation. The acceleration sensor is shown in figure 3 and its specifications represent by table 1. The acceleration sensor is fixed to the bottom of the load cell, the signal during vibration can be taken simultaneously with the load cell.

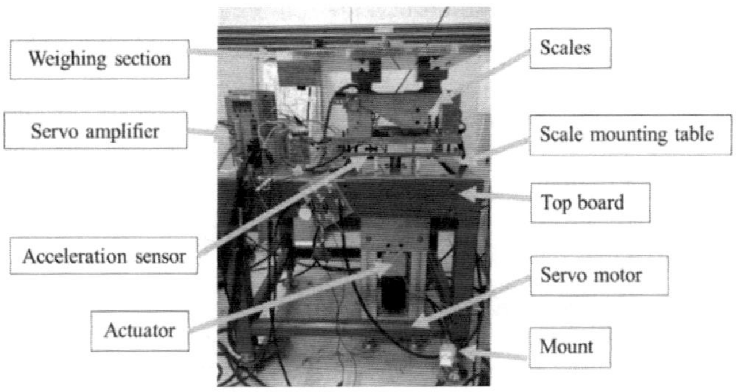

Figure 1. Vibration exciter.

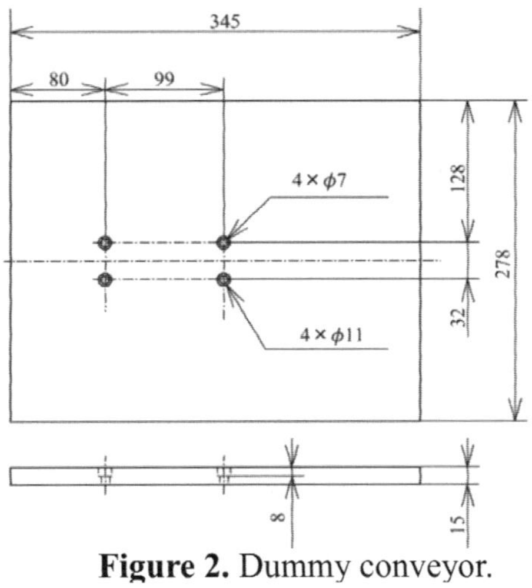

Figure 2. Dummy conveyor.

Figure 3. Acceleration sensor.

IOP Conf. Series: Materials Science and Engineering **811** (2020) 012050 doi:10.1088/1757-899X/811/1/012050

Table 1. Acceleration sensor specifications.

Manufacturer	TOKYO KEIKI INC.
Model name	TA-25J-02-1
Resolution (DC)	9.8×10^{-6} m/s^2 ±5%
Similarity measurement	0.510 V/(m/s^2)±5%
Power supply	+15~+16 VDC, -15~-16 VDC
Power consumption	Under ±25 mA
External dimensions	φ28.4 ×24.5
Mass	46 g

3. Vibration Experiment

In order to compare the response waveform of the load cell and the acceleration sensor during vibration, we make experiment with the manufactured vibration exciter. This experiment also has the meaning of confirming whether the low frequency vibration can be measured by the acceleration sensor. The experiment was operated by placing a weight on the center of gravity of the dummy conveyor shown in figure 2. Here, the weight is 0.1 kg. As the experimental conditions, the amplitude was 0.5 mm and the input frequency were changed from 1 to 2, 3 and 5 Hz.

The responses to these sinusoidal inputs are shown in figure 4(a)-(d). The response waveform is measured for 10 s with a sampling time of 1 ms. However, it is evaluated with waveform after 5 s in order to eliminate the influence of transient response. In addition, the low pass filter with 10 Hz is applied to the response waveform due to remove the noise of the measurement system.

In figure 4 (a)-(d), the left and the right axis show the voltage output of the load cell and the acceleration sensor, respectively, and the scale has been adjusted to compare both responses. From these results, it can be confirmed that the acceleration sensor can also measure low frequencies. It is reasonable that the load cell response is delayed as the frequency increases. Table 2 shows the relationship between frequencies and both voltages (peak to peak value). It can be found that the acceleration is basically proportional to the square of the frequency although the values are slightly different due to the natural frequency of the load cell etc.

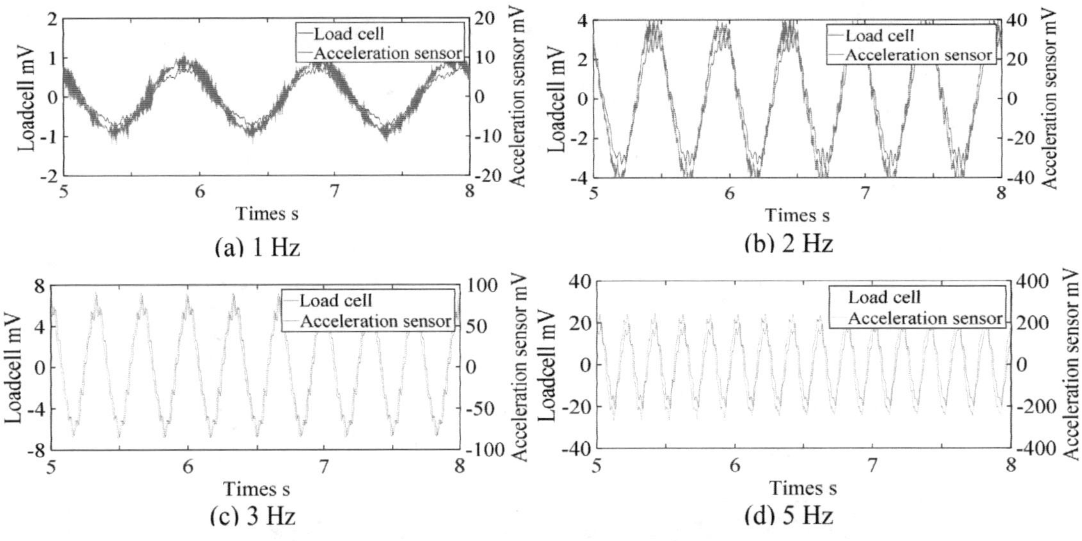

Figure 4. Response for a sinusoidal input.

Table 2. The relationship between the input frequency and the voltage value.

Input frequency [Hz]	Load cell[mV]	Acceleration sensor [mV]
1	1.53	24.2
2	6.61	87.2
3	13.6	176
5	40.5	527

4. Vibration Compensation

In this study, we call vibration compensation by removing the vibration component contained the load cell using the output of the acceleration sensor. A block diagram of vibration compensation is shown in figure 5. First, the output voltage of the load cell was passed through a low pass filter, and then converted to strain by the catalog value (K_L in figure 5). The force was calculated from the strain using the relationship between the force and strain at the load cell described by equation 1.

$$\varepsilon = \frac{3L}{2bt^2E}F \tag{1}$$

Where, b : Load cell length [mm], t : Notch thickness [mm], L : Length between notches [mm], and E : Young's modulus [GPa] in used load cell, respectively.

Next, the voltage output of the acceleration sensor was passed through a low pass filter, and then converted into acceleration by the scale factor (K_A in figure 5). The vibration is compensated by canceling out the forces obtained by these two paths and the mass variation is calculated by dividing the difference by the gravity acceleration of 9.8 m/s^2.

The comparison of mass variation before and after vibration compensation is shown in table 3. From this result, it was found that the vibration component contained in the load cell can be removed up to 65.5% by the proposed vibration compensation. The vibration removal rate is defined by equation 2.

$$\text{Removal rate} = \frac{X-Y}{Y} \times 100 \tag{2}$$

Where, X : Mass variation value after vibration compensation, Y: Mass variation value before vibration compensation.

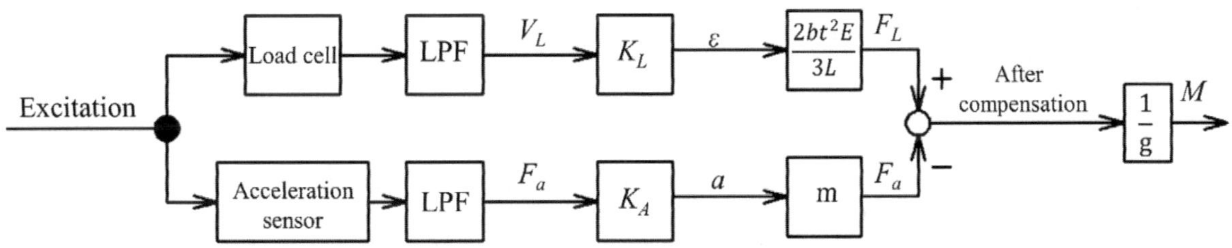

Figure 5. Block diagram of vibration compensation.

Table 3. Mass fluctuation before and after vibration compensation.

Excitation frequency [Hz]	Before vibration compensation [g]	After vibration compensation [g]	Removal rate [%]
1	37.6	22.6	39.9
2	153	57.1	62.7
3	307	106	65.5
5	942	575	39.0

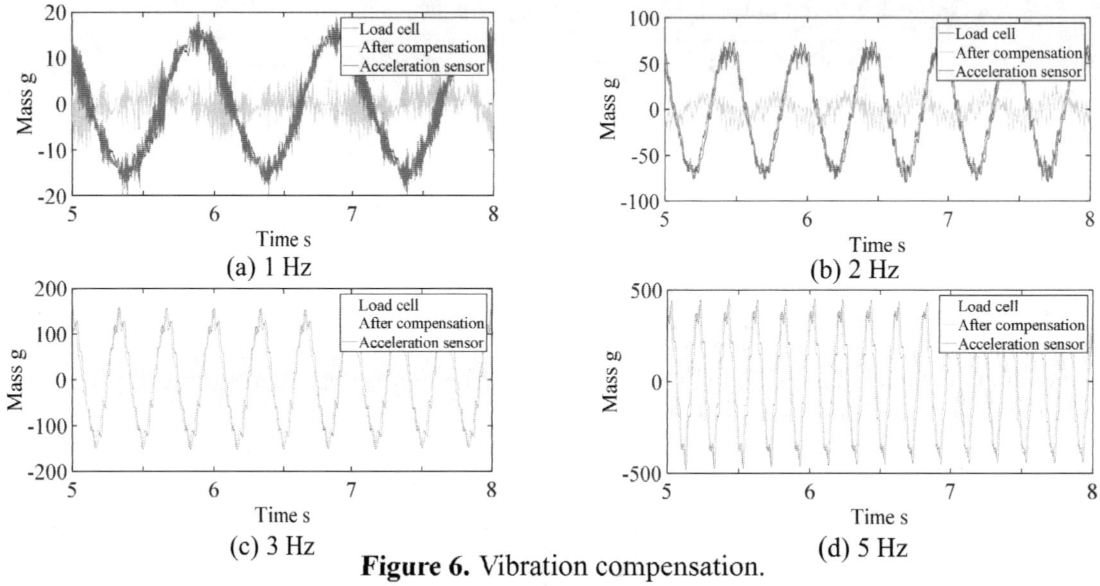

Figure 6. Vibration compensation.

The waveforms compensated for vibration using the data in figure 4 are shown in figure 6(a)-(d). Here, the compensation result is a green line, and the removal rate is as represented in table 2. From these results, it can be found that the output of the load cell is behind the acceleration sensor. The cause is the delay due to the dynamic characteristics of the load cell. In the proposed vibration compensation, it is important that the load cell and the acceleration sensor output the same waveform. The dynamic characteristics of the load cell cannot be ignored when the vibration frequency is increased.

5. Summary

In this study, we manufacture the vibration exciter that can reproduce environmental vibration. Comparing the response waveforms of the load cell and the acceleration sensor while exciting the checkweigher attached the acceleration sensor, it was confirmed that the measurement of low frequency was also possible. It is shown that environmental vibration can be removed up to 65.5% by the proposed compensation method using the output of acceleration sensor. However, the dynamic characteristics of the load cell cannot be ignored and the effect of compensation is reduced as the vibration frequency is increased. In the future, we plan to expand the application range by adding the model that takes into consideration the dynamic characteristics of the load cell.

Reference

[1] Kamimura K, Kanai H, Chubach N 1995 Design of optimum FIR filter for eliminating mechanical noise in weight measurement with load cell, *Acous Soci of Jap.* **51(11)** 845-53.

[2] Umemoto T, Kamon M, Kagawa Y 2011 Improvement of Accuracy for Continuous Mass Measurement in Checkweighers with an Adaptive Notch Filter–Processing in Measurement Time, *Trans. Soci of Inst and Con Eng.* **47(10)** 477-84.

[3] Sun J, Fujioka Y, Ono T, Nagao T, Kohashi T 1998 On the Fast and High Accurate Mass Measurement under the Conditions of Floor Vibration–On the Method of Compensating the Difference in Dynamics without On-line Identification, *Jap Soci for Preci Eng.* **64(7)** 1029-34.

[4] Sun J, Fujioka Y, Ono T, Nagao T 1999 On the Fast and High Accurate Mass Measurement under the Conditions of Floor Vibration–On the Method of Compensating the Difference in Dynamics by Using the Approximate Polynomial Model, *Jap Soci for Preci Eng.* **65(10)** 1450-5.

IOP Conf. Series: Materials Science and Engineering **811** (2020) 012051 doi:10.1088/1757-899X/811/1/012051

Damage Assessment Using Natural Frequency on Concentric Braced Frame Steel Structures

Ahmad Zaki Risadi[1, a], Josia Irwan Rastandi[1, b*], Bastian O. B. Sentosa[1, c], M. Orientilize[1, d]

[1]Program Studi Teknik Sipil, Departemen Teknik Sipil, Fakultas Teknik, Universitas Indonesia, Depok, 16424, Indonesia

[a]ahmad.zaki51@ui.ac.id, [b]jrastandi@eng.ui.ac.id, [c]bastian.sentosa@ui.ac.id, [d]mulia@eng.ui.ac.id

Abstract. At present, structural assessment using dynamic characteristics remains to be explored. In this study, a correlation between damage index and natural frequencies of Concentric Braced Frame steel structures has been observed. The study consists of two sections: the first section presents the calculation of the damage indices of structures by means of monotonic and semicyclic pushover analyses using OpenSEES software, while the second section evaluates the structures' natural frequencies by means of modal analyses using SAP2000 software. Our results show a trend in which with increasing amounts of lateral load, natural frequencies of the structures decrease while the damage indices increase. There is also an apparent 'jump' in the damage index values after the lateral loads reach their critical limits. The study results help the understanding of the structural assessment process at the global scale.

1. Introduction

Indonesia is a country in which earthquakes often happen. To anticipate the damage, structural design codes in Indonesia require structures to be able to resist lateral loads. Seismic design requires evaluation of damage in structures, which is best represented by its loss of stiffness. This makes natural frequency, which is a function of stiffness, a good indicator of structural strength degradation. Along with that, damage index is also capable of representing strength loss in structures due to it being able to quantitatively assess structures' strength.

Damage index is a function of structural deformation. A study by Park and Ang [9] in 1987 suggests that the damage index is a parameter that represents damage to a structure, where 0 indicates no damage while 1 indicates total failure. This is also true in other damage index theories, such as the one proposed by Ghobarah [1]. For this research, the damage index theory that will be used is the Park & Ang damage index which uses deformation and energy parameters, not the Ghobarah damage index which uses stiffness as a parameter. All parameters used in calculating damage index can be obtained through pushover analysis.

An earlier research by Bastian O. B. S. [3, 10] suggests that in RC frames, damage index values increase as cracks start to form. This is also accompanied by the decrease of natural frequency values, as shown in the research.

A numerous amount of research has been done to study the damage indices of structures, but only few them are done on steel structures. There are three types of lateral load resistant steel structures, namely moment-resisting frames (MRF), eccentrically braced frames (EBF) and concentric-braced frames (CBF). This research focuses on CBFs since its stiffness allows it to produce results with a high amount of disparity, allowing for better analyses. The index will be examined in relation to natural frequency, a parameter which is a function of stiffness and structural mass as stated by Chopra [2]. The stiffness decreases with increasing damage, whereas in the steel structure, the mass of the structure does not decrease with increasing damage. Therefore, natural frequencies can be a good parameter for reviewing damage to structures. The results of this study can be used as a reference to predict the damage index value of structure. The results of this study can also be used as a design reference by adjusting the target damage index and natural frequency according to the damage index.

2. Model Validation

The study starts with model validation to ensure that model built by two structural software, OpenSEES [6] and SAP2000 [8] can represent the real structure. The models were validated against experiment conducted by Wakabayashi et al [5], also shown in the research by Faggiano B. et al [4]. The CBF structure is shown in figure 1 which consist of one floor and one bay. The dimensions of the structure are 2600 mm x 5000 mm. The structure is made of SS41 steel (f_y = 235 MPa, f_u = 400MPa, strain hardening = 0.014) with different H profiles for columns, beams and braces. For columns, the profile used is H-175x175x7.5x11. For beams, the profile used is H-250x125x6x9. For braces, the profile used is H-100x50x4x6.

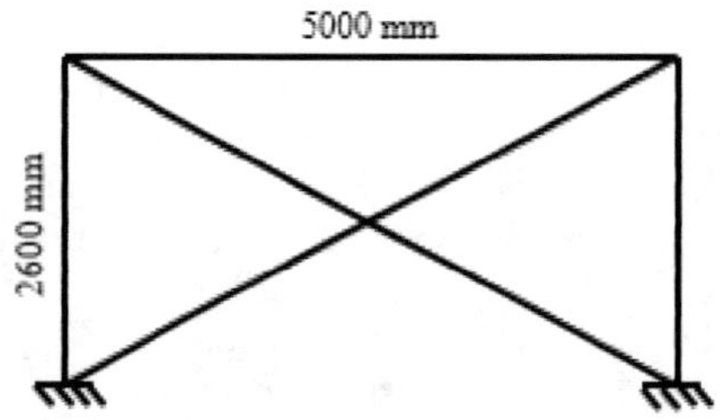

Figure 1. Model for validation.

A NonlinearBeamColumn element in OpenSEES is assigned for beam and column. The element define plastic deformation distributed along the member. The material used for the structure is Steel02, which behaviour is a bilinear curve. According to the experiment, the load is applied horizontally on top through pushover menu. In SAP2000, all members are modelled as frame element. Moment release is applied to accommodate pin connection between bracing and column. Hinge is also defined to all members, locating at the end of beam and column while braces are assigned at their ends and midpoints. Result of validation is shown in figure 2.

As can be seen, the load-displacement curves made using OpenSEES and SAP2000 has closely resemble the one shown in the research by Wakabayashi. Both curves show the same elastic behaviour, while only showing little disparity in the plastic region. This allows the assumption that both the models made in OpenSEES and SAP2000 are able to be the basis of other models made for this research.

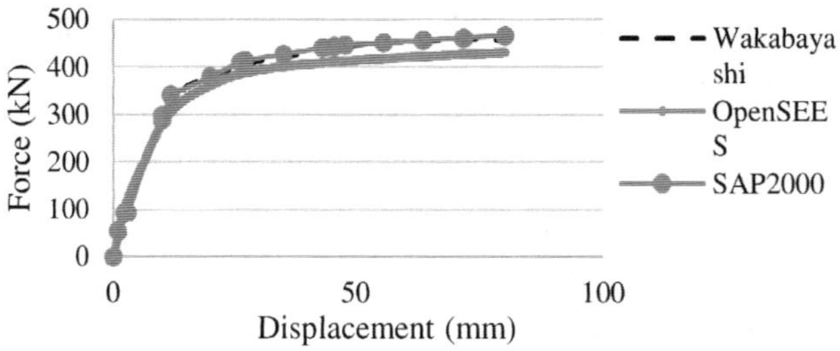

Figure 2. Validation of CBF Model.

3. Research Methodology

Since the models have been validated, then the study of damage index can be carried out on 3 different stories of CBF, one, three and four stories. The geometrics are shown in figure 1, figure 3 and figure 4. The 3 and 4-stories were based on research of CBF conducted by Ernesto G. and Alessandro R [3]. The 3 and 4-storey were made of S235 steel with yield stress and ultimate stress as 235MPa and 400MPa, respectively. The profiles used are HEA220 for the 1st and 2nd story beams and columns, HEA200 for the 3rd and 4th story beams and columns and 2L 60x6 for the braces on all floors.

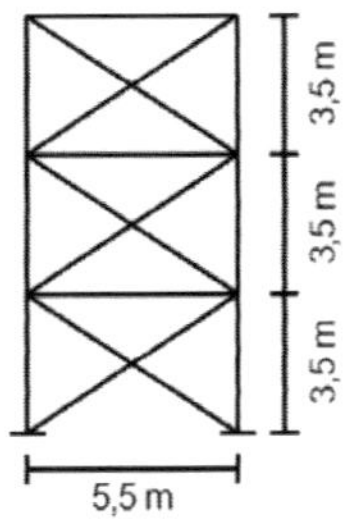

Figure 3. 3-Storey Structure.

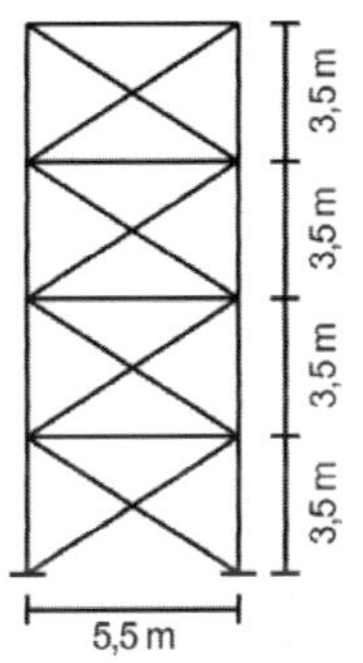

Figure 4. 4-Storey Structure.

As mention earlier, the study was conducted in two steps. First OpenSEES was used to determine the damage index. Refer to equation proposed by Park and Ang., the index can be calculated by integrate the energy below the curve obtained through semi-cyclic loading. The equation is expressed on equation 1, where D is damage index, δ_m is maximum deformation, δ_u is ultimate deformation, Q_y is yield strength, dE is dissipated hysteretic energy and β is a non-negative constant obtained through empirical data. For steel structures, $\beta = 0{,}025$.

$$D = \frac{\delta_m}{\delta_u} + \frac{\beta}{Q_y \cdot \delta_u} \int dE \quad (1)$$

The objective of the research is to investigate the correlation between damage index and natural frequency. The second part of this research revolves using SAP2000 to obtain the change of natural frequency after step-by-step loading. The geometric changes of the CBF obtained through OpenSEES was remodelled in SAP2000 to obtain its natural frequency through modal analysis. Each structure was modelled in 4 different geometric conditions to illustrate the condition of the structure after every cycle of the semi-cyclic loading, with Model 1 being its initial condition, Model 2 after the 1st cycle, Model 3 after the 2nd and Model 4 after the 3rd.

4. Results and Discussions

Monotonic and semi-cyclic pushover analysis is performed on all three structures. From the analysis, data was obtained for calculation of damage index values. Damage index is calculated at the end every cycle to represent the damage that occurs after the cyclic loading is experienced by the structure. The data for the calculation of damage index values is as follows

Table 1. Damage Index Values of Structures.

Parameter	1-Story Structure			3-Storey Structure			4-Storey Structure		
	1	2	3	1	2	3	1	2	3
Q_y (kN)	270,78			398,21			413,48		
δ_u (mm)	80,00			100,00			130,00		
δ_m (mm)	26,70	53,40	80,00	50,00	75,00	100,0	85,00	105,0	130,0
$\int dE$ (kNmm)	1021,0	2251,7	0,00	647,14	2274,3	0,00	503,30	746,25	0,00
DI_{PA}	0,33	0,67	1,00	0,50	0,75	1,00	0,62	0,81	1,00

Model 1 has a damage index value of 0 while Model 4 has a damage index value of 1. The damage index values show a trend which increases with increasing lateral load. In accordance to the research by Bastian O. B. S., the biggest factor of structural damage is structural deformation. This parameter is in the first aspect of the damage index formula, which contributes 99% of the overall damage index value by generating values between 0.10 to 1.00. The second aspect of the formula, which is an aspect involving dissipation energy, only contributes 1% by generating values between 0.001 to 0.01.

Table 2. Natural Frequencies of the Structures.

Model	Natural Frequency (Hz)		
	1-Story Structure	3-Storey Structure	4-Storey Structure
Model 1	77,717	20,588	14,803
Model 2	41,025	16,135	8,476
Model 3	27,021	13,568	7,999
Model 4	22,463	12,646	7,381

The natural frequencies of the three structures decrease with increasing lateral load. This is consistent to the theory that said that structural stiffness decreases with increasing amount of structural damage, thus resulting in lower natural frequencies. Since the structures are steel, mass loss does not factor into the changes of the natural frequencies. That is why in this research, the loss of natural frequency corresponds directly to loss of structural stiffness.

What differentiates the three structures are their initial natural frequencies, as shown in table 2. This is in accordance to the theory by Peifu Xu et al [7] which stated that the natural period of a structure increases with the increase of structural height. Natural frequency is inversely proportional to natural period, thus making it decrease as structural height increases.

As shown in figure 5, there can be seen a certain limit where excessive lateral loading causes damage index values to increase more significantly than the decrease of natural frequencies. This limit is called a critical limit in the research by Bastian O. B. S. This limit is different for the three structures. These limits are at V = 88% Vu for the 1-story structure, V = 93% Vu for the 3-storey structure and V = 97% Vu for the 4-storey structure. It can be concluded that the critical limit value of a structure increases with structural height.

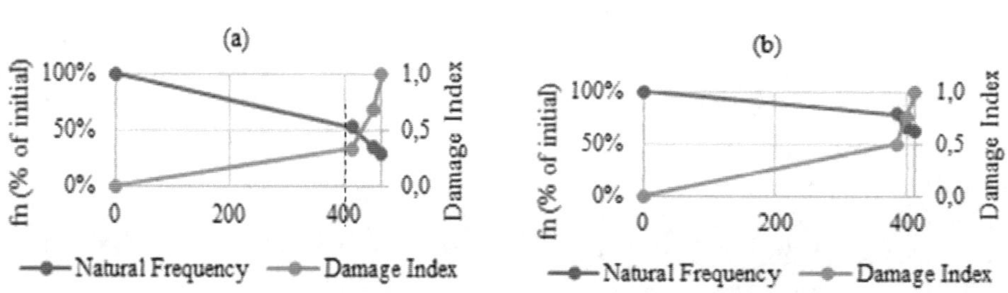

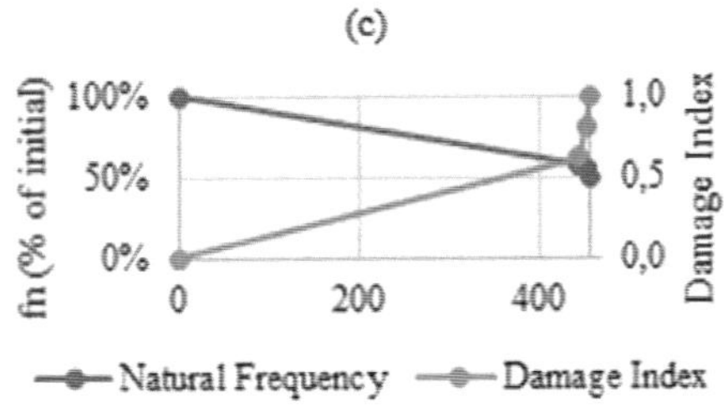

Figure 5. Damage Index and Natural Frequency of
(a) 1-Story Structure (b) 3-Storey Structure (c) 4-Storey Structure.

5. Summary

As the amount of lateral load applied to a structure increases, the corresponding structure's damage index value increases while its natural frequency decrease. This trend can be seen in all three structures. As observed, there exists a critical limit where any amount of load exceeding the limit will cause damage index values to increase much more significantly than the decrease of the structure's natural frequency. These limits are at V = 88% Vu for the 1-story structure, V = 93% Vu for the 3-storey structure and V = 97% Vu for the 4-storey structure.

Acknowledgements

This works is supported by Hibah PITTA 2019 funded by DRPM Universitas Indonesia No. 5000/UN2.R3.1/ HKP.05.00/2019.

References

[1] Ghobarah A, Abou-Elfath H, Biddah A 1999 Response-Based Damage Assessment of Structures, *Ear. Eng. Struc. Dynam.* **28(1)** 79-104.
[2] Chopra A K 2012 Dynamics of Structures 4th Ed, Prentice Hall, New Jersey.
[3] Sentosa B O B 2018 Assessing the structural elements of reinforced concrete buildings by using dynamic measurements, Doctoral dissertation, Gre. Alp.
[4] Beatrice F, Formisano A, Generoso V, Mazzolani F M 2018 Numerical Study on Concentric X Braced Frames under Monotonic and Cyclic Loads, *Key Eng. Mater.* **763** 633-41.
[5] Wakabayashi M, Matsui C, Minami K, Mitani I 1974 Inelastic Behaviour of Full-Scale Steel Frames with and without Bracings, *Bulle. Dis. Prev. Res. Ins.* **24(1)** 1-23.
[6] Open SEES version 1.7.0 2017 Open SEES Manual, Berkeley.
[7] Xu P, Xiao C, Li J 2014 Relationship between Natural Vibration Periods and Structural Heights for High-rise Buildings in China, *Int. J. High-Rise Build.* **3(1)** 49-64.
[8] SAP 2000 Adv. 2011 Analysis Reference Manual, *Comp. Struc. Inc., Berkeley.* **15.**
[9] Park Y J, Ang H S A, Wen Y K 1987 Damage-Limiting Aseismic Design of Buildings, *Ear. Spec.* **3(1)** 1-26.
[10] Sentosa B O B, Bui Q B, Plé O, Plassiard J P, Perrotin P 2019 Assessing Damage to Beam–Column Connections in Reinforced Concrete Structures from Vibrational Measurement Results, *Struc. Eng. Int.* **29(3)** 396-403.

IOP Conf. Series: Materials Science and Engineering **811** (2020) 012052 doi:10.1088/1757-899X/811/1/012052

Simulation Analysis and Optimization of a Passenger Car Ride Comfort

Daisheng Zhang[1, a], Yunqing Cai[1, b*]

[1]School of Automotive and Transportation Engineering, Hefei University of Technology, Hefei, Anhui, China

[a]daishengzhang2000@163.com, [b]caiyunqing1001@163.com

Abstract. The vehicle ride comfort research is an important part of the vehicle development process. This paper takes a passenger car as the research object, establishes the vehicle rigid-flexible coupling ride comfort simulation model based on ADAMS, and simulates the vertical vibration acceleration time domain curve at the driver's position. Through the derivation of the theoretical formula, the corresponding weighted vibration level is obtained. Next, the weighted vibration level of the front seat is used as the measurement index. The stiffness of the front suspension torsion bar spring and the damping of the front and rear suspension dampers are selected as the optimization variables to optimize ride comfort. The optimization of the performance makes the peak of vertical acceleration vibration response significantly reduced, which has certain practical value.

1. Introduction

With the advancement of China's industry and the development of new automotive technologies, automobiles have provided more and more convenience to human beings in the daily life, so that people are increasingly coming up with new demands on automobiles when using them. In the case of ensuring its safety and economy, it will consider its ride comfort more [1]. So when developing and designing a car, it requires designers to consider not only their power and fuel economy, but also take into account the ride comfort indicators. In the process of driving a car, it is inevitable to be affected by shocks and vibrations from various factors such as road surface. The research on ride comfort of the car is to reduce the impact of such shocks and vibrations on the passengers and drivers, and control them within a reasonable range to make people feel comfortable by some technical measure [4]. Therefore, the ride comfort of the car is mainly evaluated by subjective driving experience the of the members. In addition, the qualities of the car's ride comfort will directly affect the reliability of the car. The dynamic load, which generated by the vibrations and shocks of the road and the rotating parts such as the engine, the drive train and the wheels, will cause wear and tear on some parts of the car [5]. If conditions continue this way, the fatigue life of these part will be gradually reduced, which will affect the safety and reliability of the car.

Based on the theory of automobile system dynamics, this paper uses ADAMS/Car to establish the dynamic model of the vehicle and the torsion beam suspension model of the flexible body structure [2], and the simulation analysis of ride comfort under random road conditions is carried out [3]. At the same time, the RMS value of the vertical weighted acceleration at the driver's seat is taken as the optimization target, and the accuracy of the model is verified by simulation experiments. Whereafter, the stiffness and damping of the front and rear suspensions are optimized in order to improve the ride comfort [6].

IOP Conf. Series: Materials Science and Engineering **811** (2020) 012052 doi:10.1088/1757-899X/811/1/012052

2. Theoretical Analysis

2.1. Evaluation Method of Ride Comfort

When the car driving at high speed on the road, the shocks and vibrations are mainly generated by the driving speed, the unevenness of the random road, etc., passing by a complete set of vehicle vibration system which are composed of the elastic and damping elements such as tires, suspensions and cushions, and deriving from the acceleration from the seat to the human being's body. Therefore, the ride comfort of the car is judged by the response of the human body to different vibration accelerations, that is, the comfort of the ride. So the main purpose of studying ride comfort of the car is to control the dynamic characteristics of the vibration system during driving. The weighted acceleration RMS value is usually used to evaluate the impact of vibration on comfort and health of people.

When using the most basic method to determine the vibration size, firstly we should calculate the value of the RMS of the weighted acceleration of each axial vibration. In the actual research, the commonly used method is spectrum analysis method, which converts the vibration acceleration time $a(t)$ collected by the acceleration sensor into the power spectral density function $G_a(f)$, and then transforms the data through the filter grid for the weighted acceleration time a_w, the formula is as follows:

$$a_w = \left[\int_{0.5}^{80} W^2(f) G_a(f) dt\right]^{\frac{1}{2}} \tag{1}$$

Due to the convenience of experimental research, the three axial vibrations of the seat support surfaces x_s、y_s、$z_s$are usually considered together, and the total weighted acceleration value of the three axial directions is calculated as follows: Where: x_s、y_s、$z_s$represent the weighted acceleration rms values of the x, y, and z axes, respectively.

$$a_w = \left[(1.4a_{xw})^2 + (1.4a_{yw})^2 + {a_{zw}}^2\right]^{\frac{1}{2}} \tag{2}$$

In order to more intuitively display the frequency of the vibration, the weighted vibration level L_{aw} often appears in the "human vibration meter", and the conversion formula between it and a_w is as follows: Where a_0 represents the root mean square value of the reference acceleration, taking $a_0 = 10^{-6}\, m/s^2$

$$L_{aw} = 20\lg({}^{a_w}/_{a_0}) \text{ (dB)} \tag{3}$$

Table 1 below shows the correspondence between the subjective perception of vibration of the human body and the weighted acceleration root mean square value a_w and the weighted vibration level L_{aw}.

Table 1. The relationship between a_w and L_{aw} and people's subjective feelings.

$a_w(m/s^2)$	L_{aw}/dB	People's subjective feelings
<0.315	110	No discomfort
0.315~0.63	110~116	Some discomfort
0.5~1.0	114~120	Rather uncomfortable
0.8~1.6	118~124	uncomfortable
1.25~2.5	122~128	Very uncomfortable
>2.0	126	Very uncomfortable

2.2. Ride Comfort Optimization Model

Design variables X=$[K_{sf} K_{sr}]^T$ generally refer to independent parameters that need to be selected preferentially during the design process. In this paper, since the average damping coefficient of the suspension, the vertical stiffness of the seat, and the average damping coefficient are all fixed values, the vertical stiffness value of the front and rear suspensions are selected as the design variable. Through the different matching of the vertical stiffness of the front and rear suspensions, the purpose of optimal analysis of ride comfort is achieved.

2.2.1. Determination of Optimization Objective Function minσ. The purpose of optimizing the design is to optimize one or a few indicators. In this paper, the RMS of the vertical acceleration experienced by the driver when the car is driving at a speed of 60m/s on the B-class road surface is used as a single optimization goal, which optimizes ride comfort of the car to the greatest extent possible. To this end, this paper selects an eight-degree-of-freedom optimization model, the objective function minσ is defined as follows:

$$\min\sigma=\left[\int_{0.5}^{80} w^2\,(f)G_a(f)df\right]^{\frac{1}{2}} \tag{4}$$

2.2.2. Restrictions. In the process of optimizing the design, constraints are usually used to ensure the accuracy, safety, reliability, functionality and practicability of the system. Therefore, the design variables and the state variables of the system are often used as constraints. The following five aspects will be analyzed:

Vertical stiffness of the front and rear suspension: Usually, in order to ensure good ride comfort, the suspension of the car will be designed soft, so its static deflection will be relatively large. When $15 \le f_s \le 30$(cm), The limit value of the vertical stiffness of the front and rear suspensions should meet the following conditions:

$$\begin{cases} \dfrac{G\times I_r}{2(I_f+I_r)\times 0.30} \le K_{sf} \le \dfrac{G\times I_r}{2(I_f+I_r)\times 0.15} \\ \dfrac{G\times I_f}{2(I_f+I_r)\times 0.30} \le K_{sf} \le \dfrac{G\times I_f}{2(I_f+I_r)\times 0.15} \end{cases} \tag{5}$$

Suspension deflection: The relationship between the limit stroke $[f_d]$ of the suspension and the static stroke f_s is $[f_d]$=1.8~2.0$\sqrt{f_s}$, and $[f_d] = 1.9\sqrt{f_s}$ is used in this paper. It is analyzed that when the f_0 is reduced, the limit stroke $[f_d]$ of the suspension is also reduced, and ride comfort of the car becomes better, but as the f_0 decreases, the dynamic deflection $[f_d]$ of the suspension increases. The probability that the suspension may hit the limit block will increase, resulting in deteriorating the ride comfort. Therefore, it is necessary to make certain restrictions on the dynamic deflection of the car suspension during design to ensure ride comfort of the car. After calculation and verification, we found that when the RMS value of the suspension deflection $\sigma_{f_d} \le [f_d]/3$, the probability of the suspension impact limit stroke can be controlled within 0.3%, then:

$$\begin{cases} \sigma_{f_{dfL}} \le \frac{1}{3}[f_d] \\ \sigma_{f_{drL}} \le \frac{1}{3}[f_d] \\ \sigma_{f_{dfR}} \le \frac{1}{3}[f_d] \\ \sigma_{f_{drR}} \le \frac{1}{3}[f_d] \end{cases} \tag{6}$$

Natural frequency of the suspension system: In the actual design process, reducing the natural frequency of the suspension system can significantly reduce the vibration acceleration of the vehicle body. Therefore, the natural frequency of the suspension system is generally required to be between 1.1 and 1.2 Hz, so that:

$$f_0 = \frac{1}{2\pi}\sqrt{\frac{2(K_{sf}+K_{sr})}{m_{cb}}} \tag{7}$$

Relative dynamic load between wheel and road: The related experiments prove that the relative dynamic loadF_d/G between the wheel and the road surface is closely related to the safety of the car. If the relative dynamic load value of the wheel is greater than 1, the car may appear to be dragged off the road and lose road attachment. In the case of force, this may lead to dangerous situations. Therefore, we must consider the influence of the relative dynamic load of the wheel and the road. After calculation, it is found that the greater the relative dynamic load, the greater the damage to the road surface.When the mean square value of the relative dynamic load $\sigma_{F_d}/G \leq 1/3$, the probability of occurrence of the wheel leaving the ground is 0.15%, so there is:

$$\begin{cases} \sigma_{F_{dfL}}/\left(\frac{m_{cb}\times g\times I_r}{2(I_f+I_r)}\right) \leq \frac{1}{3} \\ \sigma_{F_{drL}}/\left(\frac{m_{cb}\times g\times I_f}{2(I_f+I_r)}\right) \leq \frac{1}{3} \\ \sigma_{F_{dfR}}/\left(\frac{m_{cb}\times g\times I_r}{2(I_f+I_r)}\right) \leq \frac{1}{3} \\ \sigma_{F_{drR}}/\left(\frac{m_{cb}\times g\times I_f}{2(I_f+I_r)}\right) \leq \frac{1}{3} \end{cases} \tag{8}$$

Suspension system damping ratio: In general, the damping of the suspension system is a small damping, its value $\zeta\in[0.2,0.4]$, and $\zeta=\frac{C}{2\sqrt{mK}}$, so another related constraint can be got:

$$\begin{cases} 0.2 \leq \frac{C_{sf}}{2\sqrt{K_{sf}\times m_{cb}\times I_r/(2(I_f+I_r))}} \leq 0.4 \\ 0.2 \leq \frac{C_{sf}}{2\sqrt{K_{sr}\times m_{cb}\times I_f/(2(I_f+I_r))}} \leq 0.4 \end{cases} \tag{9}$$

Through the analysis above, we combine the objective function minσ with five constraints to form a mathematical model for the optimization of the car. The model can visually reflect the various indicators of ride comfort and provide a theoretical basis for the subsequent optimization process.

3. Model Establishment

3.1. Establish an ADAMS Model

The model of the passenger car is established in ADAMS/Car, and is optimized based on the mathematical model of the optimization and related theories. The front suspension of the car studied in this paper is the McPherson independent front suspension, and the rear one is the torsion beam semi-independent rear suspension. The main parameters of the vehicle are as follows:

Table 2. The main parameters of the vehicle.

Basic parameters	value	Basic parameters	value
Dimensions(mm)	7005×2040×2645	Wheelbase (mm)	3935
Front/rear Track(mm)	1690/1490	Full load center of mass(mm)	1008
Curb quality(kg)	3550	Full load total mass(kg)	6150
No-load front/rear axle load distribution(kg)	1676/1874	Full load front/rear axle load distribution(kg)	2200/3950
Front/rear suspension unsprung mass(kg)	80/200	Centroid distance from the front axle (mm)	2024

The torsion beam suspension model of the flexible body structure is established according to the given parameters in order to obtain the ride comfort data on the driver's seat, and then the finite element model shown in figure 1 is established. The finite element mesh of the torsion beam suspension is divided, then constrained according to the condition proposed above. And finally assigning parameters to the constraints and performing analytical calculations.

Performing modal analysis in HyperMesh, calculate the modal neutral file (. MNF file) of the flexible body of the torsion beam suspension, and then import the generated neutral file into ADAMS/Car, combined with the specific hard point coordinates of the suspension, the load and constraints applied before. Finally, a simulation model of the rigid-flexible coupling of the whole vehicle is generated, as shown in figure 2 below.

Figure 1. Cross-sectional view of the beam.

Figure 2. The vehicle model.

4. Ride Comfort Analysis

4.1. Model Analysis

After the vehicle rigid-flexible coupling simulation model is established, the simulation analysis of ride comfort of the random input can be performed. The simulation was carried out at a speed of 60-100 km/h on the B-level road surface, each speed interval was 10 km/h, then test the vibration acceleration that was carried from the driver's seat position to the human body in three directions. Using the ADAMS/Post-Processor function, we can obtain the longitudinal, lateral and vertical vibration acceleration time domain curves at the driver's position. To acquire the axial RMS value, the FFT (Fast Fourier Transform) function in the ADAMS post-processing module can be used to transform each axial acceleration curve to obtain each axial acceleration power spectral density function $G_a(f)$. In figure 3 and 4, the vertical acceleration simulation curve and the vertical acceleration power spectral density curve at the driver's position when the virtual vehicle is driving on the B-level road at a speed of 80 km/h is shown, respectively.

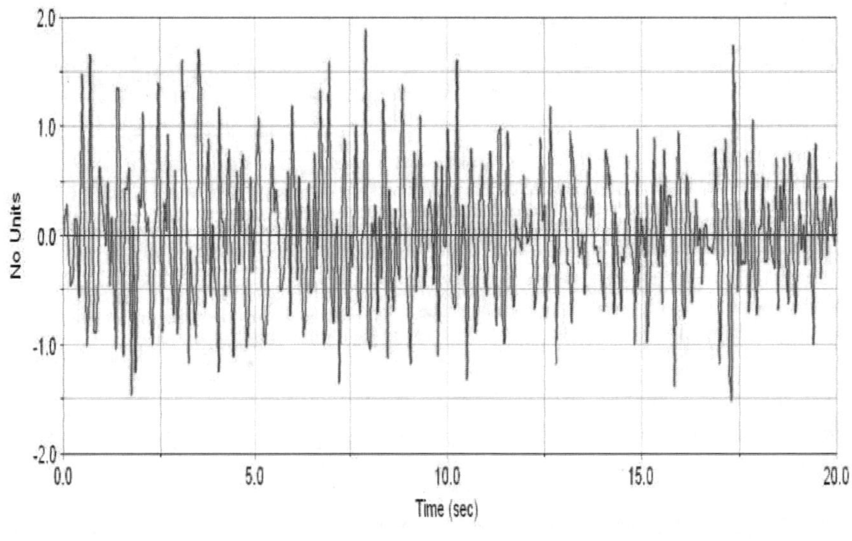

Figure 3. Vertical acceleration simulation curve.

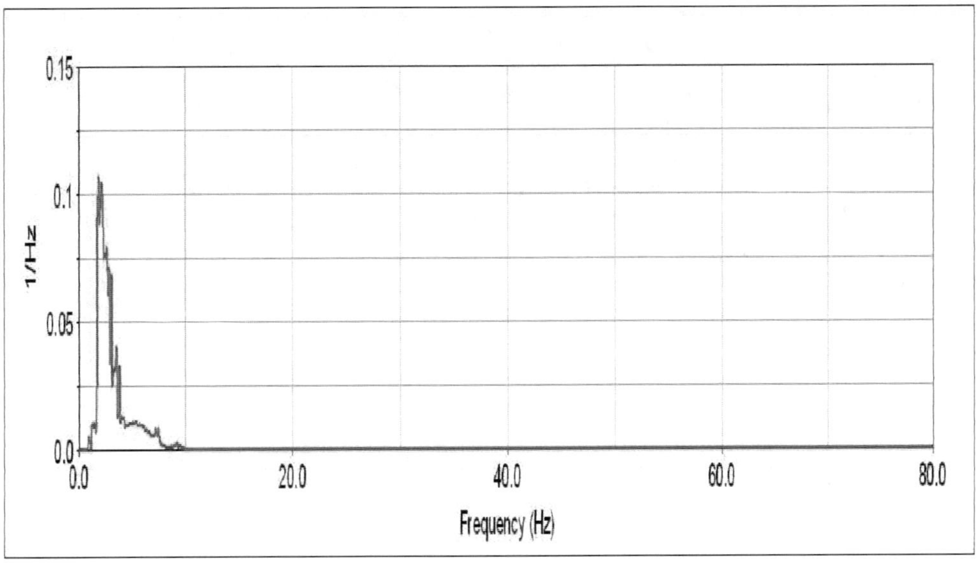

Figure 4. Vertical acceleration power spectral density curve.

From this, the RMS value of the weighted acceleration at the driver's position in each direction can be obtained, so we can calculate the weighted vibration level of the RMS and sum of the total weighted acceleration. According to the analysis method above, the above two parameter values of the vehicle driving at speeds of 60 km/h, 70 km/h, 80 km/h, 90 km/h, and 100 km/h can be calculated as shown in table 3 below.

Table 3. Weighted acceleration RMS and weighted vibration levels of vehicles at different speeds.

v (km/h)	a_w (m/s^2)	L_{aw} (dB)	People's subjective feelings
60	0.3088	109.8	No discomfort
70	0.3485	110.8	Some discomfort
80	0.4728	113.5	Some discomfort
90	0.3948	111.9	Some discomfort
100	0.4368	112.8	Some discomfort

As can be seen from the table above, the passenger may have an uncomfortable feeling when the driving speed is greater than 60 km/h, so it can be concluded that the car will be brought discomfortableness to the passenger due to vibration and other factors at high speed. Therefore, the ride comfort without optimization is not particularly desirable, and the ride comfort can be improved by changing the optimization of certain parameters.

4.2. Real Vehicle Test.

In order to explain whether the simulation model can accurately simulate the actual situation when the vehicle is running, actual test verification is required. The random input test of the real vehicle is carried out on the B-class road surface with similar simulation conditions. After comparing the simulation data with the experimental data which obtained by sensors and other experimental instruments, the feasibility of the simulation experiment and the correctness of the simulation model can be verified. This experiment uses three-way accelerometer, dynamic signal test system and data acquisition system for data acquisition and analysis, shown in figure 5.

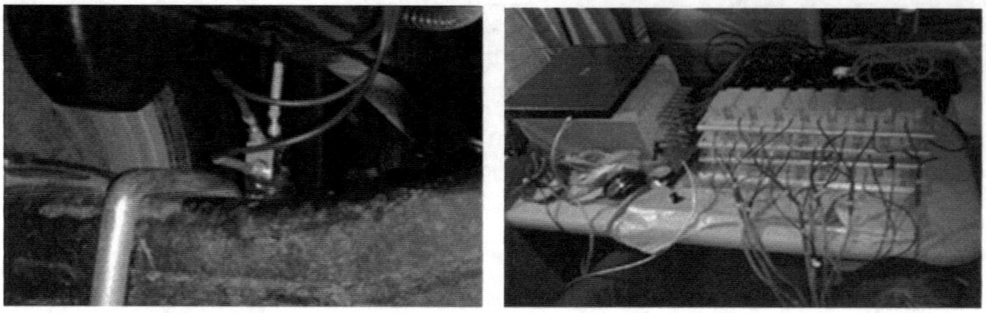

Figure 5. Three-way accelerometer and data acquisition system.

The test method is as follows: The three-way accelerometer is mounted on the seat support bracket directly below the driver's seat to measure the acceleration time in three directions. When the vehicle is started, accelerating to a speed of about 60 km / h and then drive at a constant speed. At this time, the acceleration signals of each axes can be acquired by the sensor and recorded by the data acquisition system. Similarly, recording the acceleration time of the vehicle in three directions when driving at 70km/h, 80km/h, 90km/h, and 100km/h. After analysis and calculation by dynamic signal test system and related equipment, the weighted vibration level of the front seat can be obtained as shown in table 4.

Table 4. Weighted vibration level of the front seat.

v（km/h）	60	70	80	90	100
L_{aw}（dB）	108.6	110.3	112.4	110.7	112.0

So it can be seen clearly from the table 4 above: although other factors such as the vibration caused by the sliding friction between the tire and the ground are not considered in the simulation process, but the simulation data is basically consistent with the experimental data. It shows that the vehicle model is basically consistent with the actual vehicle situation, and its correctness and accuracy can be guaranteed, which lays a foundation for the following ride comfort optimization.

5. Ride Comfort Optimization

In the optimization, considering that the human body is relatively sensitive to the vertical action, the weighted acceleration root mean square value a_{zw} at the driver's position is selected as the design target optimization function. Although many factors must be considered in the optimization of ride comfort of the vehicle, such as: structural parameters and stiffness of the suspension, damping parameters, shock absorber characteristics, cab parameters, tire stiffness and damping parameters, sprung mass and unsprung mass Parameters, etc., combined with the characteristics of the optimized platform and the size of the vehicle structure, this paper choose torsional spring stiffness of the front suspension and the damping of the front and rear suspension dampers as the optimization variables.

In the ADAMS insight module, the design parameters are arranged and combined, and select the parameters which are closest to the optimization target from them. Table 5 shows the design parameters of the front suspension stiffness and the front and rear suspension damping before and after optimization.

Table 5. Design parameters before and after optimization.

	Front suspension stiffness ($N \cdot mm/deg$)	Front suspension damping ($N \cdot s/mm$)	Rear suspension damping ($N \cdot s/mm$)
Before	175000	7.9	4.3
After	174000	8.7	3.8

The optimization result is brought into the previously established vehicle model for re-simulation, and the acceleration curve of the vertical direction at the driver's position before and after optimization at 80 km/h as shown in figure 6 is obtained.

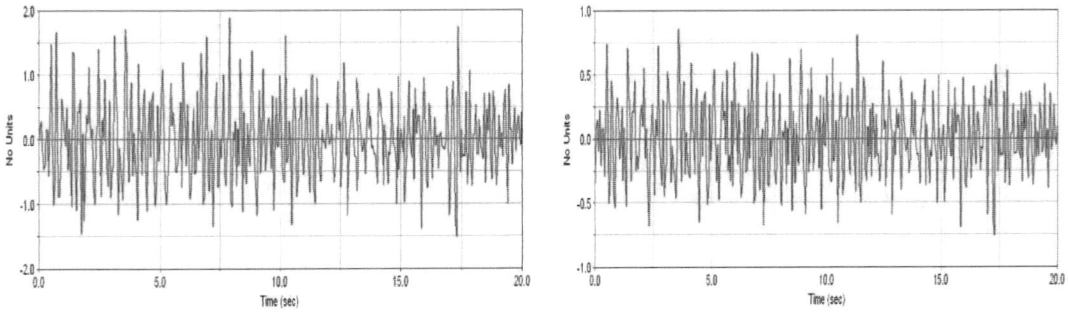

a) Before optimization b) After optimization

Figure 6. Comparison of vertical vibration acceleration curves before and after optimization.

After optimization, the vertical acceleration vibration response peak at the driver's position decreased from $0.1094m^2/s^3$ to $0.0954m^2/s^3$, which decreased by 12.8%, so the damping effect was better. After further processing, the comparison table of the acceleration root mean square value and the weighted vibration level at different speeds are obtained as shown in table 6.

Table 6. Comparison before and after optimization in acceleration RMS value and weighted vibration level.

v（km/h）	a_w（m/s^2）		L_{aw}（dB）	
	Before	After	Before	After
60	0.3088	0.1048	109.8	100.4
70	0.3485	0.1376	110.8	102.8
80	0.4728	0.2968	113.5	109.4
90	0.3948	0.1972	111.9	105.9
100	0.4368	0.2514	112.8	108.0

It can be seen clearly from the above table 6 that after optimization, the weighted vibration level of the vertical direction of the driver is reduced at each speed, so that the vibration intensity transmitted to the human body becomes smaller, and the human body feel become well. At the same time, with the improvement of the speed of the vehicle, the ride comfort and stability of the vehicle are greatly improved, so the operational reliability is.

6. Conclusion

In this paper, the RMS of the vertical direction of the driver is taken as the optimization targe. After analyzing various factors affecting ride comfort of the car, the author establishes the eight-degree-of-freedom vehicle to optimize. The accuracy of the model is verified from the perspective of theoretical calculation and actual experiment, which lays a foundation for the later optimization analysis.

The ride comfort simulation model was established by ADAMS/Car. The front suspension stiffness, front suspension damping and rear suspension damping parameters were selected for optimization design. Then comparing and analyzing the related indicators before and after optimization, the result proves the correctness of the optimization scheme.

The parameterized optimization method is used to optimize its structure. By analyzing and optimizing the structural coordinate values, the target performance can be provided under the condition of small variation, and the guiding direction for future structural optimization is provided.

References

[1] Zhang L L, zhang B 2014 Evaluation method of automobile ride comfort, *Electronic manufact.* **8** 274-5.
[2] ADAMS Reference Manual Version 12, 2002 Mechanical Dynamics, Inc.
[3] Bian A H 2019 Vehicle stability analysis based on virtual excitation method, Hubei agricultural mechanization, 2019(6): 57.
[4] Newland D E 1986 General Liner Theory of Vehicle Response to Random Road Roughness, *Stud. Appl. Mech.* **14** 303-26.
[5] Kim B S, Joo K J, et al. 2011 Ride Comfort Improvement of a Compact SUV Considering Driving Maneuver and Road Surface, *SAE Tech Paper*.
[6] Guangqiang Wu 2008 Review on Reduced Order Modeling of MBS and Its Application in Ground Vehicle, Asian Symposium on Multibody Dynamics in Shanghai, ASMD.

AUTHOR INDEX